Library of
Davidson College

Optimization and Nonsmooth Analysis

CANADIAN MATHEMATICAL SOCIETY SERIES OF MONOGRAPHS AND ADVANCED TEXTS

Monographies et Études de la Société Mathématique du Canada

EDITORIAL BOARD

Frederick V. Atkinson
Bernhard Banaschewski
Colin W. Clark
Erwin O. Kreyszig (Chairman)
John B. Walsh

Frank H. Clarke ■ *Optimization and Nonsmooth Analysis*

Optimization and Nonsmooth Analysis

FRANK H. CLARKE
Department of Mathematics
University of British Columbia
Vancouver, Canada

A Wiley-Interscience Publication
JOHN WILEY & SONS
New York · Chichester · Brisbane · Toronto · Singapore

Copyright © 1983 by John Wiley & Sons, Inc.

All rights reserved. Published simultaneously in Canada.

Reproduction or translation of any part of this work beyond that permitted by Section 107 or 108 of the 1976 United States Copyright Act without the permission of the copyright owner is unlawful. Requests for permission or further information should be addressed to the Permissions Department, John Wiley & Sons, Inc.

Library of Congress Cataloging in Publication Data:
Clarke, Frank H.
  Optimization and nonsmooth analysis.

  (Canadian Mathematical Society series in mathematics)
  "A Wiley-Interscience publication."
  Includes bibliographical references and index.
  1. Mathematical optimization. 2. Mathematical analysis. I. Title. II. Title: Nonsmooth analysis. III. Series.
QA402.5.C53 1983   519   83-1216
ISBN 0-471-87504-X

Printed in the United States of America

10 9 8 7 6 5 4 3 2 1

To my mother

Preface

"... nothing at all takes place in the universe in which some rule of maximum or minimum does not appear." So said Euler in the eighteenth century. The statement may strike us as extreme, yet it is undeniable that humankind's endeavors at least are usually associated with a quest for an optimum. This may serve to explain why there has been an enduring symbiosis between mathematical theories of optimization and the applications of mathematics, even though the forms of the problems (the "paradigms") being considered evolve in time.

The origins of analytic optimization lie in the classical calculus of variations and are intertwined with the development of the calculus. For this reason, perhaps, optimization theory has been slow to shed the strong smoothness (i.e., differentiability) hypotheses that were made at its inception. The attempts to weaken these smoothness requirements have often been ad hoc in nature and motivated by the needs of a discipline other than mathematics (e.g., engineering). In this book a general theory of nonsmooth analysis and geometry will be developed which, with its associated techniques, is capable of successful application to the spectrum of problems encountered in optimization. This leads not only to new results but to powerful versions of known ones. In consequence, the approach taken here is of interest even in the context of traditional, smooth problems in the calculus of variations, in optimal control, or in mathematical programming.

This book is meant to be useful to several types of readers. An effort is made to identify and focus upon the central issues in optimization, and results of considerable generality concerning these issues are presented. Because of this, although its scope is not encyclopedic, the work can serve as a reference text for those in any of the various fields that use optimization. An effort has been made to make the results accessible to those who are not expert in the subject. Thus the first chapter is devoted to an explanation and overview of the book's contents. Here and elsewhere the reader will find examples drawn from economics, engineering, mathematical physics, and various branches of analysis.

The reader who wishes not only to gain access to the main results in the book but also to follow all the proofs will require a graduate-level knowledge of real and functional analysis. With this prerequisite, an advanced course in optimization can be based upon the book. The remaining type of reader we have in mind is the expert, who will discover, we believe, interesting tools and techniques of nonsmooth analysis and optimization.

FRANK H. CLARKE

Vancouver, British Columbia
March 1983

Acknowledgments

Writing a book is a humbling experience. One learns (among other things) how much one depends upon others. The book could not have been written without the much appreciated support of the Killam Foundation, which awarded me a two-year Research Fellowship. I wish also to acknowledge the continued support of the Natural Sciences and Engineering Research Council of Canada. I would like to thank my colleagues at UBC and my co-workers in the field (most of them) for stimulating me over the years (wittingly or otherwise), and the Canadian Mathematical Society for choosing my book for its new series. Foremost in my personal mentions must be Terry Rockafellar who, for one thing, bears the blame for getting me interested in optimization in the first place. I am grateful also to the following colleagues, all of whom have also influenced the contents of the book: Jean-Pierre Aubin, Masako Darrough, Ivar Ekeland, Jean-Baptiste Hiriart-Urruty, Lucien Polak, Rodrigo Restrepo, Richard Vinter, Vera Zeidan. My thanks go to Gus Gassmann and Philip Loewen for courageously scouring a draft of the manuscript, to all the staff of the UBC Mathematics Department for their help and good cheer, and to the good people at John Wiley & Sons.

F. H. C.

Contents

Chapter 1 Introduction and Preview 1

1.1 Examples in Nonsmooth Analysis and Optimization 2
1.2 Generalized Gradients 9
1.3 Three Paradigms for Dynamic Optimization 13
1.4 The Theory of Optimal Control and the Calculus of Variations 18

Chapter 2 Generalized Gradients 24

2.1 Definition and Basic Properties 25
2.2 Relation to Derivatives and Subderivatives 30
2.3 Basic Calculus 38
2.4 Associated Geometric Concepts 50
2.5 The Case in Which X is Finite-Dimensional 62
2.6 Generalized Jacobians 69
2.7 Generalized Gradients of Integral Functionals 75
2.8 Pointwise Maxima 85
2.9 Extended Calculus 95

Chapter 3 Differential Inclusions 110

3.1 Multifunctions and Trajectories 111
3.2 A Control Problem 120
3.3 A Problem in Resource Economics 131
3.4 Endpoint Constraints and Perturbation 142
3.5 Normality and Controllability 147
3.6 Problems with Free Time 151
3.7 Sufficient Conditions: the Hamilton–Jacobi Equation 155

Chapter 4 The Calculus of Variations 166

4.1 The Generalized Problem of Bolza 166
4.2 Necessary Conditions 168

4.3 Sufficient Conditions 177
4.4 Finite Lagrangians 180
4.5 The Multiplier Rule for Inequality Constraints 187
4.6 Multiple Integrals 197

Chapter 5 *Optimal Control* 199

5.1 Controllability 200
5.2 The Maximum Principle 210
5.3 Example: Linear Regulator with Diode 213
5.4 Sufficient Conditions and Existence 219
5.5 The Relaxed Control Problem 222

Chapter 6 *Mathematical Programming* 227

6.1 The Lagrange Multiplier Rule 227
6.2 An Alternate Multiplier Rule 231
6.3 Constraint Qualifications and Sensitivity 234
6.4 Calmness 238
6.5 The Value Function 241
6.6 Solvability and Surjectivity 248

Chapter 7 *Topics in Analysis* 252

7.1 Inverse and Implicit Functions 252
7.2 Aumann's Theorem 256
7.3 Sets Which Are Lipschitz Epigraphs 260
7.4 Dependence on Initial Values 262
7.5 Ekeland's Theorem 265
7.6 Directional Contractions and Fixed Points 268
7.7 Hamiltonian Trajectories and Boundary-Value Problems 270
7.8 Closed Trajectories of Prescribed Period 279

Comments 284

References 291

Index 305

A Note From the Publisher

We are proud to have been chosen as the Publisher for the Canadian Mathematical Society Series of Monographs and Advanced Texts and pleased that this book, *Optimization and Nonsmooth Analysis*, represents such an outstanding start to what we anticipate will become one of the finest series in mathematics. Publishing for the Canadian Mathematical Society represents the continuation of our long history of publishing for scientific societies, and in this, our 176th year, we once again affirm our commitment to serving the scientific community in Canada and throughout the World.

June 1983

Optimization and Nonsmooth Analysis

Chapter One

Introduction and Preview

In adolescence, I hated life and was continually on the verge of suicide, from which, however, I was restrained by the desire to know more mathematics.

BERTRAND RUSSELL, *The Conquest of Happiness*

Just as "nonlinear" is understood in mathematics to mean "not necessarily linear," we intend the term "nonsmooth" to refer to certain situations in which smoothness (differentiability) of the data is not necessarily postulated. One of the purposes of this book is to demonstrate that much of optimization and analysis which have evolved under traditional smoothness assumptions can be developed in a general nonsmooth setting; another purpose is to point out the benefits of doing so. We shall make the following points:

1. Nonsmooth phenomena in mathematics and optimization occur naturally and frequently, and there is a need to be able to deal with them. We are thus led to study differential properties of nondifferentiable functions.
2. There is a recent body of theory (*nonsmooth analysis*) and associated techniques which are well suited to this purpose.
3. The interest and the utility of the tools and methods of nonsmooth analysis and optimization are not confined to situations in which nonsmoothness is present.

Our complete argument in support of these contentions is the entirety of this book. In this chapter we get under way with a nontechnical overview of the theory and some of its applications. The final sections are devoted to placing in context the book's contributions to dynamic optimization (i.e., the

1.1 EXAMPLES IN NONSMOOTH ANALYSIS AND OPTIMIZATION

For purposes of exposition, it is convenient to define five categories of examples.

1.1.1 Existing Mathematical Constructs

The first example is familiar to anyone who has had to prepare a laboratory report for a physics or chemistry class. Suppose that a set of observed data points $(x_0, y_0), (x_1, y_1), \ldots, (x_N, y_N)$ in the x-y plane is given, and consider the problem of determining the straight line in the x-y plane that best fits the data. Assuming that the given data points do not all lie on a certain line (any lab instructor would be suspicious if they did), the notion of "best" must be defined, and any choice is arbitrary. For a given line $y = mx + b$, the error e_i at the ith data point (x_i, y_i) is defined to be $|mx_i + b - y_i|$. A common definition of best approximating line requires that the slope m and the intercept b minimize $\{\sum_{i=0}^{N} e_i^2\}^{1/2}$ over all m and b (or, equivalently, $\sum_{i=0}^{N} e_i^2$). On the face of it, it seems at least as natural to ask instead that the total error $\sum_{i=0}^{N} e_i$ be minimized. The characteristics of the resulting solution certainly differ. In Figure 1.1, for example, the dashed line represents the "least total error" solution (see Example 2.3.17), and the solid line represents the "least total square error" solution. Note that the former ignores the anomalous data point which presumably corresponds to a gross measurement error. The least squares solution, in contrast, is greatly affected by that point. One or the other of these solutions may be preferable; the point we wish to make is that the function $\sum_{i=0}^{N} e_i$ is nondifferentiable as a function of m and b. Thus the usual methods for minimizing differentiable functions would be inapplicable to this function, and different methods would have to be used. Of course, the reason that the least square definition is the common one is that it leads to the minimization of a smooth function of m and b.

The two functions being minimized above are actually special cases of the L^2 and L^1 norms. The differentiability (or otherwise) of norms and of other classes of functions has been and remains a central problem in functional analysis. One of the first results in this area is due to Banach, who characterized the continuous functions x on $[0, 1]$ at which the supremum norm

$$\|x\| := \max_{0 \leq t \leq 1} |x(t)|$$

is differentiable. (His result is rederived in Section 2.8.)

1.1 Examples in Nonsmooth Analysis and Optimization

An interesting example of a nondifferentiable function is the *distance function* d_C of a nonempty closed subset C of R^n. This is the function defined by

$$d_C(x) := \min\{|x - c| : c \in C\},$$

where $|\cdot|$ refers to the Euclidean norm. (It is a consequence of the results of Section 2.5 that when C is convex, for example, d_C must fail to be differentiable at any point on the boundary of C.) The distance function has been a useful tool in the geometrical theory of Banach spaces; it will serve us as well, acting as a bridge between the analytic and the geometric concepts developed later. As an illustration, consider the natural attempt to define directions of tangency to C, at a point x lying in C, as the vectors v satisfying $d'_C(x; v) = 0$, where the notation refers to the customary one-sided directional derivative. Since such directional derivatives do not necessarily exist (unless extra smoothness or convexity hypotheses are made), this approach is only feasible (see Section 2.4) when an appropriate nonsmooth calculus exists.

As a final example in this category, consider the initial-value problem

$$\frac{d}{dt}x(t) = f(t, x(t)), \qquad x(0) = u.$$

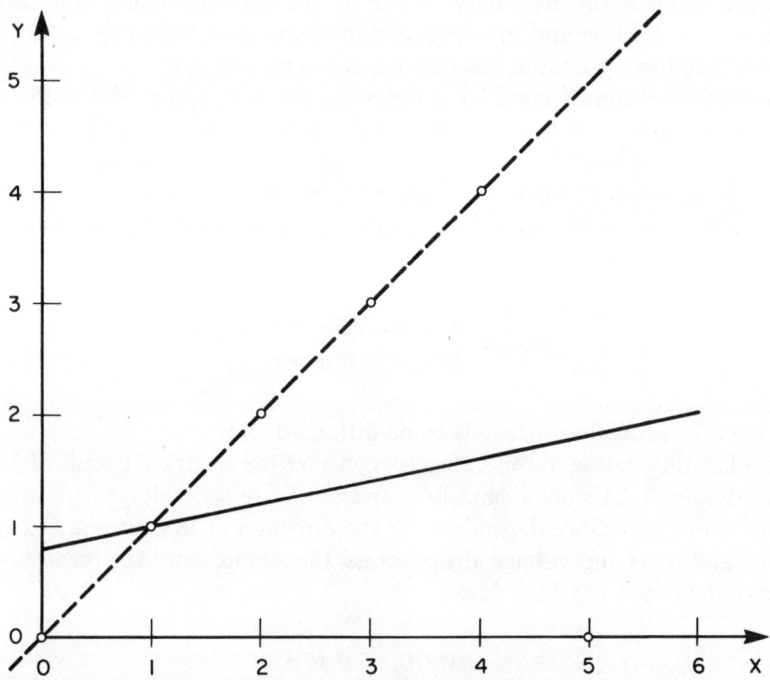

Figure 1.1 Two lines fitted to six data points (indicated by circles).

It is well known that the natural framework for studying existence and uniqueness of solutions is that of functions f which satisfy a Lipschitz condition in the x variable. It is desirable, then, to be able to study in this same framework the closely related issue of how solutions depend on the initial value u. The classical theory, however, hinges upon the resolvent, which is defined in terms of derivatives of f. This confines the analysis to smooth functions f. In Section 7.4 we extend the theory to the Lipschitz setting.

1.1.2 Direction Phenomena

Consider an elastic band whose upper end is fixed, and whose lower end is tied to a unit point mass. When the band is stretched a positive amount x, it exerts an upward (restoring) force proportional to x (Hooke's Law). When unstretched, no force is exerted. (This contrasts to a spring, which also exerts a restoring force when compressed.) If the mass is oscillating vertically, and if $x(t)$ measures the (positive or negative) amount by which the distance from the mass to the upper end of the band exceeds the natural (unstretched) length of the band, Newton's Law yields $\ddot{x}(t) = f(x(t))$, where f is given by

$$f(x) = \begin{cases} g - kx & \text{if } x \geq 0 \\ g & \text{if } x \leq 0. \end{cases}$$

(g is the acceleration of gravity, and k is the proportionality constant for Hooke's Law; friction and the weight of the band have been neglected.) Note that the function f is continuous but not differentiable at 0.

As another example, consider a flat solar panel in space. When the sun's rays meet its surface, the energy produced is proportional to $\cos \alpha$, where α is the (positive) angle of incidence (see Figure 1.2). When the panel's back is turned to the sun (i.e., when α exceeds $\pi/2$), no energy is produced. It follows then that the energy produced is proportional to the quantity $f(\alpha)$, where f is given by

$$f(\alpha) = \begin{cases} \cos \alpha & \text{if } \alpha \leq \pi/2 \\ 0 & \text{if } \alpha \geq \pi/2. \end{cases}$$

This again is a function that fails to be differentiable.

As a last illustration in this category, consider the electrical circuit of Figure 1.3 consisting of a diode, a capacitor, and an impressed voltage. A diode is a resistor whose resistance depends upon the direction of the current. If I is the current and V is the voltage drop across the diode, one has the following nonsmooth version of Ohm's Law:

$$I = \begin{cases} V/R_+ & \text{if } V \geq 0 \\ V/R_- & \text{if } V \leq 0, \end{cases}$$

1.1 Examples in Nonsmooth Analysis and Optimization

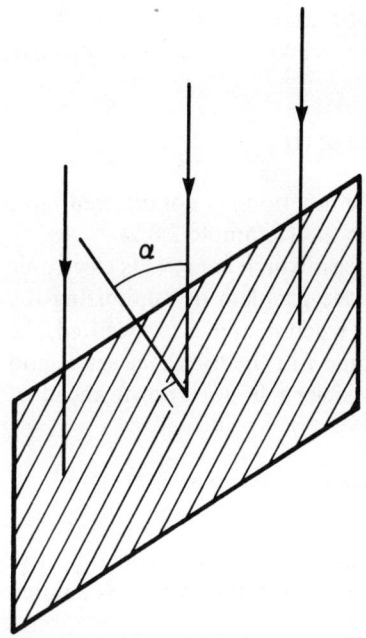

Figure 1.2 The angle of incidence between the sun's rays and a solar panel.

where R_+ and R_- are different positive constants. If x is the voltage across the capacitor, application of Kirchhoff's Laws yields

$$\frac{dx}{dt} = \begin{cases} \alpha(u - x) & \text{if } x \leq u \\ -\beta(x - u) & \text{if } x \geq u, \end{cases}$$

where u is the impressed voltage. In Section 5.3 we shall formulate and solve an optimal control problem with these nonsmooth dynamics.

1.1.3 Objective Functions

The first example is drawn from engineering design in which, in the study of physical systems, an appealing criterion is to minimize the greatest eigenvalue

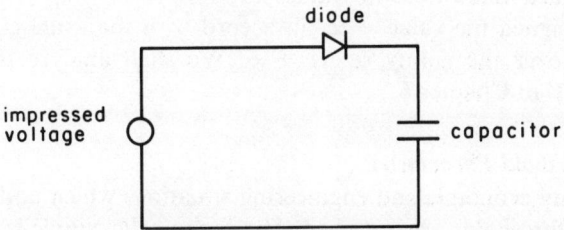

Figure 1.3 An electrical network.

of the system; that is, the greatest eigenvalue of a matrix A associated with it. Suppose that a vector x of parameters is open to choice, so that the objective function f to be minimized becomes

$$f(x) := \text{greatest eigenvalue of } A(x).$$

Even if the matrix A depends smoothly on x, the function f is not differentiable in general. We shall pursue this topic in Section 2.8 (Example 2.8.7).

As a second example, suppose we have a process which constructs electronic parts within a certain tolerance, and consider the problem of minimizing the error in manufacture. Specifically, suppose that when a state x is specified, the process actually produces the state $x + \tau$ for some τ in the tolerance set T, and let $\theta(x + \tau)$ measure the resulting distortion. Since τ is not known ahead of time, the worst-case distortion is

$$f(x) := \max_{\tau \in T} \theta(x + \tau).$$

We seek to minimize f over some feasible set of x values. The objective function f, which is not differentiable in general, is of a type which is studied in Section 2.8.

These two examples are minimization problems in which the possible constraints on x are not specified. In most cases such constraints exist and are themselves dependent upon parameters. Consider, for example, the problem of minimizing $f(x)$ subject to the constraints

(1) $$g(x) + p \leq 0, \quad h(x) + q = 0, \quad x \in C,$$

where C is a given closed subset of some Euclidean space X, and where $g: X \to R^n$, $h: X \to R^m$ are given functions. Here p and q are fixed vectors in R^n and R^m, respectively. The value (minimum) of this prototypical mathematical programming problem clearly depends upon p and q; let us designate it $V(p, q)$. The function V is of fundamental interest in the analysis of the problem, from both the theoretical and the computational points of view. Not only does V fail to be differentiable in general, it even fails to be finite everywhere. For example, our main interest could be in the case $p = 0$, $q = 0$, and $V(0, 0)$ could be finite. Yet there may easily be values of p and q arbitrarily near 0 for which there exist no values of x satisfying Eq. (1). In that case, $V(p, q)$ is assigned the value $+\infty$, in accord with the usual convention that the infimum over the empty set is $+\infty$. We shall analyze the differential properties of V in Chapter 6.

1.1.4 Threshold Phenomena

There are many economic and engineering situations which undergo a change when certain thresholds are breached. The canonical example is that of a dam holding a reserve of water whose level fluctuates with time (due, for example, to tidal or other action). If $h(t)$ is the reservoir's natural water level at time t and if h_0 is the height of the dam, the amount of water that will have to be coped with downstream from the dam is proportional to $f(t)$, where

$$f(t) := \max\{h(t) - h_0, 0\}.$$

Of course, f is nondifferentiable in general, even if $h(\cdot)$ is smooth.

Let us turn to an economic example (of which there are a great number). Suppose that a firm is faced with a *demand function* $q(p)$ for its product. This means that the quantity demanded (in some units) is $q(p)$ when the asking price is set at p. Let Q be the largest quantity of the product that the firm can produce. Then, corresponding to any price p, the quantity that the firm can actually sell is

$$\min\{q(p), Q\},$$

so that effectively any firm with bounded output has a nonsmooth demand function.

One of the growth laws of the classical theory of capital growth is known as *fixed proportions*. It refers to situations in which inputs to production can only be used in certain proportions (e.g., one worker, one shovel). If, for example, L and K measure in some units two inputs that must be used in equal proportions, then the effective quantity of either input that is available to the production process is $\min\{L, K\}$. Most dynamic problems in capital growth or resource theory are treated by continuous models (i.e., neither time nor input levels are discrete) which incorporate some production law. If the fixed-proportions law is used, the resulting optimal control problem has nonsmooth dynamics. In Section 3.3 we study and solve such a problem in the theory of nonrenewable resources.

1.1.5 Mathematical Techniques

There exist mathematical methods which involve nonsmoothness in a fundamental way, and which can be very useful even in situations which seem smooth at first glance. We shall illustrate this with three such techniques, all of which will be used subsequently. The first is called *exact penalization*. To begin, let us consider the admittedly trivial problem of minimizing the function $f(x) = x$ over the reals, subject to the constraint $x = 0$. The minimum is 0, attained at the only point x satisfying the constraint (i.e., $x = 0$). The technique of exact penalization transforms constrained problems to unconstrained ones by adding to the original objective function a term that penalizes violation of the constraint. For example, as the reader may verify, the function $x + 2|x|$ attains its unique unconstrained minimum at $x = 0$ (which is the solution to the original constrained problem). Note that the penalty term $2|x|$ is nonsmooth. This is not due simply to the choice that was made; no smooth penalty function (i.e., a nonnegative function vanishing at 0) will ever produce a problem whose minimum occurs at 0. For example, the function $x + kx^2$ does not attain a minimum at 0, no matter how large k is.

This simple example illustrates that exact penalization hinges upon nonsmoothness. We shall use the technique in Chapter 3 to replace a constraint $\dot{x} \in F(x)$, where x is a function on $[a, b]$, by a penalty term $\int_a^b \rho(x, \dot{x})\, dt$, where ρ is defined by

$$\rho(x, v) := \text{Euclidean distance from } v \text{ to } F(x).$$

This is reminiscent of the distance function cited earlier. In fact, distance functions frequently arise in connection with exact penalization (see Proposition 2.4.3).

Our second technique is the method which classically is known as the variational principle. It consists of obtaining some desired conclusion by constructing a functional F and finding a point u which minimizes it. Applying the stationarity conditions (e.g., $F'(u) = 0$) yields the required conclusion. Many of the initial applications of this method took place within the calculus of variations (hence the name). (A famous example is Dirichlet's Principle, which seeks to find solutions of Laplace's equation $u_{xx} + u_{yy} = 0$ by minimizing the functional $\iint \{u_x^2 + u_y^2\}\, dx\, dy$.) Now it is clear that the method can only become more powerful as the functionals that can be considered become more general. In particular, the admissibility of nonsmooth functionals is useful, as we shall see.

Suppose that in applying the method of variational principles, or through some other means, we are led to consider a functional F which does not attain its infimum. (This is more likely to happen in infinite than in finite dimensions.) Further suppose that u is a point which "almost minimizes" F. Ekeland's Theorem, which is presented in Chapter 7, is a useful tool which says something about this situation. It asserts, roughly speaking, that there is a functional $\tilde{F}$ which is a slight perturbation of F and which attains a minimum at a point $\tilde{u}$ near u. The perturbed functional $\tilde{F}$ is of the form $\tilde{F}(v) = F(v) + k\|v - \tilde{u}\|$ for a positive constant k (where $\|\cdot\|$ is the norm on the space in question), so that the theorem leads to nonsmooth functionals.

The final example of how nonsmoothness can intervene in new mathematical techniques is provided by a *dual principle of least action* due to the author, and which we shall describe here only briefly. (The principle is used in Sections 7.7 and 7.8.) Hamilton's Principle of least action in classical mechanics amounts to the assertion that a physical system evolves in such a way as to minimize (or, more precisely, render stationary) the *action*, which is the variational functional

(2) $$\int \{\langle p, \dot{x}\rangle - H(x, p)\}\, dt.$$

This leads to the Hamiltonian equations $-\dot{p} = \nabla_x H$, $\dot{x} = \nabla_p H$, which are basic to classical mechanics. The functional (2) suffers from the serious mathematical defect of being indefinite; that is, it admits no local maxima or minima. This has limited its role, for example, in the qualitative theory of existence of solutions to Hamilton's equations.

Consider now the function G which is conjugate to H in the sense of convex analysis:

$$
(3) \qquad G(u, v) := \sup_{x, p} \{\langle u, x \rangle + \langle v, p \rangle - H(x, p)\},
$$

and let us define the *dual action* to be the functional

$$
(4) \qquad \int \{\langle \dot{p}, x \rangle + G(-\dot{p}, \dot{x})\}\, dt.
$$

The function G, and hence the functional (4), are nonsmooth in general. In Chapter 4 we shall define and derive stationarity conditions for such functionals. It turns out that when H is convex, these stationarity conditions for the functional (4) almost coincide with the Hamiltonian equations. (To be precise, if (x, p) satisfies the generalized stationarity conditions for Eq. (4), then there exist translates of x and p by constants which satisfy the Hamiltonian equations.)

The dual action, Eq. (4), in contrast to the classical action, Eq. (2), can be shown to attain a minimum for certain classes of Hamiltonians H. In consequence, the dual action, which is a nonsmooth and nonconvex functional, has proven to be a valuable tool in the study of classical Hamiltonian systems, notably in the theory of periodic solutions.

1.2 GENERALIZED GRADIENTS

The generalized gradient is a replacement for the derivative. It can be defined for a very general class of functions (and will be in Chapter 2). Our purpose here is to give a nontechnical summary of the main definitions for those whose primary interest lies in the results of later chapters. We begin the discussion with the simplest setting, that of a locally Lipschitz real-valued function defined on R^n (n-dimensional Euclidean space).

Let $f: R^n \to R$ be a given function, and let x be a given point in R^n. The function f is said to be *Lipschitz near* x if there exist a scalar K and a positive number ε such that the following holds:

$$
|f(x'') - f(x')| \leq K |x'' - x'| \quad \text{for all } x'', x' \text{ in } x + \varepsilon B.
$$

(Here B signifies the open unit ball in R^n, so that $x + \varepsilon B$ is the open ball of radius ε about x.) This is also referred to as a Lipschitz condition of rank K.

We wish to study differential properties of f. Note to begin with that f need not be differentiable at x. Indeed, it is not hard to produce a function f which is Lipschitz near x and which fails to admit even one-sided directional derivatives at x.

We define instead the *generalized directional derivative* of f which, when evaluated at x in the direction v, is given by

$$(1) \qquad f^\circ(x; v) := \limsup_{\substack{y \to x \\ \lambda \downarrow 0}} \frac{f(y + \lambda v) - f(y)}{\lambda}.$$

The difference quotient whose upper limit is being taken is bounded above by $K|v|$ (for y sufficiently near x and λ sufficiently near 0) in light of the Lipschitz condition, so that $f^\circ(x; v)$ is a well-defined finite quantity. One might conceive of using other expressions in defining a generalized directional derivative. What makes f° above so useful is that, as a function of v, $f^\circ(x; v)$ is positively homogeneous and subadditive. This fact allows us to define a nonempty set $\partial f(x)$, the *generalized gradient* of f at x, as follows:

$$(2) \qquad \partial f(x) := \{\zeta \in R^n : f^\circ(x; v) \geq \langle v, \zeta \rangle \text{ for all } v \text{ in } R^n\}.$$

By considering the properties of f°, it is possible to show that $\partial f(x)$ is a nonempty convex compact subset of R^n. One has, for any v,

$$(3) \qquad f^\circ(x; v) = \max\{\langle \zeta, v \rangle : \zeta \in \partial f(x)\},$$

so that knowing f° is equivalent to knowing $\partial f(x)$.

When f is smooth (continuously differentiable), $\partial f(x)$ reduces to the singleton set $\{\nabla f(x)\}$, and when f is convex, then $\partial f(x)$ coincides with what is called the subdifferential of convex analysis; that is, the set of vectors ζ in R^n satisfying

$$f(x + u) - f(x) \geq \langle u, \zeta \rangle \quad \text{for all } u \text{ in } R^n.$$

The computation of $\partial f(x)$ from this definition may appear to be a formidable task. In fact, it is something that one seeks to avoid, just as in differential calculus, where one rarely computes derivatives from the definition. Instead, one appeals to a body of theory that characterizes generalized gradients of certain kinds of functions, and to rules which relate the generalized gradient of some compound function (e.g., sum, composition) to those of its simpler components. This calculus of generalized gradients is developed in detail in Chapter 2.

We shall see that there are several equivalent ways of defining the generalized gradient. One alternate to the route taken above hinges upon Rademacher's Theorem, which asserts that a locally Lipschitz function is differentiable almost everywhere (in the sense of Lebesgue measure). Let Ω_f be the set of points in $x + \varepsilon B$ at which f fails to be differentiable, and let S be any other set of measure zero. We shall obtain in Section 2.5 the following characterization of the generalized gradient:

(4) $$\partial f(x) = \operatorname{co}\left\{\lim_{i \to \infty} \nabla f(x_i) : x_i \to x, x_i \notin S, x_i \notin \Omega_f\right\}.$$

In words; $\partial f(x)$ is the convex hull of all points of the form $\lim \nabla f(x_i)$, where $\{x_i\}$ is any sequence which converges to x while avoiding $S \cup \Omega_f$.

Normals and Tangents

Let C be a nonempty closed subset of R^n. An interesting, nonsmooth, Lipschitz function related to C is its *distance function* d_C, defined by

$$d_C(x) = \min\{|x - c| : c \in C\}.$$

The generalized directional derivative defined earlier can be used to develop a notion of tangency that does not require C to be smooth or convex. The *tangent cone* $T_C(x)$ to C at a point x in C is defined as follows:

(5) $$T_C(x) := \{v \in R^n : d_C^\circ(x; v) = 0\}.$$

It can be shown that this condition does in fact specify a closed convex cone.

Having defined a tangent cone, the likely candidate for the normal cone is the one obtained from $T_C(x)$ by polarity. Accordingly, we define $N_C(x)$, the *normal cone* to C at x, as follows:

(6) $$N_C(x) := \{\zeta : \langle \zeta, v \rangle \leq 0 \text{ for all } v \text{ in } T_C(x)\}.$$

It follows then that $N_C(x)$ is the closed convex cone generated by $\partial d_C(x)$.

It is natural to ask whether T_C and N_C can be defined directly without recourse to the distance function. We shall prove in Section 2.4 that a vector v belongs to $T_C(x)$ if and only if (iff) it satisfies the following property:

(7) $\begin{cases} \text{For every sequence } x_i \text{ in } C \text{ converging to } x \text{ and every sequence } t_i \text{ in} \\ (0, \infty) \text{ converging to } 0, \text{ there is a sequence } v_i \text{ converging to } v \text{ such} \\ \text{that } x_i + t_i v_i \text{ belongs to } C \text{ for all } i. \end{cases}$

The direct definition of $N_C(x)$ requires the notion of *perpendicular*. A nonzero vector v is *perpendicular* to C at a point x in C (this is denoted $v \perp C$ at x) if $v = x' - x$, where the point x' has unique closest point x in C. (Equivalently, $v = x' - x$, where there is a closed ball centered at x' which

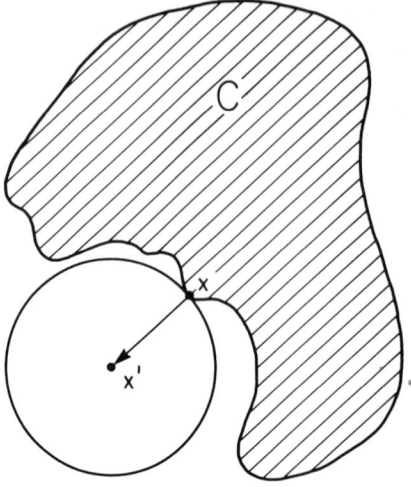

Figure 1.4 The vector $x' - x$ is perpendicular to C at x.

meets C only at x; see Figure 1.4.) It will be shown that one has

$$(8) \quad N_C(x) = \overline{\text{co}}\left\{ \lambda \lim \frac{v_i}{|v_i|} : \lambda \geq 0,\, v_i \perp C \text{ at } x_i,\, x_i \to x,\, v_i \to 0 \right\}.$$

(An overbar denotes closure.) This is probably the easiest way to see how the normal cones of Figure 1.5 are obtained; tangents follow by polarity.

Everything Is One

We now have two analytic notions for functions (f° and ∂f) and two geometric notions for sets (T_C and N_C). Each of these pairs is dual: f° and ∂f are obtainable one from the other via Eq. (3), while T_C and N_C are each related to the other by polarity. It turns out moreover that these analytic and geometric concepts can be linked to one another. This is an important factor in their utility, just as chess pieces gain in strength when used cooperatively. The key to the relationship is the *epigraph* of a function f, denoted epi f, which is

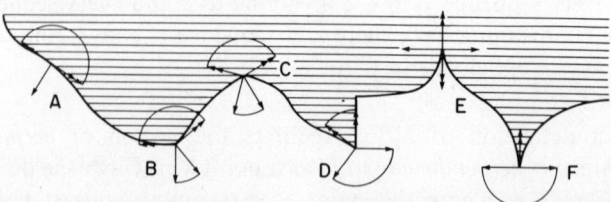

Figure 1.5 Five normal (single-arrow) and tangent (double-arrow) cones to a set.

the set

$$\{(x, r) \in R^n \times R : f(x) \leqslant r\}$$

(i.e., the set of points on or above the graph of f). In Chapter 2 we shall prove the following formulas:

(9) $$T_{\text{epi} f}(x, f(x)) = \text{epi} f^\circ(x; \cdot)$$

(10) $$\partial f(x) = \{\zeta : (\zeta, -1) \in N_{\text{epi} f}(x, f(x))\}.$$

(This last relationship reflects the familiar calculus result that the vector $[f'(x), -1]$ is normal to the graph of f at the point $(x, f(x))$.) It follows that T_C, N_C, ∂f, and f° form a circle of concepts so that any of them can be defined (by Eq. (7), (8), (4), or (1), respectively) as the nucleus from which the others evolve.

Thus far the functions we have been considering have been locally Lipschitz. This turns out to be a useful class of functions, and not only because a very satisfactory theory can be developed for it. As we shall see, the local Lipschitz property is a reasonable and verifiable hypothesis in a wide range of applications, and one that is satisfied in particular when smoothness or convexity is present. Locally Lipschitz functions possess the important property of being closed under the main functional operations (sums, compositions, etc.).

Nonetheless, there are good reasons for wanting to develop the theory for non-Lipschitz functions as well, and we shall do so in Chapter 2. For now, let us mention only that the key is to use Eq. (10) as a *definition* of ∂f in the extended setting, rather than as a derived result. When f is not necessarily Lipschitz near x, $\partial f(x)$ may not be compact, and may be empty.

If the set in Figure 1.5 is interpreted as the epigraph of a function f, then (using Eq. (10)) we find that at the x value corresponding to point A we have $\partial f = \{-1\}$, and f is smooth there. The function is Lipschitz near the points corresponding to B and C, and one has $\partial f = [0, 1]$ and $[-\frac{1}{2}, 1]$, respectively. At D there is a discontinuity, and $\partial f = [-1, \infty)$. At E, ∂f is empty, while at F it coincides with the real line.

Chapter 2 goes beyond the theory of this section in two further directions, in considering vector-valued functions, and in allowing the domain of f to be a (possibly infinite-dimensional) Banach space.

1.3 THREE PARADIGMS FOR DYNAMIC OPTIMIZATION

In later chapters we shall be dealing with three distinct though overlapping problems in optimal control and the calculus of variations. We propose now to describe them and discuss their origins and relative merits.

The Problem of Bolza: P_B

The calculus of variations will soon be celebrating its three-hundredth birthday (some would say posthumously). It is a beautiful body of mathematics in which the origins of several of today's mathematical subdisciplines can be found, and to which most of mathematics' historic names have contributed. The basic problem in the calculus of variations is the following: to minimize, over all functions x in some class mapping $[a, b]$ to R^n, the integral functional

$$\text{(1)} \qquad \int_a^b L(x(t), \dot{x}(t))\, dt.$$

In the main developments of the theory (which, along with other matters, will be discussed in the following section), the given function L is smooth (four times continuously differentiable will do). A common version of the problem is that in which the functions x admitted above must satisfy $x(a) = x_0$, $x(b) = x_1$, where x_0 and x_1 are given (the *fixed endpoint* problem). Another classical version of the problem is the so-called *problem of Bolza*, in which the functional to be minimized is given by

$$\text{(2)} \qquad l(x(a), x(b)) + \int_a^b L(x(t), \dot{x}(t))\, dt.$$

An *arc* is an absolutely continuous function mapping $[a, b]$ to R^n. We shall adopt the problem of minimizing the functional (2) over all arcs x as our first paradigm; we label it P_B. The problem as we shall treat it, however, is much more general than the classical problem of Bolza, due to the fact we shall allow l and L to be extended-valued. For example, the following choice for l is permissible:

$$l(u, v) = \begin{cases} 0 & \text{if } u = x_0 \text{ and } v = x_1 \\ +\infty & \text{otherwise.} \end{cases}$$

With this choice of l, minimizing (2) is equivalent to minimizing (1) under the constraint $x(a) = x_0$, $x(b) = x_1$. Thus the fixed endpoint problem is a special case of P_B. In similar fashion, we may define L to take account of such constraints as, say,

$$g(x(t), \dot{x}(t)) \leq 0, \qquad a \leq t \leq b.$$

While the extension of the problem of Bolza just described takes us beyond the pale of the classical calculus of variations, the formal resemblance that we take with us is a valuable guide in building an extended theory, as we shall see presently. Let us first define the second paradigm.

1.3 Three Paradigms for Dynamic Optimization

The Optimal Control Problem: P_C

The optimal control problem, which is generally thought to have originated in the engineering literature of the 1940s and 1950s, can be formulated as follows. Let the word *control* signify any member of some given class of functions $u(\cdot)$ mapping $[a, b]$ to a given subset U of R^m. We associate to each control u an arc x (called the *state*) by means of the differential equation and initial condition

(3) $$\dot{x}(t) = \phi(x(t), u(t)), \qquad x(a) = x_0,$$

where ϕ is given, and where the point x_0 is free to be chosen within a given set C_0. The problem P_C consists of choosing u and x_0 (and hence x) so as to minimize a given functional of the type

(4) $$f(x(b)) + \int_a^b F(x(t), u(t))\, dt,$$

subject to the condition that $x(b)$ lie in a given set C_1. (In Chapter 5 we allow U, F, and ϕ to depend on t, and we admit a further *state constraint* $g(t, x(t)) \leq 0$, but in this chapter we suppress these possibilities in order to unburden the exposition.)

A large number of physical systems can be modeled by differential equations in which some parameter values can be chosen within a certain range. This is precisely the framework of Eq. (3), which explains in part why the optimal control problem has found such wide application. From the strictly mathematical point of view, the optimal control problem subsumes the classical calculus of variations. For example, the fixed-endpoint problem of minimizing (1) subject to $x(a) = x_0$, $x(b) = x_1$ is seen to be the special case of P_C in which

$$\phi(x, u) \equiv u, \qquad U = R^n, \qquad F = L,$$

$$C_0 = \{x_0\}, \qquad C_1 = \{x_1\}, \qquad f \equiv 0.$$

Conversely, it is also the case that many (but not all) optimal control problems can be rephrased as standard variational problems with inequality and equality constraints (which are admitted in the classical theory).

Accordingly, it seems that optimal control theory can simply be viewed as the modern face of the calculus of variations. Yet we would argue that this fails to reflect the actual state of affairs, and that in fact a "paradigm shift" has taken place which outweighs in importance the mathematical links between the two. There are significant distinctions in outlook; many of the basic questions differ, and different sorts of answers are required. We have found that by and large the methodology and the areas of application of the two disciplines have little in common. We believe that this is due in large part to the fact that [as Young (1969) has pointed out] the smoothness requirements of the classical variational calculus are inappropriate in optimal control theory. Indeed, we would argue that in establishing a bridge between the two, nondifferentiability intervenes fundamentally; more anon.

We now complete our triad of central problems.

The Differential Inclusion Problem: P_D

A *multifunction F* from R^n to itself is a mapping such that, for each x in R^n, $F(x)$ is a subset of R^n. Given such a multifunction F, along with two subsets C_0 and C_1 of R^n and a function f on R^n, we define the problem P_D to consist of minimizing $f(x(b))$ over all arcs x satisfying

(5) $$\dot{x}(t) \in F(x(t)), \quad a \leq t \leq b$$

$$x(a) \in C_0, \quad x(b) \in C_1.$$

(We refer to (5) as a *differential inclusion*.) In contrast to P_B and P_C, this problem has very little history or tradition associated with it. It is in a sense the antithesis of P_B, which suppresses all explicit constraints by incorporating them (via extended values) in the objective functional. The only explicit minimization in P_D, on the other hand, involves one endpoint; all else is constraint. The optimal control problem P_C is a hybrid from this point of view.

We now turn to the relationships among these three problems and their roles in what follows.

Equivalence: Why Three?

Some relationships between the problems P_B, P_C, and P_D are well known. For example, let us suppose that a problem P_C is given, and let us attempt to recast it in the mold of P_D. To begin with, we may suppose that F in P_C is identically zero, by using the following device: let $\bar{x}$ signify the point (x°, x) in $R \times R^n$, and define

$$\bar{\phi}(\bar{x}, u) = [F(x, u), \phi(x, u)], \quad \bar{f}(\bar{x}) = f(x) + x^\circ$$

$$\bar{C}_0 = \{0\} \times C_0, \quad \bar{C}_1 = R \times C_1.$$

Then the original P_C is equivalent to the problem of minimizing $\bar{f}(\bar{x}(b))$ over the arcs $\bar{x}$ and controls u, satisfying

$$\dot{\bar{x}}(t) = \bar{\phi}(\bar{x}(t), u(t))$$

$$\bar{x}(a) \in \bar{C}_0, \quad \bar{x}(b) \in \bar{C}_1.$$

1.3 Three Paradigms for Dynamic Optimization

This continues to have the form of P_C, but with F identically zero. Assuming then that F in P_C is zero, we define the data of a problem P_D as follows: f, C_0, C_1 remain as they are, and the multifunction F is defined by $F(x) := \phi(x, U)$. Under very mild assumptions on ϕ, it follows that the arc x satisfies $\dot{x} \in F(x)$ iff there is a control u such that $\dot{x} = \phi(x, u)$. (This fact is known as Filippov's Lemma.) We deduce therefore that the original P_C is equivalent to the resulting special case of P_D, in the sense that x solves the latter iff there is a control u, which, along with x, solves the former.

In turn, it is easy to see that any problem of the form P_D can be viewed as a special case of P_B. We have merely to define l and L as follows:

$$l(u, v) = \begin{cases} f(v) & \text{if } u \in C_0 \text{ and } v \in C_1 \\ +\infty & \text{otherwise} \end{cases}$$

$$L(s, v) = \begin{cases} 0 & \text{if } v \in F(s) \\ +\infty & \text{otherwise.} \end{cases}$$

Together with the foregoing, then, we can summarize notationally as follows:

$$(6) \qquad P_C \subset P_D \subset P_B.$$

The relationship $P_C \subset P_B$ can also be obtained directly without first reducing to $F = 0$ in the following way: define l as above, and define L by

$$(7) \qquad L(s, v) = \inf\{F(s, u) : u \in U, v = \phi(s, u)\}.$$

(Note that in keeping with custom, the infimum over the empty set is taken to be $+\infty$.)

Thus far in the discussion we have been neglecting an important factor: the hypotheses under which each of the three problems is analyzed. For example, the function L produced by Eq. (7) will in general be extended-valued and not even continuous where it is finite, so that the transformation of P_C to P_B which uses Eq. (7) is of little use if we are not prepared to deal with such functions (rest assured, we are). Thus the practical import of such reductions depends upon hypotheses, and these in turn depend upon the issue under discussion (e.g., existence, necessary conditions, etc.). By and large, the relations embodied in (6) accurately reflect the hierarchy that exists. An example of a formal reduction that is incompatible with our hypotheses in general is the one in which P_B is phrased in the form of P_C by taking

$$\phi(x, u) \equiv u, \qquad U = R^n, \qquad F(x, u) = L(x, u).$$

(We ignore boundary terms.)

Why treat three problems if one of them subsumes the other two? One reason is that the differing structure of the problems (together with some prodding from established custom) leads to certain results (e.g., necessary conditions) of differing nature according to the problem, and this can be advantageous. Many problems encountered in applications can be phrased as any one of the three central types we have defined. As we shall illustrate, the methods which correspond to one problem may yield more precise information than those of another.

Each of the three paradigms has advantages specific to it. P_B, by its very form, facilitates our drawing inspiration from the calculus of variations. It is possible within the framework of P_B to achieve a unification of the classical calculus of variations and optimal control theory. Indeed, the extension to P_B of classical variational methods, which requires consideration of nonsmoothness, turns out to be a powerful tool in optimal control theory. A disadvantage of P_B is that a rather high level of technical detail is required to deal with the full generality of the integrands L that may arise.

The standard optimal control problem P_C is a paradigm that has proven itself natural and useful in a wide variety of modeling situations. It is the only one of the three in which consideration of nondifferentiability can be circumvented to a great extent by certain smoothness hypotheses. As we shall see in the next section and in Section 5.3, however, the other formulations can be better in some cases.

The main advantage of P_D derives from the fact that its structure is the simplest of the three. For this reason, most of the results that we shall obtain for the problems P_B and P_C are derived as consequences of the theory developed for P_D in Chapter 3. P_D, being of a novel form, has of course not been used very much in modeling applications. It is our belief that some problems can most naturally be modeled this way; an example is analyzed in Section 3.3.

We turn now to a discussion of the main issues that arise in connection with variational and control problems, and the existing theory of such problems. We shall then be able to place in context the results of the three chapters on dynamic optimization.

1.4 THE THEORY OF OPTIMAL CONTROL AND THE CALCULUS OF VARIATIONS

The three central themes common to any optimization problem are (1) existence—is there a solution to the problem?; (2) necessary conditions—what clues are there to help identify solutions?; (3) sufficient conditions—how can a candidate be confirmed to be a solution? There are of course other issues that depend on the particular structure of the problem (such as controllability and sensitivity, which will also be studied), but we shall organize the present discussion around the three themes given above. We shall lead off with necessary conditions; we begin at the beginning.

Necessary Conditions in the Calculus of Variations

Consider the basic problem in the calculus of variations: that of minimizing the functional

(1) $$\int_a^b L(x(t), \dot{x}(t))\, dt$$

over a class of arcs x with prescribed values at a and b, where $(s, v) \to L(s, v)$ is a given function. When L is sufficiently smooth, any (local) solution to this problem satisfies the *Euler–Lagrange equation*

(2) $$\frac{d}{dt}\{\nabla_v L(x(t), \dot{x}(t))\} = \nabla_s L(x(t), \dot{x}(t)).$$

Another condition that a solution x must satisfy is known as the *Weierstrass condition*. This is the assertion that for each t, the following holds:

(3)

$$L(x(t), \dot{x}(t) + w) - L(x(t), \dot{x}(t)) \geq \langle w, \nabla_v L(x(t), \dot{x}(t))\rangle \quad \text{for all } w.$$

The other main necessary condition is known as the *Jacobi condition*; it is described in Section 4.3.

Classical mechanics makes much use of a function called the *Hamiltonian*. It is the function H derived from L (the *Lagrangian*) via the Legendre transform, as follows. Suppose that (at least locally) the equation $p = \nabla_v L(x, v)$ defines v as a (smooth) function of (x, p). Then we define

(4) $$H(x, p) = \langle p, v(x, p)\rangle - L(x, v(x, p)).$$

It follows that if x satisfies the Euler–Lagrange equation (2), then the function $p(t) := \nabla_v L(x(t), \dot{x}(t))$ satisfies

(5) $$-\dot{p}(t) = \nabla_x H(x(t), p(t)), \qquad \dot{x}(t) = \nabla_p H(x(t), p(t)),$$

a celebrated system of differential equations called *Hamilton's equations*.

In order to facilitate future comparison, let us depart from classical variational theory by defining the *pseudo-Hamiltonian* H_P:

(6) $$H_P(x, p, v) := \langle p, v\rangle - L(x, v).$$

The reader may verify that the Euler–Lagrange equation (2) and the Weierstrass condition (3) can be summarized in terms of H_P as follows:

$$-\dot{p} = \nabla_x H_P(x, p, \dot{x}), \qquad \dot{x} = \nabla_p H_P(x, p, \dot{x})$$
(7)
$$H_P(x, p, \dot{x}) = \max_v H_P(x, p, v).$$

Now suppose that side constraints of the type, say, $g(x, \dot{x}) \leq 0$ are added to the problem. Inspired by the technique of Lagrange multipliers, we may seek to prove that for an appropriate multiplier function λ, a solution x satisfies the necessary conditions above for L replaced by $L + \langle \lambda, g \rangle$ (see Section 4.5). The proof of this *multiplier rule* (in its fullest generality) was the last great quest of the classical school of the calculus of variations; it was completed in the work of Bliss (1930) and McShane (1939). (See Hestenes (1966) for a thorough treatment of the approach.) It has been eclipsed in prominence, however, by the work of Pontryagin et al. (1962), who proved the *maximum principle*, which we shall now describe.

Necessary Conditions for P_C, P_D, and P_B

Suppose that we now constrain the arcs x under consideration to satisfy

(8)
$$\dot{x}(t) = \phi(x(t), u(t)),$$

where $u(t)$ belongs to U. Then $L(x, \dot{x})$ can be expressed in the form $F(x, u)$, and the basic problem of minimizing (1), but subject now to (8), is seen to be the optimal control problem P_C described in the preceding section. If in the definition (6) of the pseudo-Hamiltonian H_P we take account of Eq. (8) by replacing v (which stands in for $\dot{x}$) with the term $\phi(x, u)$, we obtain H_P as a function of (x, p, u):

(9)
$$H_P(x, p, u) = \langle p, \phi(x, u) \rangle - F(x, u).$$

The maximum principle is simply the assertion that the classical necessary conditions (7) continue to hold in this situation.

We have seen two approaches to giving necessary conditions for constrained problems: the multiplier rule (which is a Lagrangian approach) and the maximum principle (which employs the pseudo-Hamiltonian, and which is therefore essentially Lagrangian in nature also). Clearly, neither of these is a direct descendant of the Hamiltonian theory. What in fact is the true Hamiltonian H of a constrained problem?

To answer this question, we first note the extension of the Legendre transform defined by Fenchel. Applied to L, it expresses the Hamiltonian H as follows:

(10)
$$H(x, p) = \sup_v \{\langle p, v \rangle - L(x, v)\}.$$

1.4 The Theory of Optimal Control and the Calculus of Variations

The advantage of this version of the transform as opposed to the classical procedure (see Eq. (4)) is that H can be defined without referring to derivatives of L. Now we have seen in the preceding section that the control problem P_C is equivalent to minimizing $\int L(x, \dot{x})\, dt$, where L is defined by

(11) $$L(s, v) = \inf\{F(s, u) : u \in U,\, v = \phi(s, u)\}.$$

If this Lagrangian is substituted in Eq. (10), a simple calculation gives

(12) $$H(x, p) = \sup_{u \in U} \{\langle p, \phi(x, u)\rangle - F(x, u)\}.$$

This is the true Hamiltonian for P_C; observe that (in light of Eq. (9))

$$H(x, p) = \sup_{u \in U} H_P(x, p, u).$$

Why has the true Hamiltonian been neglected? Because it is nondifferentiable in general, in contrast to H_P, which is as smooth as the data of the problem.

This causes no difficulty, however, if the nonsmooth calculus described in Section 1.2 can be brought to bear. We shall prove that if x solves the optimal control problem, then an arc p exists such that one has

(13) $$\begin{bmatrix} -\dot{p}(t) \\ \dot{x}(t) \end{bmatrix} \in \partial H(x(t), p(t)),$$

which reduces to the classical Hamiltonian system (5) when H is smooth. We shall confirm in discussing existence and sufficiency that H is heir to the rich Hamiltonian theory of the calculus of variations.

The Hamiltonian H for the more general problem P_B is simply the function defined by Eq. (10); it shall figure prominently in Chapter 4. As noted earlier, the differential inclusion problem P_D is the case of P_B in which

$$L(x, v) = \begin{cases} 0 & \text{if } v \in F(x) \\ +\infty & \text{otherwise.} \end{cases}$$

When this Lagrangian is substituted in Eq. (10), we obtain

$$H(x, p) = \sup\{\langle p, v\rangle : v \in F(x)\},$$

a function which lies at the heart of things in Chapter 3. Thus, the three main problems P_C, P_D, and P_B all admit necessary conditions featuring (13); one need only interpret the Hamiltonian appropriately. (We are omitting in this section other components of the necessary conditions that deal with boundary terms.) The approach just outlined above is different than, for example, that of the maximum principle; the resulting necessary conditions are not equivalent.

The problem treated in Section 5.3 illustrates that (13) can in some cases provide more precise information than the maximum principle; the opposite is also possible.

Existence and Sufficient Conditions

The classical existence theorem of Tonelli in the calculus of variations postulates two main conditions under which the basic problem of minimizing (1) admits a solution. These are: first, the convexity of the function $v \to L(s, v)$ for each s, and second, certain growth conditions on L, a simple example being

(14) $\qquad L(s, v) \geq \alpha + \beta |v|^2 \quad \text{for all } s \text{ and } v \qquad (\beta > 0).$

When L is convex in v, L and its Hamiltonian H (see Eq. (10)) are mutually conjugate functions, each obtainable from the other by the Fenchel transform. It follows that growth conditions on L can be expressed in terms of H. For example, (14) is equivalent to

$$H(s, p) \leq \frac{p^2}{4\beta} - \alpha \quad \text{for all } s \text{ and } p.$$

In Section 4.1 we state an extension in Hamiltonian terms due to R. T. Rockafellar (1975) of the classical existence theory to the generalized problem of Bolza P_B. Since P_C is a special case of P_B, this result applies to the optimal control problem as well (see Section 5.4).

Let us now turn to the topic of sufficient conditions. There are four main techniques in the calculus of variations to prove that a given arc actually does solve the basic problem. We examine these briefly with an eye to how they extend to constrained problems.

1. *Elimination.* This refers to the situation in which we know that a solution exists, and in which we apply the necessary conditions to eliminate all other candidates. This method is common to all optimization problems. Note the absolute necessity of an appropriate existence theorem, and of knowing that the necessary conditions apply.

2. *Convexity.* Every optimization problem has its version of the general rule that for "convex problems," the necessary conditions are also sufficient. It is easy to prove that if $L(s, v)$ is convex (jointly in (s, v)), then any (admissible) solution x of the Euler–Lagrange equation (2) is a solution of the basic problem. (As simple and useful as it is, this fact is not known in the classical theory.) As for P_B, we may use the fact that convexity of L is equivalent to the Hamiltonian $H(s, p)$ being concave in s (it is always convex in p) in order to state sufficient conditions in Hamiltonian terms (see Section 4.3). This approach can be extended to derive sufficient conditions for the optimal control problem (see Section 5.4).

1.4 The Theory of Optimal Control and the Calculus of Variations

3. *Conjugacy and Fields.* There is a beautiful chapter in the calculus of variations that ties together certain families of solutions to the Euler–Lagrange equation (*fields*), the zeros of a certain second-order differential equation (*conjugate points*), and solutions of a certain partial differential equation (the *Hamilton–Jacobi equation*), and uses these ingredients to derive sufficient conditions. There has been only partial success in extending this theory to constrained problems (e.g., P_C; see the method of geodesic coverings (Young, 1969), which is based on fields). Recently Zeidan (1982) has used the Hamiltonian approach and the classical technique of canonical transformations to develop a conjugacy approach to sufficient conditions for P_B (see Section 4.3).

4. *Hamilton–Jacobi Methods.* The classical Hamilton–Jacobi equation is the following partial differential equation for a function $W(t, x)$:

$$W_t(t, x) + H(x, W_x(t, x)) = 0.$$

It has long been known in classical settings that this equation is closely linked to optimality. A discretization of the equation leads to the useful numerical technique known as dynamic programming. Our main interest in it stems from its role as a verification technique. Working in the context of the differential inclusion problem P_D, we shall use generalized gradients and the true Hamiltonian to define a generalized Hamilton–Jacobi equation (see Section 3.7). To a surprising extent, it turns out that the existence of a certain solution to this extended Hamilton–Jacobi equation is both a necessary and a sufficient condition for optimality.

Chapter Two

Generalized Gradients

What is now proved was once only imagined.

WILLIAM BLAKE, *The Marriage of Heaven and Hell*

I was losing my distrust of generalizations.

THORNTON WILDER, *Theophilus North*

In this chapter we gather a basic toolkit that will be used throughout the rest of the book. The theory and the calculus of generalized gradients are developed in detail, beginning with the case of a real-valued "locally Lipschitz" function defined on a Banach space. We also develop an associated geometric theory of normal and tangent cones, and explore the relationship between all of these concepts and their counterparts in smooth and in convex analysis. Later in the chapter special attention is paid to the case in which the underlying space is finite-dimensional, and also to the generalized gradients of certain important kinds of functionals. Examples are given, and extensions of the theory to non-Lipschitz and vector-valued functions are also carried out.

At the risk of losing some readers for one of our favorite chapters, we will concede that the reader who wishes only to have access to the statements of certain results of later chapters may find the introduction to generalized gradients given in Chapter 1 adequate for this purpose. Those who wish to follow all the details of the proofs (in this and later chapters) will require a background in real and functional analysis. In particular, certain standard constructs and results of the theory of Banach spaces are freely invoked. (Both the prerequisites and the difficulty of many proofs are greatly reduced if it is assumed that the Banach space is finite-dimensional.)

2.1 DEFINITION AND BASIC PROPERTIES

We shall be working in a Banach space X whose elements x we shall refer to as vectors or points, whose norm we shall denote $\|x\|$, and whose open unit ball is denoted B; the closed unit ball is denoted $\bar{B}$.

The Lipschitz Condition

Let Y be a subset of X. A function $f: Y \to R$ is said to satisfy a *Lipschitz condition* (on Y) provided that, for some nonnegative scalar K, one has

$$(1) \qquad |f(y) - f(y')| \leq K\|y - y'\|$$

for all points y, y' in Y; this is also referred to as a Lipschitz condition of *rank K*. We shall say that f is Lipschitz (of rank K) *near x* if, for some $\varepsilon > 0$, f satisfies a Lipschitz condition (of rank K) on the set $x + \varepsilon B$ (i.e., within an ε-neighborhood of x).

For functions of a real variable, a Lipschitz condition is a requirement that the graph of f not be "too steep." It is easy to see that a function having this property near a point need not be differentiable there, nor need it admit directional derivatives in the classical sense.

The Generalized Directional Derivative

Let f be Lipschitz near a given point x, and let v be any other vector in X. The *generalized directional derivative* of f at x in the direction v, denoted $f^\circ(x; v)$, is defined as follows:

$$f^\circ(x; v) = \limsup_{\substack{y \to x \\ t \downarrow 0}} \frac{f(y + tv) - f(y)}{t},$$

where of course y is a vector in X and t is a positive scalar. Note that this definition does not presuppose the existence of any limit (since it involves an upper limit only), that it involves only the behavior of f near x, and that it differs from traditional definitions of directional derivatives in that the base point (y) of the difference quotient varies. The utility of f° stems from the following basic properties.

2.1.1 Proposition

Let f be Lipschitz of rank K near x. Then

(a) *The function $v \to f^\circ(x; v)$ is finite, positively homogeneous, and subadditive on X, and satisfies*

$$|f^\circ(x; v)| \leq K\|v\|.$$

(b) $f°(x; v)$ *is upper semicontinuous as a function of* (x, v) *and, as a function of* v *alone, is Lipschitz of rank* K *on* X.

(c) $f°(x; -v) = (-f)°(x; v)$.

Proof. In view of the Lipschitz condition, the absolute value of the difference quotient in the definition of $f°(x; v)$ is bounded by $K\|v\|$ when y is sufficiently near x and t sufficiently near 0. It follows that $|f°(x; v)|$ admits the same upper bound. The fact that $f°(x; \lambda v) = \lambda f°(x; v)$ for any $\lambda > 0$ is immediate, so let us now turn to the subadditivity. With all the upper limits below understood to be taken as $y \to x$ and $t \downarrow 0$, we calculate:

$$f°(x; v + w) = \limsup \frac{f(y + tv + tw) - f(y)}{t}$$

$$\leq \limsup \frac{f(y + tv + tw) - f(y + tw)}{t} + \limsup \frac{f(y + tw) - f(y)}{t}$$

(since the upper limit of a sum is bounded above by the sum of the upper limits). The first upper limit in this last expression is $f°(x; v)$, since the term $y + tw$ represents in essence just a dummy variable converging to x. We conclude

$$f°(x; v + w) \leq f°(x; v) + f°(x; w),$$

which establishes (a).

Now let $\{x_i\}$ and $\{v_i\}$ be arbitrary sequences converging to x and v, respectively. For each i, by definition of upper limit, there exist y_i in X and $t_i > 0$ such that

$$\|y_i - x_i\| + t_i < \frac{1}{i},$$

$$f°(x_i; v_i) - \frac{1}{i} \leq \frac{f(y_i + t_i v_i) - f(y_i)}{t_i}$$

$$= \frac{f(y_i + t_i v) - f(y_i)}{t_i} + \frac{f(y_i + t_i v_i) - f(y_i + t_i v)}{t_i}.$$

Note that the last term is bounded in magnitude by $K\|v_i - v\|$ (in view of the Lipschitz condition). Upon taking upper limits (as $i \to \infty$), we derive

$$\limsup_{i \to \infty} f°(x_i; v_i) \leq f°(x; v),$$

which establishes the upper semicontinuity.

Finally, let any v and w in X be given. We have

$$f(y + tv) - f(y) \leq f(y + tw) - f(y) + K\|v - w\|t$$

for y near x, t near 0. Dividing by t and taking upper limits as $y \to x$, $t \downarrow 0$, gives

$$f°(x; v) \leq f°(x; w) + K\|v - w\|.$$

Since this also holds with v and w switched, (b) follows. To prove (c), we calculate:

$$f°(x; -v) := \limsup_{\substack{x' \to x \\ t \downarrow 0}} \frac{f(x' - tv) - f(x')}{t}$$

$$= \limsup_{\substack{u \to x \\ t \downarrow 0}} \frac{(-f)(u + tv) - (-f)(u)}{t}, \quad \text{where } u := x' - tv$$

$$= (-f)°(x; v), \quad \text{as stated.} \quad \square$$

The Generalized Gradient

The Hahn–Banach Theorem asserts that any positively homogeneous and subadditive functional on X majorizes some linear functional on X. Under the conditions of Proposition 2.1.1, therefore, there is at least one linear functional $\zeta: X \to R$ such that, for all v in X, one has $f°(x; v) \geq \zeta(v)$. It follows also that ζ is bounded, and hence belongs to the dual space X^* of continuous linear functionals on X, for which we adopt the convention of using $\langle \zeta, v \rangle$ or $\langle v, \zeta \rangle$ for $\zeta(v)$. We are led to the following definition: the *generalized gradient* of f at x, denoted $\partial f(x)$, is the subset of X^* given by

$$\{\zeta \in X^* : f°(x; v) \geq \langle \zeta, v \rangle \text{ for all } v \text{ in } X\}.$$

We denote by $\|\zeta\|_*$ the norm in X^*:

$$\|\zeta\|_* := \sup\{\langle \zeta, v \rangle : v \in X, \|v\| \leq 1\},$$

and B_* denotes the open unit ball in X^*. The following summarizes some basic properties of the generalized gradient.

2.1.2 Proposition

Let f be Lipschitz of rank K near x. Then

(a) $\partial f(x)$ is a nonempty, convex, weak*-compact subset of X^* and $\|\zeta\|_* \leq K$ for every ζ in $\partial f(x)$.

(b) For every v in X, one has

$$f°(x; v) = \max\{\langle \zeta, v \rangle : \zeta \in \partial f(x)\}.$$

Proof. Assertion (a) is immediate from our preceding remarks and Proposition 2.1.1. (The weak*-compactness follows from Alaoglu's Theorem.) Assertion (b) is simply a restatement of the fact that $\partial f(x)$ is by definition the weak*-closed convex set whose support function (see Section 1.2 and below) is $f^\circ(x; \cdot)$. To see this independently, suppose that for some v, $f^\circ(x; v)$ exceeded the given maximum (it can't be less, by definition of $\partial f(x)$). According to a common version of the Hahn–Banach Theorem, there is a linear functional ζ majorized by $f^\circ(x; \cdot)$ and agreeing with it at v. It follows that ζ belongs to $\partial f(x)$, whence $f^\circ(x; v) > \langle \zeta, v \rangle = f^\circ(x; v)$, a contradiction which establishes (b). □

2.1.3 Example — The Absolute-Value Function

As in the case of the classical calculus, one of our goals is to have to resort rarely to the definition in order to calculate generalized gradients in practice. For illustrative purposes, however, let us nonetheless so calculate the generalized gradient of the absolute-value function on the reals. In this case, $X = R$ and $f(x) = |x|$ (f is Lipschitz by the triangle inequality). If x is strictly positive, we calculate

$$f^\circ(x; v) = \lim_{\substack{y \to x \\ t \downarrow 0}} \frac{y + tv - y}{t} = v,$$

so that $\partial f(x)$, the set of numbers ζ satisfying $v \geq \zeta v$ for all v, reduces to the singleton $\{1\}$. Similarly, $\partial f(x) = \{-1\}$ if $x < 0$. The remaining case is $x = 0$. We find

$$f^\circ(0; v) = \begin{cases} v & \text{if } v \geq 0 \\ -v & \text{if } v < 0; \end{cases}$$

that is, $f^\circ(0; v) = |v|$. Thus $\partial f(0)$ consists of those ζ satisfying $|v| \geq \zeta v$ for all v; that is, $\partial f(0) = [-1, 1]$.

Support Functions

As Proposition 2.1.2 makes clear, it is equivalent to know the set $\partial f(x)$ or the function $f^\circ(x; \cdot)$; each is obtainable from the other. This is an instance of a general fact: closed convex sets are characterized by their support functions. Recall that the *support function* of a nonempty subset C of X is the function $\sigma_C : X^* \to R \cup \{+\infty\}$ defined by

$$\sigma_C(\zeta) := \sup\{\langle \zeta, x \rangle : x \in C\}.$$

If Σ is a subset of X^*, its support function is defined on X^{**}. If we view X as a subset of X^{**}, then for $x \in X$, one has

$$\sigma_\Sigma(x) = \sup\{\langle \zeta, x \rangle : \zeta \in \Sigma\}.$$

The following facts are known (see Hörmander, 1954).

2.1.4 Proposition

Let C, D be nonempty closed convex subsets of X, and let Σ, Δ be nonempty weak*-closed convex subsets of X^*. Then

(a) $C \subset D$ iff $\sigma_C(\zeta) \leq \sigma_D(\zeta)$ for all $\zeta \in X^*$.
(b) $\Sigma \subset \Delta$ iff $\sigma_\Sigma(x) \leq \sigma_\Delta(x)$ for all $x \in X$.
(c) Σ is weak*-compact iff $\sigma_\Sigma(\cdot)$ is finite-valued on X.
(d) A given function $\sigma: X \to R \cup \{+\infty\}$ is positively homogeneous, subadditive, lower semicontinuous (strong or weak) and not identically $+\infty$ iff there is a nonempty weak*-closed convex subset Σ of X^* such that $\sigma = \sigma_\Sigma$. Any such Σ is unique.

A *multifunction* $\Gamma: X \to Y$ is a mapping from X to the subsets of Y. Its *graph* is the set

$$\operatorname{Gr} \Gamma := \{(x, y) : x \in X, y \in \Gamma(x)\}.$$

Γ is *closed* if $\operatorname{Gr} \Gamma$ is closed in $X \times Y$ (relative to a given topology). When X and Y are Banach spaces, we define *upper semicontinuity* of Γ at x to be the following property: for all $\varepsilon > 0$, there exists $\delta > 0$ such that

$$\Gamma(x') \subset \Gamma(x) + \varepsilon B_Y \quad \text{for all } x' \in x + \delta B_X.$$

We continue to assume that f is Lipschitz near x. The first assertion below reiterates that $f^\circ(x; \cdot)$ is the support function of $\partial f(x)$.

2.1.5 Proposition

(a) $\zeta \in \partial f(x)$ iff $f^\circ(x; v) \geq \langle \zeta, v \rangle$ for all v in X.
(b) Let x_i and ζ_i be sequences in X and X^* such that $\zeta_i \in \partial f(x_i)$. Suppose that x_i converges to x, and that ζ is a cluster point of ζ_i in the weak* topology. Then one has $\zeta \in \partial f(x)$. (That is, the multifunction ∂f is weak*-closed.)
(c)
$$\partial f(x) = \bigcap_{\delta > 0} \bigcup_{y \in x + \delta B} \partial f(y).$$

(d) If X is finite-dimensional, then ∂f is upper semicontinuous at x.

Proof. Assertion (a) is evident from Propositions 2.1.2 and 2.1.4. In order to prove (b), let any v in X be given. There is a subsequence of the numerical sequence $\langle \zeta_i, v \rangle$ which converges to $\langle \zeta, v \rangle$ (we do not relabel). One has $f^\circ(x_i; v) \geq \langle \zeta_i, v \rangle$ by property (a), which implies $f^\circ(x; v) \geq \langle \zeta, v \rangle$ by the upper semicontinuity of f° (see Proposition 2.1.1(b)). Since v is arbitrary, ζ belongs to $\partial f(x)$ by assertion (a).

Part (c) of the proposition is an immediate consequence of (b), so we turn now to (d). If ∂f fails to be upper semicontinuous at x, then we can construct a sequence x_i converging to x and a sequence ζ_i converging to ζ such that ζ_i belongs to $\partial f(x_i)$ for each i, yet ζ does not belong to $\partial f(x)$. This contradicts assertion (b). □

2.2 RELATION TO DERIVATIVES AND SUBDERIVATIVES

We pause to recall some terminology before pursuing the study of generalized gradients. The main results of this section will be that ∂f reduces to the derivative if f is C^1, and to the subdifferential of convex analysis if f is convex.

Classical Derivatives

Let F map X to another Banach space Y. The usual (one-sided) directional derivative of F at x in the direction v is

$$F'(x; v) := \lim_{t \downarrow 0} \frac{F(x + tv) - F(x)}{t}$$

when this limit exists. F is said to admit a *Gâteaux derivative* at x, an element in the space $\mathcal{L}(X, Y)$ of continuous linear functionals from X to Y denoted $DF(x)$, provided that for every v in X, $F'(x; v)$ exists and equals $\langle DF(x), v \rangle$. Let us note that this is equivalent to saying that the difference quotient converges for each v, that one has

$$\lim_{t \downarrow 0} \frac{F(x + tv) - F(x)}{t} = \langle DF(x), v \rangle,$$

and that the convergence is uniform with respect to v in finite sets (the last is automatically true). If the word "finite" in the preceding sentence is replaced by "compact," the derivative is known as *Hadamard*; for "bounded" we obtain the *Fréchet* derivative. In general, these are progressively more demanding requirements. When $X = R^n$, Hadamard and Fréchet differentiability are equivalent; when F is Lipschitz near x (i.e., when for some constant K one has $\|F(x') - F(x'')\|_Y \leq K\|x' - x''\|_X$ for all x', x'' near x) then Hadamard and Gâteaux differentiabilities coincide.

Strict Differentiability

It turns out that the differential concept most naturally linked to the theory of this chapter is that of *strict differentiability* (Bourbaki: "strictement dérivable"). We shall say that F admits a strict derivative at x, an element of $\mathcal{L}(X, Y)$ denoted $D_s F(x)$, provided that for each v, the following holds:

$$\lim_{\substack{x' \to x \\ t \downarrow 0}} \frac{F(x' + tv) - F(x')}{t} = \langle D_s F(x), v \rangle,$$

and provided the convergence is uniform for v in compact sets. (This last condition is automatic if F is Lipschitz near x). Note that ours is a "Hadamard-type strict derivative."

2.2.1 Proposition

Let F map a neighborhood of x to Y, and let ζ be an element of $\mathcal{L}(X, Y)$. The following are equivalent:

(a) *F is strictly differentiable at x and $D_s F(x) = \zeta$.*
(b) *F is Lipschitz near x, and for each v in X one has*

$$\lim_{\substack{x' \to x \\ t \downarrow 0}} \frac{F(x' + tv) - F(x')}{t} = \langle \zeta, v \rangle.$$

Proof. Assume (a). The equality in (b) holds by assumption, so to prove (b) we need only show that F is Lipschitz near x. If this is not the case, there exist sequences $\{x_i\}$ and $\{x'_i\}$ converging to x such that x_i, x'_i lie in $x + (1/i)B$ and

$$\|F(x'_i) - F(x_i)\|_Y > i\|x'_i - x_i\|_X.$$

Let us define $t_i > 0$ and v_i via $x'_i = x_i + t_i v_i$ and $\|v_i\| = i^{-1/2}$. It follows that $t_i \to 0$.

Let V consist of the points in the sequence $\{v_i\}$ together with 0. Note that V is compact, so that by definition of $D_s F(x)$ for any $\varepsilon > 0$ there exists n_ε such that, for all $i \geq n_\varepsilon$, for all v in V, one has

$$\left\| \frac{F(x_i + t_i v) - F(x_i)}{t_i} - \langle DF(x), v \rangle \right\|_Y < \varepsilon.$$

But this is impossible since when $v = v_i$, the term $[F(x_i + t_i v) - F(x_i)]/t_i$ has norm exceeding $i^{1/2}$ by construction. Thus (b) holds.

We now posit (b). Let V be any compact subset of X and ε any positive number. In view of (b), there exists for each v in V a number $\delta(v) > 0$ such that

(1) $$\left\| \frac{F(x' + tv) - F(x')}{t} - \langle \zeta, v \rangle \right\|_Y < \varepsilon$$

for all $x' \in x + \delta B$ and $t \in (0, \delta)$. Since the norm of

$$\frac{F(x' + tv') - F(x')}{t} - \frac{F(x' + tv) - F(x')}{t}$$

is bounded above by $K\|v - v'\|$ (where K is a Lipschitz constant for F and where x', t are sufficiently near x and 0), we deduce from (1) that for a suitable redefinition of $\delta(v)$, one has

(2) $$\left\| \frac{F(x' + tv') - F(x')}{t} - \langle \zeta, v' \rangle \right\|_Y < 2\varepsilon$$

for all x' in $x + \delta B$, v' in $v + \delta B$, and t in $(0, \delta)$. A finite number of the open sets $\{v + \delta(v)B : v \in V\}$ will cover V, say, those that correspond to $v_1, v_2, \ldots, v_n$. If we set $\delta' = \min_{1 \leq i \leq n} \delta(v_i)$, it follows then that (1) holds (with 2ε for ε) for any v in V, for all x' in $x + \delta' B$ and t in $(0, \delta')$. Thus ζ is the strict derivative of F at x, and the proof is complete. □

We shall call F *continuously (Gâteaux) differentiable* at x provided that on a neighborhood of x the Gâteaux derivative DF exists and is continuous as a mapping from X to $\mathcal{L}(X, Y)$ (with its operator norm topology).

Corollary

If F is continuously differentiable at x, then F is strictly differentiable at x and hence Lipschitz near x.

Proof. To prove that $DF(x)$ is actually the strict derivative, it suffices by the proposition to prove the equality in (b) for each v, for it follows easily from the vector mean value theorem (see McLeod, 1965) that F is Lipschitz near x. In turn, it suffices to show that, for any trio of sequences $\{x_i\}, \{v_i\}, \{t_i\}$ (with $t_i > 0$) converging to $x, v, 0$, respectively, one has

$$\lim_{i \to \infty} \sup_{\|\theta\|_{Y^*} \leq 1} \theta \left[\frac{F(x_i + t_i v_i) - F(x_i)}{t_i} - \langle DF(x), v_i \rangle \right] = 0.$$

By the mean-value theorem, there is x_i^* between x_i and $x_i + t_i v_i$ such that the last expression equals

$$\langle \theta DF(x_i^*), v_i \rangle - \langle \theta DF(x), v_i \rangle = \langle \theta [DF(x_i^*) - DF(x)], v_i \rangle.$$

This goes to 0 (uniformly in θ) as $i \to \infty$ by the continuity of DF. □

We are now ready to explore the relationship between the various derivatives defined above and the generalized gradient.

2.2.2 Proposition

Let f be Lipschitz near x and admit a Gâteaux (or Hadamard, or strict, or Fréchet) derivative $Df(x)$. Then $Df(x) \in \partial f(x)$.

Proof. By definition, $f'(x; v)$ exists for each v and equals $\langle Df(x), v \rangle$. Clearly one has $f' \leq f°$ from the definition of the latter, so one has $f°(x; v) \geq$

$\langle Df(x), v \rangle$ for all v in X. The required conclusion now follows from Proposition 2.1.5(a). □

2.2.3 Example

That $\partial f(x)$ can contain points other than $Df(x)$ is illustrated by the familiar example on $R: f(x) := x^2 \sin(1/x)$. This function is Lipschitz near 0, and it is easy to show that $f^\circ(0; v) = |v|$. It follows that $\partial f(0) = [-1, 1]$, a set which contains the (nonstrict) derivative $Df(0) = 0$.

2.2.4 Proposition

If f is strictly differentiable at x, then f is Lipschitz near x and $\partial f(x) = \{D_s f(x)\}$. Conversely, if f is Lipschitz near x and $\partial f(x)$ reduces to a singleton $\{\zeta\}$, then f is strictly differentiable at x and $D_s f(x) = \zeta$.

Proof. Suppose first that $D_s f(x)$ exists (so that f is Lipschitz near x by Proposition 2.2.1). Then, by definition of f°, one has $f^\circ(x; v) = \langle D_s f(x), v \rangle$ for all v, and it follows from Proposition 2.1.5(a) that $\partial f(x)$ reduces to $\{D_s f(x)\}$. To prove the converse, it suffices to show that the condition of Proposition 2.2.1(b) holds for each v in X. We begin by showing that $f^\circ(x; v) = \langle \zeta, v \rangle$ for each v. (Note that $f^\circ(x; v) \geq \langle \zeta, v \rangle$ by Proposition 2.1.2(b).) By the Hahn–Banach Theorem there exists $\zeta' \in X^*$ majorized by $f^\circ(x; \cdot)$ and agreeing with $f^\circ(x; \cdot)$ at v. It follows that $\zeta' \in \partial f(x)$, and we have $f^\circ(x; v) = \langle \zeta', v \rangle \geq \langle \zeta, v \rangle$. If $\langle \zeta, v \rangle$ were less than $f^\circ(x; v)$, then ζ, ζ' would be distinct elements of $\partial f(x)$, contrary to hypothesis. Thus $f^\circ(x; v) = \langle \zeta, v \rangle$ for all v.

We now calculate:

$$\liminf_{\substack{x' \to x \\ t \downarrow 0}} \frac{f(x' + tv) - f(x')}{t} = -\limsup_{\substack{x' \to x \\ t \downarrow 0}} \frac{f(x') - f(x' + tv)}{t}$$

$$= -\limsup_{\substack{x' \to x \\ t \downarrow 0}} \frac{f(x' + tv - tv) - f(x' + tv)}{t}$$

$$= -f^\circ(x; -v) = -\langle \zeta, -v \rangle = \langle \zeta, v \rangle$$

$$= f^\circ(x; v) = \limsup_{\substack{x' \to x \\ t \downarrow 0}} \frac{f(x' + tv) - f(x')}{t}.$$

This establishes the limit condition of Proposition 2.2.1(b) and completes the proof. □

Corollary

If f is Lipschitz near x and X is finite-dimensional, then $\partial f(x')$ reduces to a singleton for every x' in $x + \varepsilon B$ iff f is continuously differentiable on $x + \varepsilon B$.

Proof. Invoke the corollary to Proposition 2.2.1 together with Proposition 2.1.5(d); note that for a point-valued map, continuity and upper semicontinuity coincide. □

2.2.5 Example (Indefinite Integrals)

Let $\phi: [0, 1] \to R$ belong to $L^\infty[0, 1]$, and define a (Lipschitz) function $f: [0, 1] \to R$ via $f(x) = \int_0^x \phi(t)\, dt$. Let us calculate $\partial f(x)$. We know that f is differentiable for almost all x, with $f'(x) = \phi(x)$; for any such x, we have $\phi(x) \in \partial f(x)$ by Proposition 2.2.2. It follows from this and from upper semicontinuity (see Proposition 2.1.5) that all essential cluster points of ϕ at x (i.e., those that persist upon the removal of any set of measure zero) belong to $\partial f(x)$.

Let $\phi^+(x)$ and $\phi^-(x)$ denote the essential supremum and essential infimum of ϕ at x. The remarks above and the fact that $\partial f(x)$ is convex imply that $\partial f(x)$ contains the interval $[\phi^-(x), \phi^+(x)]$. From the equality

$$f(y + t) - f(y) = \int_y^{y+t} \phi(s)\, ds$$

one easily deduces that $f^\circ(x; 1)$ is bounded above by $\phi^+(x)$. Any ζ in $\partial f(x)$ satisfies $f^\circ(x; 1) \geq 1\zeta$ (by Proposition 2.1.5(a)), so we derive $\zeta \leq \phi^+(x)$. Similarly, $\zeta \geq \phi^-(x)$. We arrive at the conclusion

$$\partial f(x) = [\phi^-(x), \phi^+(x)].$$

Convex Functions

Let U be an open convex subset of X. Recall that a function $f: U \to R$ is said to be *convex* provided that, for all u, u' in U and λ in $[0, 1]$, one has

$$f(\lambda u + (1 - \lambda) u') \leq \lambda f(u) + (1 - \lambda) f(u').$$

As we now see, convex functions are Lipschitz except in pathological cases.

2.2.6 Proposition

Let f above be bounded above on a neighborhood of some point of U. Then, for any x in U, f is Lipschitz near x.

Proof (Roberts and Varberg, 1974). We begin by proving that f is bounded on a neighborhood of x. Without loss of generality, let us suppose that f is bounded above by M on the set $\varepsilon B \subset U$. Choose $\rho > 1$ so that $y = \rho x$ is in U. If $\lambda = 1/\rho$, then the set

$$V = \{v : v = (1 - \lambda) x' + \lambda y, x' \in \varepsilon B\}$$

2.2 Relation to Derivatives and Subderivatives

is a neighborhood of $x = \lambda y$ with radius $(1 - \lambda)\varepsilon$. For all v in V, one has by convexity

$$f(v) \leq (1 - \lambda)f(x') + \lambda f(y) \leq M + \lambda f(y),$$

so that f is bounded above on a neighborhood of x. If z is any point in $x + (1 - \lambda)\varepsilon B$, there is another such point z' such that $x = (z + z')/2$, whence

$$f(x) \leq \tfrac{1}{2} f(z) + \tfrac{1}{2} f(z').$$

It follows that

$$f(z) \geq 2f(x) - f(z') \geq 2f(x) - M - \lambda f(y),$$

so that now f is also seen to be bounded below near x; we have established that f is bounded near x.

Let N be a bound on $|f|$ on the set $x + 2\delta B$, where $\delta > 0$. For distinct x_1, x_2 in $x + \delta B$, set $x_3 = x_2 + (\delta/\alpha)(x_2 - x_1)$ where $\alpha = \|x_2 - x_1\|$, and note that x_3 is in $x + 2\delta B$. Solving for x_2 gives

$$x_2 = \frac{\delta}{\alpha + \delta} x_1 + \frac{\alpha}{\alpha + \delta} x_3,$$

and so by convexity

$$f(x_2) \leq \frac{\delta}{\alpha + \delta} f(x_1) + \frac{\alpha}{\alpha + \delta} f(x_3).$$

Then

$$f(x_2) - f(x_1) \leq \frac{\alpha}{\alpha + \delta} [f(x_3) - f(x_1)] \leq \frac{\alpha}{\delta} |f(x_3) - f(x_1)|,$$

which combined with $|f| \leq N$ and $\alpha = \|x_2 - x_1\|$ yields

$$f(x_2) - f(x_1) \leq \frac{2N}{\delta} \|x_2 - x_1\|.$$

Since the roles of x_1 and x_2 may be interchanged, we conclude that f is Lipschitz near x. □

The proof showed:

Corollary

Let f be convex with $|f| \leq N$ on an open convex set U which contains a δ-neighborhood of a subset V. Then f satisfies a Lipschitz condition of rank $2N/\delta$ on V.

Recall that the *subdifferential* of the (convex) function f at x is defined to be the set of those ζ in X^* satisfying

$$f(x') - f(x) \geq \langle \zeta, x' - x \rangle \quad \text{for all } x' \text{ in } U.$$

Since ∂f is the established notation for the subdifferential, the following asserts that (fortunately) $\partial f = \partial f$.

2.2.7 Proposition

When f is convex on U and Lipschitz near x, then $\partial f(x)$ coincides with the subdifferential at x in the sense of convex analysis, and $f^\circ(x; v)$ coincides with the directional derivative $f'(x; v)$ for each v.

Proof. It is known from convex analysis that $f'(x; v)$ exists for each v and that $f'(x; \cdot)$ is the support function of the subdifferential at x. It suffices therefore to prove that for any v, $f^\circ(x; v) = f'(x; v)$. The former can be written as

$$\lim_{\varepsilon \downarrow 0} \sup_{\|x' - x\| < \varepsilon \delta} \sup_{0 < t < \varepsilon} \frac{f(x' + tv) - f(x')}{t},$$

where δ is any fixed positive number. It follows readily from the definition of convex function that the function

$$t \to \frac{f(x' + tv) - f(x')}{t}$$

is nondecreasing, whence

$$f^\circ(x; v) = \lim_{\varepsilon \downarrow 0} \sup_{\|x' - x\| < \varepsilon \delta} \frac{f(x' + \varepsilon v) - f(x')}{\varepsilon}.$$

Now by the Lipschitz condition, for any x' in $x + \varepsilon \delta B$, one has

$$\left| \frac{f(x' + \varepsilon v) - f(x')}{\varepsilon} - \frac{f(x + \varepsilon v) - f(x)}{\varepsilon} \right| \leq 2\delta K,$$

so that

$$f^\circ(x; v) \leq \lim_{\varepsilon \downarrow 0} \frac{f(x + \varepsilon v) - f(x)}{\varepsilon} + 2\delta K = f'(x; v) + 2\delta K.$$

Since δ is arbitrary, we deduce $f^\circ(x; v) \leq f'(x; v)$. Of course equality follows, and the proof is complete. □

2.2.8 Example

Let us determine the generalized gradient of the function $f: R^n \to R$ defined by

$$f(x_1, x_2, \ldots, x_n) = \max\{x_i : i = 1, 2, \ldots, n\}.$$

Observe first that f, as a maximum of linear functions, is convex; let us calculate $f'(x; v)$. Let $I(x)$ denote the set of indices i for which $f_i(x) = f(x)$ (i.e., the indices at which the maximum defining f is attained). We find

$$f'(x; v) := \lim_{t \downarrow 0} \max_{i} \frac{\{x_i + tv_i\} - f(x)}{t}$$

$$= \lim_{t \downarrow 0} \max_{i \in I(x)} \frac{\{x_i + tv_i\} - f(x)}{t}$$

(since for t small enough, any index i not in $I(x)$ can be ignored in the maximum)

$$= \lim_{t \downarrow 0} \max_{i \in I(x)} \frac{\{x_i + tv_i - x_i\}}{t}$$

$$= \max_{i \in I(x)} v_i.$$

Since f° and f' coincide (Proposition 2.2.7), $\partial f(x)$ consists of those vectors ζ in R^n satisfying

$$\max_{i \in I(x)} v_i \geq \zeta \cdot v \quad \text{for all } v \text{ in } R^n.$$

It follows that $\partial f(x)$ consists of all vectors $(\zeta_1, \zeta_2, \ldots, \zeta_n)$ such that $\zeta_i \geq 0$, $\Sigma \zeta_i = 1$, $\zeta_i = 0$ if $i \notin I(x)$.

We conclude the discussion by giving a criterion for convexity in terms of the generalized gradient. The proof, which can be based upon the mean value theorem, Theorem 2.3.7, is left as an exercise.

2.2.9 Proposition

Let f be Lipschitz near each point of an open convex subset U of X. Then f is convex on U iff the multifunction ∂f is monotone on U; that is, iff

$$\langle x - x', \zeta - \zeta' \rangle \geq 0 \quad \text{for all } x, x' \in U, \zeta \in \partial f(x), \zeta' \in \partial f(x').$$

2.3 BASIC CALCULUS

We now proceed to derive an assortment of formulas that facilitate greatly the calculation of ∂f when (as is often the case) f is "built up" from simple functionals through linear combination, maximization, composition, and so on. We assume that a function f is given which is Lipschitz near a given point x.

2.3.1 Proposition (Scalar Multiples)

For any scalar s, one has

$$\partial(sf)(x) = s\partial f(x).$$

Proof. Note that sf is also Lipschitz near x. When s is nonnegative, $(sf)^\circ = sf^\circ$, and it follows easily that $\partial(sf)(x) = s\partial f(x)$. It suffices now to prove the formula for $s = -1$. An element ζ of X^* belongs to $\partial(-f)(x)$ iff $(-f)^\circ(x; v) \geq \langle \zeta, v \rangle$ for all v. By Proposition 2.1.1(c), this is equivalent to: $f^\circ(x; -v) \geq \langle \zeta, v \rangle$ for all v, which is equivalent to $-\zeta$ belonging to $\partial f(x)$ (by Proposition 2.1.5(a)). Thus $\zeta \in \partial(-f)(x)$ iff $\zeta \in -\partial f(x)$, as claimed. □

2.3.2 Proposition (Local Extrema)

If f attains a local minimum or maximum at x, then $0 \in \partial f(x)$.

Proof. In view of the formula $\partial(-f) = -\partial f$, it suffices to prove the proposition when x is a local minimum. But in this case it is evident that for any v in X, one has $f^\circ(x; v) \geq 0$. Thus $\zeta = 0$ belongs to $\partial f(x)$ (by Proposition 2.1.5(a)). □

If f_i ($i = 1, 2, \ldots, n$) is a finite family of functions each of which is Lipschitz near x, it follows easily that their sum $f = \Sigma f_i$ is also Lipschitz near x.

2.3.3 Proposition (Finite Sums)

$$\partial\left(\sum f_i\right)(x) \subset \sum \partial f_i(x).$$

Proof. Note that the right-hand side denotes the (weak* compact) set of all points ζ obtainable as a sum $\Sigma \zeta_i$ (sum 1 to n), where each ζ_i belongs to $\partial f_i(x)$. It suffices to prove the formula for $n = 2$; the general case follows by induction.

The support functions (on X) of the left- and right-hand sides (evaluated at v) are, respectively, $(f_1 + f_2)^\circ(x; v)$ and $f_1^\circ(x; v) + f_2^\circ(x; v)$ (by Proposition 2.1.2(b)). In view of Proposition 2.1.4(b), it suffices therefore (in fact, it is equivalent) to prove the general inequality

$$(f_1 + f_2)^\circ(x; v) \leq f_1^\circ(x; v) + f_2^\circ(x; v).$$

2.3 Basic Calculus

This follows readily from the definition; we leave the verification to the reader. □

Corollary 1

Equality holds in Proposition 2.3.3 if all but at most one of the functions f_i are strictly differentiable at x.

Proof. By adding all the strictly differentiable functions together to get a single strictly differentiable function, we can reduce the proof to the case where we have two functions f_1 and f_2, with f_1 strictly differentiable. In this case, as is easily seen, we actually have

$$(f_1 + f_2)^\circ(x; v) = f_1'(x; v) + f_2^\circ(x; v) = f_1^\circ(x; v) + f_2^\circ(x; v),$$

so that the two sets in the statement of the proposition have support functions that coincide, and hence are equal. □

Corollary 2

For any scalars s_i, one has

$$\partial\left(\sum_{i=1}^n s_i f_i\right)(x) \subset \sum_{i=1}^n s_i \partial f_i(x),$$

and equality holds if all but at most one of the f_i are strictly differentiable at x.

Proof. Invoke Proposition 2.3.1. □

Regularity

It is often the case that calculus formulas for generalized gradients involve inclusions, such as in Proposition 2.3.3. The addition of further hypotheses can serve to sharpen such rules by turning the inclusions to equalities. For instance, equality certainly holds in Proposition 2.3.3 if all the functions in question are continuously differentiable, since then the generalized gradient is essentially the derivative, a linear operator. However, one would wish for a less extreme condition, one that would cover the convex (nondifferentiable) case, for example (in which equality does hold). A class of functions that proves useful in this connection is the following.

2.3.4 Definition

f is said to be *regular* at x provided

(i) For all v, the usual one-sided directional derivative $f'(x; v)$ exists.
(ii) For all v, $f'(x; v) = f^\circ(x; v)$.

To see how regularity contributes, let us note the following addendum to Proposition 2.3.3.

Corollary 3

If each f_i is regular at x, equality holds in Proposition 2.3.3. Equality then holds in Corollary 2 as well, if in addition each s_i is nonnegative.

Proof. Since, as we shall see below, a positive linear combination of regular functions is regular, it suffices again (in both cases) to consider the case $n = 2$. The proof made clear that equality would hold provided the two sets had the same support functions. The support function of the one on the left is
$(f_1 + f_2)^\circ(x; \cdot) = (f_1 + f_2)'(x; \cdot) = f_1'(x; \cdot) + f_2'(x; \cdot) = f_1^\circ(x; \cdot) + f_2^\circ(x; \cdot)$,
which is the support function of the one on the right. □

2.3.5 Remark

In view of the formula $\partial(-f) = -\partial f$, equality also holds in Proposition 2.3.3 if, for each i, $-f_i$ is regular at x. (In the future we shall often leave implicit such dual results.) To see that the inclusion can be strict in general, it suffices to take any function f such that $\partial f(x)$ is not a singleton, and set $f_1 = f$, $f_2 = -f$.

Here are some first observations about regular functions.

2.3.6 Proposition

Let f be Lipschitz near x.
- (a) *If f is strictly differentiable at x, then f is regular at x.*
- (b) *If f is convex, then f is regular at x.*
- (c) *A finite linear combination (by nonnegative scalars) of functions regular at x is regular at x.*
- (d) *If f admits a Gâteaux derivative $Df(x)$ and is regular at x, then $\partial f(x) = \{Df(x)\}$.*

Proof. (a) follows from Propositions 2.1.5(a) and 2.2.4, since $f^\circ(x; v)$ is the support function of $\partial f(x) = \{D_s f(x)\}$, and one has the equality $D_s f(x)(v) = f'(x; v)$. To prove (b) we invoke convex analysis, which asserts that $f'(x; \cdot)$ exists, and is the support function of the subdifferential; in view of Proposition 2.2.7, we derive $f'(x; \cdot) = f^\circ(x; \cdot)$ as required. As for (c), once again only the case of two functions f_1 and f_2 need be treated, and since sf is clearly regular when f is regular and $s \geq 0$, it suffices to prove $(f_1 + f_2)' = (f_1 + f_2)^\circ$ when f_1 and f_2 are regular at x (the existence of $(f_1 + f_2)'$ being evident). We have $(f_1 + f_2)' = f_1' + f_2' = f_1^\circ + f_2^\circ \geq (f_1 + f_2)^\circ$ (as in the proof of Proposition 2.3.3). Since we always have the opposite inequality $(f_1 + f_2)^\circ \geq (f_1 + f_2)'$, we arrive at (c). The assertion (d) is evident from Proposition 2.1.5(a). □

We shall extend (c) in Section 2.7 to sums which are not necessarily finite, as part of a more general study of integral functionals. Given x and y in X, the notation $[x, y]$ signifies the closed line segment consisting of all points $tx + (1 - t)y$ for $t \in [0, 1]$; (x, y) signifies the open line segment.

Mean-Value Theorem

2.3.7 Theorem (Lebourg)

Let x and y be points in X, and suppose that f is Lipschitz on an open set containing the line segment $[x, y]$. Then there exists a point u in (x, y) such that

(1) $$f(y) - f(x) \in \langle \partial f(u), y - x \rangle.$$

We shall need the following special chain rule for the proof. We denote by x_t the point $x + t(y - x)$.

Lemma. The function $g: [0, 1] \to R$ defined by $g(t) = f(x_t)$ is Lipschitz on $(0, 1)$, and we have

(2) $$\partial g(t) \subset \langle \partial f(x_t), y - x \rangle.$$

Proof. The fact that g is Lipschitz is plain. The two closed convex sets appearing in Eq. (2) are in fact intervals in R, so it suffices to prove that for $v = \pm 1$, we have

$$\max\{\partial g(t) v\} \leq \max\{\langle \partial f(x_t), y - x \rangle v\}.$$

Now the left-hand side is just $g°(t; v)$; that is,

$$\limsup_{\substack{s \to t \\ \lambda \downarrow 0}} \frac{g(s + \lambda v) - g(s)}{\lambda}$$

$$= \limsup_{\substack{s \to t \\ \lambda \downarrow 0}} \frac{f(x + [s + \lambda v](y - x)) - f(x + s(y - x))}{\lambda}$$

$$\leq \limsup_{\substack{y' \to x_t \\ \lambda \downarrow 0}} \frac{f(y' + \lambda v(y - x)) - f(y')}{\lambda}$$

$$= f°(x_t; v(y - x))$$

$$= \max\langle \partial f(x_t), v(y - x) \rangle.$$

Now to the proof of the theorem. Consider the function θ on $[0, 1]$ defined by

$$\theta(t) = f(x_t) + t[f(x) - f(y)].$$

Note $\theta(0) = \theta(1) = f(x)$, so that there is a point t in $(0, 1)$ at which θ attains a local minimum or maximum (by continuity). By Proposition 2.3.2, we have

$0 \in \partial\theta(t)$. We may calculate the latter by appealing to Propositions 2.3.1, 2.3.3, and the above lemma. We deduce

$$0 \in [f(x) - f(y)] + \langle \partial f(x_t), y - x \rangle,$$

which is the assertion of the theorem (take $u = x_t$). □

2.3.8 Example

We shall find the mean-value theorem very useful in what is to come. As an immediate illustration of its use, consider the function

$$f(x) = \int_0^x \phi(t)\, dt$$

of Example 2.2.5. If we combine the mean-value theorem with the expression for ∂f, we arrive at the following. For any $\varepsilon > 0$, there are points x and y in $(0, 1)$ such that $|x - y| < \varepsilon$ and such that

$$\phi(x) - \varepsilon < \int_0^1 \phi(t)\, dt < \phi(y) + \varepsilon.$$

Chain Rules

We now turn to a very useful chain rule that pertains to the following situation: $f = g \circ h$, where $h: X \to R^n$ and $g: R^n \to R$ are given functions. The component functions of h will be denoted h_i ($i = 1, 2, \ldots, n$). We assume that each h_i is Lipschitz near x and that g is Lipschitz near $h(x)$; this implies that f is Lipschitz near x (as usual). If we adopt the convention of identifying $(R^n)^*$ with R^n, then an element α of ∂g can be considered an n-dimensional vector: $\alpha = (\alpha_1, \alpha_2, \ldots, \alpha_n)$. (All sums below are from 1 to n.)

2.3.9 Theorem (Chain Rule I)
One has

$$\partial f(x) \subset \overline{\operatorname{co}}\left\{\sum \alpha_i \zeta_i : \zeta_i \in \partial h_i(x), \alpha \in \partial g(h(x))\right\}$$

(where $\overline{\operatorname{co}}$ denotes weak*-closed convex hull), and equality holds under any one of the following additional hypotheses:

(i) g is regular at $h(x)$, each h_i is regular at x, and every element α of $\partial g(h(x))$ has nonnegative components. (In this case it follows that f is regular at x.)
(ii) g is strictly differentiable at $h(x)$ and $n = 1$. (In this case the $\overline{\operatorname{co}}$ is superfluous.)
(iii) g is regular at $h(x)$ and h is strictly differentiable at x. (In this case it follows that f is regular at x, and the $\overline{\operatorname{co}}$ is superfluous.)

2.3 Basic Calculus

Proof. The set S whose convex hull is taken in the formula is weak*-compact, and therefore its convex hull has compact closure (and is in fact closed if X is finite-dimensional, so that $\overline{\text{co}}$ may be replaced by co in that case). The support function (of either S or $\overline{\text{co}}\, S$) evaluated at a point v in X is easily seen to be given by the quantity

$$(3) \qquad q_0 := \max\{\sum \alpha_i \langle \zeta_i, v \rangle : \zeta_i \in \partial h_i(x), \alpha \in \partial g(h(x))\}.$$

It suffices (by Proposition 2.1.4(b)) to prove that q_0 majorizes $f^\circ(x; v)$ for any v. To show this, it suffices in turn to show that for any $\varepsilon > 0$, the following quantity q_ε majorizes $f^\circ(x; v) - \varepsilon$:

$$\max\{\sum \alpha_i \langle \zeta_i, v \rangle : \zeta_i \in \partial h_i(x_i), \alpha \in \partial g(u), x_i \in x + \varepsilon B, u \in h(x) + \varepsilon B\}.$$

The reason for this is that, in light of a lemma that we shall provide later, q_ε decreases to q_0 as $\varepsilon \downarrow 0$.

From the definition of $f^\circ(x; v)$ it follows that we can find a point x' near x and a positive t near 0 such that

$$(4) \qquad f^\circ(x; v) \leq \frac{f(x' + tv) - f(x')}{t} + \varepsilon.$$

The degree of nearness is chosen to guarantee

$$x' \in x + \varepsilon B, \qquad x' + tv \in x + \varepsilon B, \qquad h(x') \in h(x) + \varepsilon B,$$

$$h(x' + tv) \in h(x) + \varepsilon B.$$

By the mean-value theorem 2.3.7 we may write

$$f(x' + tv) - f(x') = g(h(x' + tv)) - g(h(x'))$$

$$= \sum \alpha_i [h_i(x' + tv) - h_i(x')],$$

where $\alpha \in \partial g(u)$ and u is a point in the segment $[h(x' + tv), h(x')]$ (and hence in $h(x) + \varepsilon B$). By another application of the mean-value theorem, the last term above can be expressed as

$$\sum \alpha_i \langle \zeta_i, tv \rangle,$$

where $\zeta_i \in \partial h_i(x_i)$ and x_i is a point in the segment $[x' + tv, x']$ (and hence in $x + \varepsilon B$). These observations combine with Eq. (4) to imply

$$f^\circ(x; v) \leq \sum \alpha_i \langle \zeta_i, v \rangle + \varepsilon,$$

which clearly yields $f^\circ(x; v) \leq q_\varepsilon + \varepsilon$ as required.

Here is the missing link we promised:

Lemma. $\lim_{\varepsilon \downarrow 0} q_\varepsilon = q_0$. To see this, let any $\delta > 0$ be given, and let K be a common local Lipschitz constant for the functions h_i. We shall show that q_ε is bounded above by $q_0 + n\delta(1 + K|v|)$ for all ε sufficiently small, which yields the desired result (since $q_\varepsilon \geq q_0$).

Choose ε so that each h_i is Lipschitz of rank K on $x + \varepsilon B$, and so that, for each index i, for any x_i in $x + \varepsilon B$, one has

$$h_i^\circ(x_i; \pm v) \leq h_i^\circ(x; \pm v) + \frac{\delta}{K}.$$

Multiplying across by $|\zeta_i|$, where ζ_i is any element of $\partial h_i(x_i)$, gives

$$h_i^\circ(x_i; \zeta_i v) \leq h_i^\circ(x; \zeta_i v) + \delta$$

since $|\zeta_i| \leq K$. By Proposition 2.1.5(d), we can also choose ε small enough to guarantee that $\partial g(h(x) + \varepsilon B)$ is contained in $\partial g(h(x)) + \delta B$. We may now calculate as follows:

$$q_\varepsilon \leq$$

$$\max\left\{ \sum_i \max[\alpha_i \langle \zeta_i, v \rangle : \zeta_i \in \partial h_i(x_i), x_i \in x + \varepsilon B] : \alpha \in \partial g(x) + \delta B \right\}$$

$$\leq \max\left\{ \sum_i (h_i^\circ(x; \alpha_i v) + \delta) : \alpha \in \partial g(x) + \delta B \right\}$$

$$\leq \max\left\{ \sum_i \max[\alpha_i \langle \zeta_i, v \rangle : \zeta_i \in \partial h_i(x)] : \alpha \in \partial g(x) + \delta B \right\} + n\delta$$

$$\leq q_0 + n\delta K |v| + n\delta,$$

which completes the proof of the lemma.

Having completed the proof of the general formula, we turn now to the additional assertions, beginning with the situation depicted by (i).

Consider the quantity q_0 defined by Eq. (3). We have, since the α are all nonnegative,

$$q_0 = \max\{\sum \alpha_i \max\{\langle \zeta_i, v \rangle : \zeta_i \in \partial h_i(x)\} : \alpha \in \partial g(h(x))\}$$

$$= \max\{\sum \alpha_i h_i'(x; v) : \alpha \in \partial g(h(x))\}$$

$$= g'(h(x); w), \quad \text{where } w_i := h_i'(x; v)$$

$$= \lim_{t \downarrow 0} \frac{g(h(x) + tw) - g(h(x))}{t}$$

$$= \lim_{t \downarrow 0} \left\{ \frac{g(h(x + tv)) - g(h(x))}{t} + \frac{g(h(x) + tw) - g(h(x + tv))}{t} \right\}.$$

(note that the second term under the limit goes to 0 because g is Lipschitz near $h(x)$ and $w - [h(x + tv) - h(x)]/t$ goes to 0)

$$= \lim_{t \downarrow 0} \frac{g(h(x + tv)) - g(h(x))}{t} = f'(x; v).$$

Stepping back, let us note that we have just proven $q_0 \leq f'(x; v)$, and that earlier we proved $f°(x; v) \leq q_0$. Since $f'(x; v) \leq f°(x; v)$ always, we deduce that $q_0 = f'(x; v) = f°(x; v)$ (thus f is regular). Recall that q_0 is the support function (evaluated at v) of the right-hand side of the general formula in the theorem, and that $f°(x; \cdot)$ is that of $\partial f(x)$. The equality follows, and (i) is disposed of. The reader may verify that the preceding argument will adapt to case (iii) as well. This leaves case (ii), in which $D_s g(h(x))$ is a scalar α. We may assume $\alpha \geq 0$. We find, much as above,

$$q_0 = \alpha h_1°(x; v)$$

$$= \limsup_{\substack{x' \to x \\ t \downarrow 0}} \frac{\alpha[h(x' + tv) - h(x')]}{t}$$

$$= \limsup_{\substack{x' \to x \\ t \downarrow 0}} \frac{g(h(x' + tv)) - g(h(x'))}{t}$$

(an immediate consequence of the strict differentiability of g at $h(x)$)

$$= f°(x; v).$$

As before, this yields the equality in the theorem. The $\overline{co}$ is superfluous in cases (ii) and (iii), since when $\partial g(h(x))$ or else each $\partial h_i(x)$ consists of a single point, the set whose convex hull is being taken is already convex (it fails to be so in general) and closed. □

2.3.10 Theorem (Chain Rule II)
Let F be a map from X to another Banach space Y, and let g be a real-valued function on Y. Suppose that F is strictly differentiable at x and that g is Lipschitz near $F(x)$. Then $f = g \circ F$ is Lipschitz near x, and one has

(5) $$\partial f(x) \subset \partial g(F(x)) \circ D_s F(x).$$

Equality holds if g (or $-g$) is regular at $F(x)$, in which case f (or $-f$) is also regular at x. Equality also holds if F maps every neighborhood of x to a set which is dense in a neighborhood of $F(x)$ (for example, if $D_s F(x)$ is onto).

2.3.11 Remark
The meaning of (5), of course, is that every element z of $\partial f(x)$ can be represented as a composition of a map ζ in $\partial g(F(x))$ and $D_s F(x)$: $\langle z, v \rangle =$

$\langle \zeta, D_s F(x)(v)\rangle$ for all v in X. If * denotes adjoint, we may write (5) in the equivalent form

$$\partial f(x) \subset [D_s F(x)]^* \partial g(F(x)).$$

Proof. The fact that f is Lipschitz at x is straightforward. Once again (5) is an inclusion between convex, weak*-compact sets. In terms of support functions, therefore, (5) is equivalent to the following inequality being true for any v (we let $A = D_s F(x)$):

(6) $$f^\circ(x; v) \leq \max\{\langle z, Av\rangle : z \in \partial g(F(x))\}$$

$$= g^\circ(F(x); Av).$$

The proof of this is quite analogous to that of Theorem 2.3.9, and we omit it.

Now suppose that g is regular. (The case in which $-g$ is regular is then handled by looking at $-f$ and invoking $\partial(-f) = -\partial f$.) Then the last term in (6) coincides with $g'(F(x); Av)$, which is

$$\lim_{t \downarrow 0} \frac{g(F(x) + tAv) - g(F(x))}{t} = \lim_{t \downarrow 0} \frac{g(F(x+tv)) - g(F(x))}{t}$$

(since $[F(x + tv) - F(x) - tAv]/t$ goes to 0 with t and g is Lipschitz)

$$= \lim_{t \downarrow 0} \frac{f(x + tv) - f(x)}{t}$$

$$= f'(x; v) \leq f^\circ(x; v).$$

This shows that f' exists and establishes the opposite inequality to that of (6). Thus equality actually holds above (so f is regular) and in (5).

Finally, suppose that F maps every neighborhood of x to a set dense in a neighborhood of $F(x)$. This permits us to write the final term in (6) as follows:

$$g^\circ(F(x); Av) = \limsup_{\substack{y \to F(x) \\ t \downarrow 0}} \frac{g(y + tAv) - g(y)}{t}$$

$$= \limsup_{\substack{x' \to x \\ t \downarrow 0}} \frac{g(F(x') + tAv) - g(F(x'))}{t}$$

$$= \limsup_{\substack{x' \to x \\ t \downarrow 0}} \frac{g(F(x' + tv)) - g(F(x'))}{t}$$

(since $[F(x' + tv) - F(x') - tAv]/t$ goes to zero as $x' \to x$ and $t \downarrow 0$, and g is

Lipschitz)
$$= f^\circ(x; v).$$

As before, this establishes the reverse inequality to (6), and equality in (5) ensues (but not, this time, that f is regular). □

Corollary

Let $g: Y \to R$ be Lipschitz near x, and suppose the space X is continuously imbedded in Y, is dense in Y, and contains the point x. Then the restriction f of g to X is Lipschitz near x, and $\partial f(x) = \partial g(x)$, in the sense that every element z of $\partial f(x)$ admits a unique extension to an element of $\partial g(x)$.

Proof. Apply the theorem with F being the imbedding map from X to Y. □

Suppose now that $\{f_i\}$ is a finite collection of functions ($i = 1, 2, \ldots, n$) each of which is Lipschitz near x. The function f defined by

$$f(x') = \max\{f_i(x') : i = 1, 2, \ldots, n\}$$

is easily seen to be Lipschitz near x as well. For any x' we let $I(x')$ denote the set of indices i for which $f_i(x') = f(x')$ (i.e., the indices at which the maximum defining f is attained).

2.3.12 Proposition (Pointwise Maxima)

$$\partial f(x) \subset \operatorname{co}\{\partial f_i(x) : i \in I(x)\},$$

and if f_i is regular at x for each i in $I(x)$, then equality holds and f is regular at x.

Proof. Define $g: R^n \to R$ via

$$g(u_1, u_2, \ldots, u_n) = \max\{u_i : i = 1, 2, \ldots, n\},$$

and define $h: X \to R^n$ via

$$h(x) = [f_1(x), f_2(x), \ldots, f_n(x)].$$

Observe that $f = g \circ h$. It suffices now to apply Theorem 2.3.9 and the characterization of ∂g from Example 2.2.8. Because g is convex, it is regular at $h(x)$ (see Proposition 2.3.6). The assertions regarding equality and regularity follow from Theorem 2.3.9(i). (We may assume that $I(x) = \{1, 2, \ldots, n\}$, for dropping any f_i for which $i \notin I(x)$ has no bearing on f locally; the closure operation is superfluous here.) □

2.3.13 Proposition (Products)

Let f_1, f_2 be Lipschitz near x. Then $f_1 f_2$ is Lipschitz near x, and one has

$$\partial(f_1 f_2)(x) \subset f_2(x)\partial f_1(x) + f_1(x)\partial f_2(x).$$

If in addition $f_1(x) \geq 0$, $f_2(x) \geq 0$ and if f_1, f_2 are both regular at x, then equality holds and $f_1 f_2$ is regular at x.

Proof. Let $g: R^2 \to R$ be the function $g(u_1, u_2) = u_1 \cdot u_2$, and let $h: X \to R^2$ be the function

$$h(x) = [f_1(x), f_2(x)].$$

Note that $f_1 f_2 = g \circ h$. It suffices now to apply Theorem 2.3.9; condition (i) of that result applies to yield the additional assertions. □

The next result is proven very similarly:

2.3.14 Proposition (Quotients)

Let f_1, f_2 be Lipschitz near x, and suppose $f_2(x) \neq 0$. Then f_1/f_2 is Lipschitz near x, and one has

$$\partial\left(\frac{f_1}{f_2}\right)(x) \subset \frac{f_2(x)\partial f_1(x) - f_1(x)\partial f_2(x)}{f_2^2(x)}.$$

If in addition $f_1(x) \geq 0$, $f_2(x) > 0$ and if f_1 and $-f_2$ are regular at x, then equality holds and f_1/f_2 is regular at x.

Partial Generalized Gradients

Let $X = X_1 \times X_2$, where X_1, X_2 are Banach spaces, and let $f(x_1, x_2)$ on X be Lipschitz near (x_1, x_2). We denote by $\partial_1 f(x_1, x_2)$ the (partial) generalized gradient of $f(\cdot, x_2)$ at x_1, and by $\partial_2 f(x_1, x_2)$ that of $f(x_1, \cdot)$ at x_2. The notation $f_1^\circ(x_1, x_2; v)$ will represent the generalized directional derivative at x_1 in the direction $v \in X_1$ of the function $f(\cdot, x_2)$. It is a fact that in general neither of the sets $\partial f(x_1, x_2)$ and $\partial_1 f(x_1, x_2) \times \partial_2 f(x_1, x_2)$ need be contained in the other; an example will be given in Section 2.5. For regular functions, however, a general relationship does hold between these sets.

2.3.15 Proposition

If f is regular at $x = (x_1, x_2)$, then

$$\partial f(x_1, x_2) \subset \partial_1 f(x_1, x_2) \times \partial_2 f(x_1, x_2).$$

Proof. Let $z = (z_1, z_2)$ belong to $\partial f(x_1, x_2)$. It suffices to prove that z_1 belongs to $\partial_1 f(x_1, x_2)$, which in turn is equivalent to proving that for any v in

2.3 Basic Calculus

X_1, one has $\langle z_1, v \rangle \leq f_1^\circ(x_1, x_2; v)$. But the latter coincides with $f_1'(x_1, x_2; v)$ $= f'(x_1, x_2; v, 0) = f^\circ(x_1, x_2; v, 0)$, which majorizes $\langle z_1, v \rangle$ (by Proposition 2.1.2(b)). □

To obtain a relationship in the nonregular case, define the projection $\pi_1 \partial f(x_1, x_2)$ as the set

$$\{z_1 \in X_1^* : \text{for some } z_2 \in X_2^*, (z_1, z_2) \in \partial f(x_1, x_2)\},$$

and analogously for $\pi_2 \partial f(x_1, x_2)$.

2.3.16 Proposition

$$\partial_1 f(x_1, x_2) \subset \pi_1 \partial f(x_1, x_2).$$

Proof. Fix x_2, and let f_1 be the function $f_1(x) = f(x, x_2)$ on X_1. Let $F: X_1 \to X_1 \times X_2$ be defined by $F(x) = (x, x_2)$, and note that $D_s F(x)$ is given by $\langle D_s F(x), v \rangle = (v, 0)$. Since $f_1 = f \circ F$, the result now follows by applying Theorem 2.3.10, with $x = x_1$. □

Corollary

$$\partial_1 f(x_1, x_2) \times \partial_2 f(x_1, x_2) \subset \pi_1 \partial f(x_1, x_2) \times \pi_2 \partial f(x_1, x_2).$$

2.3.17 Example

Let us prove the assertion made in Section 1.1 (see Figure 1.1) about the best L^1-approximating straight line. Specifically, let the $N + 1$ data points be $(0,0), (1,1), \ldots, (N-1, N-1)$ together with $(N, 0)$. Recall that the problem becomes that of minimizing the following function on R^2:

$$f(\alpha, \beta) = |\alpha N + \beta| + \sum_{i=1}^{N-1} |\alpha i + \beta - i|.$$

Note first that the function $f_{c,k}$ defined by

$$f_{c,k}(\alpha, \beta) = |\alpha c + \beta - k|$$

is the composition of g and F, where

$$g(y) = |y|, \quad F(\alpha, \beta) = \alpha c + \beta - k.$$

Note that F is strictly differentiable, with $D_s F(\alpha, \beta) = [c, 1]$. Since g is regular (in fact, convex), it follows from Theorem 2.3.10 that one has

$$\partial f_{c,k}(\alpha, \beta) = \begin{cases} \{[c, 1]\} & \text{if } \alpha c + \beta - k > 0 \\ \{[-c, -1]\} & \text{if } \alpha c + \beta - k < 0 \\ \{\lambda[c, 1] : |\lambda| \leq 1\} & \text{if } \alpha c + \beta - k = 0. \end{cases}$$

There is a point (α, β) at which f is minimized, and it must satisfy $0 \in \partial f(\alpha, \beta)$ by Proposition 2.3.2. The preceding remarks and Proposition 2.3.3 allow us to interpret this as follows:

(7) $$0 = \lambda_N[N, 1] + \sum_{i=0}^{N-1} \lambda_i[i, 1].$$

If this necessary condition holds for the line $y = x$ (i.e., for $\alpha = 1, \beta = 0$), then $\lambda_N = 1$, and $|\lambda_i| \leq 1$ ($i = 0, \ldots, N - 1$). It is easy to see that if $N \geq 3$, then there exist such λ_i's satisfying Eq. (7). Since f is convex, the condition $0 \in \partial f(1, 0)$ is also sufficient to imply that the line $y = x$ is a solution to the problem. Any other candidate would pass through at most two of the data points; the reader is invited to show that Eq. (7) cannot hold in such a case if N exceeds 4.

2.4 ASSOCIATED GEOMETRIC CONCEPTS

The Distance Function

Let C be a nonempty subset of X, and consider its *distance function*; that is, the function $d_C(\cdot): X \to R$ defined by

$$d_C(x) = \inf\{\|x - c\| : c \in C\}.$$

If C happens to be closed (which we do not assume), then $x \in C$ iff $d_C(x) = 0$. The function d_C is certainly not differentiable in any of the standard senses; it is, however, globally Lipschitz, as we are about to prove. We shall use the generalized gradient of d_C to lead us to new concepts of tangents and normals to an arbitrary set C. Subsequently, we shall characterize these normals and tangents topologically, thus making it clear that they do not depend on the particular norm (or distance function) that we are using. We shall prove that the new tangents and normals defined here reduce to the known ones in the smooth or convex cases. Finally, we shall indicate how these geometrical ideas lead to an extended definition of the generalized gradient ∂f of a function f which is not necessarily locally Lipschitz, and possibly extended-valued.

2.4.1 Proposition

The function d_C satisfies the following global Lipschitz condition on X:

$$|d_C(x) - d_C(y)| \leq \|x - y\|.$$

2.4 Associated Geometric Concepts

Proof. Let any $\varepsilon > 0$ be given. By definition, there is a point c in C such that $d_C(y) \geq \|y - c\| - \varepsilon$. We have

$$d_C(x) \leq \|x - c\| \leq \|x - y\| + \|y - c\|$$
$$\leq \|x - y\| + d_C(y) + \varepsilon.$$

Since ε is arbitrary, and since the argument can be repeated with x and y switched, the result follows. □

Tangents

Suppose now that x is a point in C. A vector v in X is *tangent* to C at x provided $d_C^\circ(x; v) = 0$. The set of all tangents to C at x is denoted $T_C(x)$. Of course, only the local nature of C near x is involved in this definition.

It is an immediate consequence of Proposition 2.1.1 that $T_C(x)$ is a closed convex cone in X (in particular, $T_C(x)$ always contains 0).

Normals

We define the *normal cone* to C at x by polarity with $T_C(x)$:

$$N_C(x) = \{\zeta \in X^* : \langle \zeta, v \rangle \leq 0 \text{ for all } v \text{ in } T_C(x)\}.$$

We have the following alternate characterization of $N_C(x)$ in terms of generalized gradients:

2.4.2 Proposition

$$N_C(x) = \text{cl}\left\{ \bigcup_{\lambda \geq 0} \lambda \, \partial d_C(x) \right\},$$

where cl *denotes weak* closure.*

Proof. The definition of $T_C(x)$, along with Proposition 2.1.2(b), implies that v belongs to $T_C(x)$ iff $\langle v, \zeta \rangle \leq 0$ for every ζ in $\partial d_C(x)$. It follows that the cone polar to $T_C(x)$ is the weak*-closed, convex cone generated by $\partial d_C(x)$, which is the statement of the proposition. □

The distance function figures in later chapters on optimization largely because of its role in "exact penalization." Here is a first such result:

2.4.3 Proposition

Let f be Lipschitz of rank K on a set S. Let x belong to a set $C \subset S$ and suppose that f attains a minimum over C at x. Then for any $\hat{K} \geq K$, the function

$g(y) = f(y) + \hat{K}d_C(y)$ attains a minimum over S at x. If $\hat{K} > K$ and C is closed, then any other point minimizing g over S must also lie in C.

Proof. Let us prove the first assertion by supposing the contrary. Then there is a point y in S and $\varepsilon > 0$ such that $f(y) + \hat{K}d_C(y) < f(x) - \hat{K}\varepsilon$. Let c be a point in C such that $\|y - c\| \leq d_C(y) + \varepsilon$. Then

$$f(c) \leq f(y) + \hat{K}\|y - c\| \leq f(y) + \hat{K}(d_C(y) + \varepsilon) < f(x),$$

which contradicts the fact that x minimizes f over C. Now let $\hat{K} > K$, and let y also minimize g over S. Then

$$f(y) + \hat{K}d_C(y) = f(x) \leq f(y) + (K + \hat{K})d_C(y)/2$$

(by the first assertion applied to $(\hat{K} + K)/2$), which implies $d_C(y) = 0$, and hence that y belongs to C. □

Corollary

Suppose that f is Lipschitz near x and attains a minimum over C at x. Then $0 \in \partial f(x) + N_C(x)$.

Proof. Let S be a neighborhood of x upon which f is Lipschitz (of rank K). We may suppose $C \subset S$ (since C and $C \cap S$ have the same normal cones at x), so by the proposition we deduce that x minimizes $f(y) + Kd_C(y)$ locally. Thus

$$0 \in \partial(f + Kd_C)(x) \subset \partial f(x) + K\partial d_C(x).$$

The result now follows from Proposition 2.4.2. □

When C is convex, there is a well-known concept of normal vector: $\zeta \in X^*$ is said to be normal to C at x (in the sense of convex analysis) provided $\langle \zeta, x - c \rangle \geq 0$ for all $c \in C$.

2.4.4 Proposition

If C is convex, $N_C(x)$ coincides with the cone of normals in the sense of convex analysis.

Proof. Let ζ be normal to C at x as per convex analysis. Then the point $c = x$ minimizes $f(c) = \langle \zeta, x - c \rangle$ over C, so that by the corollary to Proposition 2.4.3 (and since f is continuously, or strictly, differentiable) we obtain

$$0 \in -\zeta + N_C(x),$$

which shows that ζ belongs to $N_C(x)$.

2.4 Associated Geometric Concepts

To complete the proof, it will suffice, in view of Proposition 2.4.2, to prove that any element ζ of $\partial d_C(x)$ is normal to C at x in the sense of convex analysis (since the set of such normals is a weak*-closed convex cone).

Lemma. $d_C(\cdot)$ is convex.

Let x, y in X and λ in $(0, 1)$ be given. For any $\varepsilon > 0$, pick c_x, c_y in C such that
$$\|c_x - x\| \leq d_C(x) + \varepsilon, \qquad \|c_y - y\| \leq d_C(y) + \varepsilon,$$
and let c in C be given by $c = \lambda c_x + (1 - \lambda) c_y$. Then
$$d_C(\lambda x + (1 - \lambda) y) \leq \|c - \lambda x - (1 - \lambda) y\|$$
$$\leq \lambda \|c_x - x\| + (1 - \lambda)\|c_y - y\|$$
$$\leq \lambda d_C(x) + (1 - \lambda) d_C(y) + \varepsilon.$$

Since ε is arbitrary, the lemma is proved.

We now resume the proof of the proposition. By Proposition 2.2.7, ζ is a subgradient of d_C at x; that is,
$$d_C(y) - d_C(x) \geq \langle \zeta, y - x \rangle \qquad \text{for all } y \text{ in } X.$$

Thus $\langle \zeta, c - x \rangle \leq 0$ for any c in C. □

Corollary
If C is convex, then $v \in T_C(x)$ iff $d_C^0(x; v) = d_C'(x; v) = 0$.

Proof. The proof showed that d_C is convex, and therefore regular by Proposition 2.3.6. Thus d_C^0 and d_C' coincide, and the result follows. □

An Intrinsic Characterization of $T_C(x)$

We now show that the tangency concept defined above is actually independent of the norm (and hence distance function) used on X. Knowing this allows us to choose in particular circumstances a distance function which makes calculating tangents (or normals) more convenient.

2.4.5 Theorem
An element v of X is tangent to C at x iff, for every sequence x_i in C converging to x and sequence t_i in $(0, \infty)$ decreasing to 0, there is a sequence v_i in X converging to v such that $x_i + t_i v_i \in C$ for all i.

Proof. Suppose first that $v \in T_C(x)$, and that sequences $x_i \to x$ (with $x_i \in C$), $t_i \downarrow 0$ are given. We must produce the sequence v_i alluded to in the statement of the theorem. Since $d_C^0(x; v) = 0$ by assumption, we have

$$(1) \quad \lim_{i \to \infty} \frac{d_C(x_i + t_i v) - d_C(x_i)}{t_i} = \lim_{i \to \infty} \frac{d_C(x_i + t_i v)}{t_i} = 0.$$

Let c_i be a point in C which satisfies

$$(2) \quad \|x_i + t_i v - c_i\| \leq d_C(x_i + t_i v) + \frac{t_i}{i}$$

and let us set

$$v_i = \frac{c_i - x_i}{t_i}.$$

Then (1) and (2) imply that $\|v - v_i\| \to 0$; that is, that v_i converges to v. Furthermore, $x_i + t_i v_i = c_i \in C$, as required.

Now for the converse. Let v have the stated property concerning sequences, and choose a sequence y_i converging to x and t_i decreasing to 0 such that

$$(3) \quad \lim_{i \to \infty} \frac{d_C(y_i + t_i v) - d_C(y_i)}{t_i} = d_C^0(x; v).$$

Our purpose is to prove this quantity nonpositive, for then $v \in T_C(x)$ by definition. Let c_i in C satisfy

$$(4) \quad \|c_i - y_i\| \leq d_C(y_i) + \frac{t_i}{i}.$$

It follows that c_i converges to x. Thus there is a sequence v_i converging to v such that $c_i + t_i v_i \in C$. But then, since d_C is Lipschitz,

$$d_C(y_i + t_i v) \leq d_C(c_i + t_i v_i) + \|y_i - c_i\| + t_i \|v - v_i\|$$

$$\leq d_C(y_i) + t_i\left(\|v - v_i\| + \frac{1}{i}\right) \quad \text{(by (4))}.$$

We deduce that the limit (3) is nonpositive, which completes the proof. □

Corollary

Let $X = X_1 \times X_2$, where X_1, X_2 are Banach spaces, and let $x = (x_1, x_2) \in C_1 \times C_2$, where C_1, C_2 are subsets of X_1, X_2, respectively. Then

$$T_{C_1 \times C_2}(x) = T_{C_1}(x_1) \times T_{C_2}(x_2)$$

$$N_{C_1 \times C_2}(x) = N_{C_1}(x_1) \times N_{C_2}(x_2).$$

2.4 Associated Geometric Concepts

Proof. The first equality follows readily from the characterization of the tangent cone given in the theorem, and the second then follows by polarity. □

Regularity of Sets

In order to establish the relationship between the geometric concepts defined above and the previously known ones in smooth contexts, we require a notion of regularity for sets which will play the role that regularity for functions played in Section 2.3. We recall first the *contingent cone* $K_C(x)$ of tangents to a set C at a point x. A vector v in X belongs to $K_C(x)$ iff, for all $\varepsilon > 0$, there exist t in $(0, \varepsilon)$ and a point w in $v + \varepsilon B$ such that $x + tw \in C$ (thus $x \in \operatorname{cl} C$ necessarily). It follows immediately from Theorem 2.4.5 that $T_C(x)$ is always contained in $K_C(x)$. The latter may not be convex, however.

2.4.6 Definition

The set C is *regular* at x provided $T_C(x) = K_C(x)$.

Any convex set is regular at each of its points by the corollary to Proposition 2.4.4. The second corollary to the following theorem will confirm the fact that N_C and T_C reduce to the classical notions when C is a "smooth" set.

2.4.7 Theorem

Let f be Lipschitz near x, and suppose $0 \notin \partial f(x)$. If C is defined as $\{y \in X: f(y) \leq f(x)\}$, then one has

$$(5) \qquad \{v \in X: f^\circ(x; v) \leq 0\} \subset T_C(x).$$

If f is regular at x, then equality holds, and C is regular at x.

Proof. We observe first that there is a point $\hat{v}$ in X such that $f^\circ(x; \hat{v}) < 0$, since $f^\circ(x; \cdot)$ is the support function of a set (i.e., $\partial f(x)$) not containing zero. If v belongs to the left-hand side of (5), then for any $\varepsilon > 0$, $f^\circ(x; v + \varepsilon \hat{v}) < 0$, since $f^\circ(x; \cdot)$ is subadditive (Proposition 2.1.1(a)). In consequence, it suffices to prove that any v for which $f^\circ(x; v) < 0$ belongs to $T_C(x)$, as we now proceed to do.

It follows from the definition of $f^\circ(x; v)$ that there are ε and $\delta > 0$ such that, for all y within ε of x and t in $(0, \varepsilon)$, we have

$$f(y + tv) - f(y) \leq -\delta t.$$

Now let x_i be any sequence in C converging to x, and t_i any sequence decreasing to 0. By definition of C, we have $f(x_i) \leq f(x)$, and for all i sufficiently large,

$$f(x_i + t_i v) \leq f(x_i) - \delta t_i$$
$$\leq f(x) - \delta t_i.$$

It follows that $x_i + t_i v$ (for i large) belongs to C, and this establishes that $v \in T_C(x)$ (Theorem 2.4.5).

Now suppose that f is regular at x. In order to derive the extra assertions in this case, it will suffice to prove that any member v of $K_C(x)$ belongs to the left-hand side of (5). Since we always have $T_C(x) \subset K_C(x)$, it will then follow that the three sets coincide.

So let v belong to $K_C(x)$. Then by definition

$$\liminf_{t \downarrow 0} \frac{d_C(x + tv)}{t} = 0.$$

For any $\varepsilon > 0$, we may therefore choose a sequence t_i decreasing to 0 such that, for all i sufficiently large, we have

$$d_C(x + t_i v) \leq \varepsilon t_i.$$

Consequently there is a point x_i in C satisfying

$$\|x + t_i v - x_i\| \leq 2\varepsilon t_i,$$

and of course $f(x_i) \leq f(x)$. We deduce

$$\frac{f(x + t_i v) - f(x)}{t_i} \leq \frac{f(x_i) + 2\varepsilon K t_i - f(x)}{t_i} \leq 2\varepsilon K,$$

where K is the Lipschitz constant for f near x. Taking limits, and recalling that ε is arbitrary, we arrive at $f'(x; v) = f^\circ(x; v) \leq 0$, as required. □

Corollary 1

$$N_C(x) \subset \bigcup_{\lambda \geq 0} \lambda \, \partial f(x).$$

If f is regular at x, then equality holds.

Proof. The cone polar to the left-hand side of (5) is precisely the weak*-closed convex cone generated by $\partial f(x)$; that is, the right-hand side of the above inclusion (which is already closed because $\partial f(x)$ is a weak*-compact set not containing 0). Since taking polars reverses inclusions, the result follows immediately from (5). □

Corollary 2

Let C be given as follows:

$$\{y \in X: f_1(y) \leq 0, f_2(y) \leq 0, \ldots, f_n(y) \leq 0\},$$

and let x be such that $f_i(x) = 0$ ($i = 1, 2, \ldots, n$). Then, if each f_i is strictly differentiable at x, and if $D_s f_i(x)$, $i = 1, 2, \ldots, n$, are positively linearly independent, it follows that C is regular at x, and one has

$$N_C(x) = \left\{ \sum_{i=1}^{n} \lambda_i D_s f_i(x) : \lambda_i \geq 0, i = 1, 2, \ldots, n \right\}.$$

Proof. Define $f(y)$ via

$$f(y) = \max\{f_i(y) : i = 1, 2, \ldots, n\}.$$

Then (by Propositions 2.3.6(a) and 2.3.12) f is Lipschitz near x and regular at x. C is the set $\{y : f(y) \leq 0\}$, and $f(x) = 0$, so that C is regular and the assertion of Corollary 1 holds with equality. By Proposition 2.3.12, $\partial f(x)$ is the convex hull of the n points $D_s f_i(x)$, which leads to the desired formula for $N_C(x)$. □

Hypertangents

A vector v in X is said to be *hypertangent* to the set C at the point x in C if, for some $\varepsilon > 0$,

$$y + tw \in C \quad \text{for all } y \in (x + \varepsilon B) \cap C, w \in v + \varepsilon B, t \in (0, \varepsilon).$$

It follows easily that any vector v hypertangent to C at x belongs to $T_C(x)$. It is possible, however, for there to be no hypertangents at all.

2.4.8 Theorem (Rockafellar)

Suppose there is at least one vector v hypertangent to C at x. Then the set of all hypertangents to C at x coincides with int $T_C(x)$.

Proof (Rockafellar, 1980). Let K denote the set of all hypertangents to C at x; it follows easily that K is an open set containing all positive multiples of its elements, and that $K \subset T_C(x)$. It will therefore suffice to prove int $T_C(x) \subset K$. This inclusion can be established by verifying

(6) $$K + T_C(x) \subset K,$$

for if this holds and $v \in \text{int } T_C(x)$, then for arbitrary w in K there exists $\lambda > 0$ with $v - \lambda w \in \text{int } T_C(x)$. Since also $\lambda w \in K$ and $\lambda w + (v - \lambda w) = v$, it follows from (6) that $v \in K$. We therefore establish (6).

Let $v_1 \in K$ and $v_2 \in T_C(x)$. In order to verify (6), we must demonstrate that $v_1 + v_2 \in K$; that is, that there exists an $\varepsilon > 0$ such that

(7) $\quad C \cap (x + \varepsilon B) + t(v_1 + v_2 + \varepsilon B) \subset C \quad$ for all t in $(0, \varepsilon)$.

Since $v_1 \in K$, we know there exists an $\varepsilon_1 > 0$ such that

(8) $\quad C \cap (x + \varepsilon_1 B) + t(v_1 + \varepsilon_1 B) \subset C \quad$ for all t in $(0, \varepsilon_1)$.

Next, because $v_2 \in T_C(x)$, there exists $\varepsilon_2 > 0$ such that

(9) $\quad C \cap (x + \varepsilon_2 B) + t v_2 \subset C + t\left(\dfrac{\varepsilon_1}{2}\right) B \quad$ for $t \in (0, \varepsilon_2)$,

a consequence of the characterization of T_C given in Theorem 2.4.5. Let $\varepsilon < \min(\varepsilon_2, \varepsilon_1/2, \varepsilon_1/[1 + \varepsilon_1 + \|v_2\|])$. We claim that (7) is valid for this ε.

To see this, let v be any element of the left-hand side of (7). Then $v = y + t(v_1 + v_2 + \varepsilon u)$ where $y \in C \cap (x + \varepsilon B)$ and $u \in B$. Since $\varepsilon \leq \varepsilon_2$, we derive from (9) the fact that $y + t v_2 - t\varepsilon_1 w/2 \in C$ for some vector w of norm at most 1. The vector $y + t(v_2 - \varepsilon_1 w/2)$ also belongs to $x + \varepsilon_1 B$:

$$\left| x - y - t\left(v_2 - \dfrac{\varepsilon_1 w}{2}\right) \right| \leq \varepsilon + \varepsilon(\|v_2\| + \varepsilon_1) < \varepsilon_1.$$

Consequently $y + t(v_2 - \varepsilon_1 w/2) \in C \cap (x + \varepsilon_1 B)$, and by (8) we conclude that

(10) $\quad y + t\left(v_2 - \dfrac{\varepsilon_1 w}{2}\right) + t v_1 + t \varepsilon_1 B \subset C.$

If we now write v as follows:

$$v = y + t\left(v_2 - \dfrac{\varepsilon_1 w}{2}\right) + t v_1 + t\left(\varepsilon u + \dfrac{\varepsilon_1 w}{2}\right)$$

and observe $|\varepsilon u + \varepsilon_1 w/2| \leq \varepsilon + \varepsilon_1/2 < \varepsilon_1$, we deduce from (10) that v belongs to C. □

Corollary

Let $x \in C$, and suppose that a hypertangent to C at x exists. Then the multifunction N_C is closed at x; that is, if $\zeta_i \in N_C(x_i)$ and $\zeta_i \to \zeta$ (weak*), $x_i \to x$, then it follows that $\zeta \in N_C(x)$.

Proof. Consider first any v hypertangent to C at x. It follows easily that $v \in T_C(y)$ for all y in C near x, whence $0 \geq \lim_{i \to \infty} \langle \zeta_i, v \rangle = \langle \zeta, v \rangle$. Since

2.4 Associated Geometric Concepts

int $T_C(x)$ consists of hypertangents by the theorem, we have proved

$$\langle \zeta, v \rangle \leq 0 \quad \text{for all } v \in \text{int } T_C(x).$$

But $T_C(x)$ is convex with nonempty interior, so $T_C(x)$ is the closure of its interior, hence $\langle \zeta, v \rangle \leq 0$ for all $v \in T_C(x)$; that is, $\zeta \in N_C(x)$. □

Epigraphs and Non-Lipschitz Functions

As we have seen, the distance function d_C serves as a bridge between the analytical theory of generalized gradients and the geometric theory of the preceding section. A different link can be forged through the notion of *epigraph*. The epigraph of a real-valued function f defined on X is the following subset of $X \times R$:

$$\text{epi } f := \{(x, r) \in X \times R : f(x) \leq r\}.$$

Clearly epi f captures all information about f. The following confirms that tangency is consistent with the generalized directional derivative. (Note that only the local nature of epi f near the point $(x, f(x))$ is involved.)

2.4.9 Theorem

Let f be Lipschitz near x. Then

(i) The epigraph of $f^\circ(x; \cdot)$ is $T_{\text{epi} f}(x, f(x))$; that is, (v, r) lies in $T_{\text{epi} f}(x, f(x))$ iff $r \geq f^\circ(x; v)$.
(ii) f is regular at x iff epi f is regular at $(x, f(x))$.

Proof. Suppose first that (v, r) lies in $T_{\text{epi} f}(x, f(x))$. Choose sequences $y_i \to x$, $t_i \downarrow 0$ such that

$$(11) \qquad \lim_{i \to \infty} \frac{f(y_i + t_i v) - f(y_i)}{t_i} = f^\circ(x; v).$$

Note that $(y_i, f(y_i))$ is a sequence in epi f converging to $(x, f(x))$. By Theorem 2.4.5, therefore, there exists a sequence (v_i, r_i) converging to (v, r) such that $(y_i, f(y_i)) + t_i(v_i, r_i) \in \text{epi } f$. Thus

$$f(y_i) + t_i r_i \geq f(y_i + t_i v_i).$$

We rewrite this as

$$\frac{f(y_i + t_i v_i) - f(y_i)}{t_i} \leq r_i.$$

Taking limits, and recalling Eq. (11), we obtain $f^\circ(x; v) \leq r$ as desired.

It suffices now to prove that for any v, for any $\delta \geq 0$, the point $(v, f^\circ(x; v) + \delta)$ lies in $T_{\text{epi } f}(x, f(x))$. Accordingly, let (x_i, r_i) be any sequence in epi f converging to $(x, f(x))$, and let $t_i \downarrow 0$. We are to produce a sequence (v_i, s_i) converging to $(v, f^\circ(x; v) + \delta)$ with the property that $(x_i, r_i) + t_i(v_i, s_i)$ lies in epi f for each i; that is, such that

(12) $$r_i + t_i s_i \geq f(x_i + t_i v_i).$$

Let us define $v_i = v$ and

$$s_i = \max\left\{ f^\circ(x; v) + \delta, \frac{f(x_i + t_i v) - f(x_i)}{t_i} \right\}.$$

Observe first that $s_i \to f^\circ(x; v) + \delta$, since

$$\limsup_{i \to \infty} \frac{f(x_i + t_i v) - f(x_i)}{t_i} \leq f^\circ(x; v).$$

We need only verify (12) to complete the proof of (i). We have

$$r_i + t_i s_i \geq r_i + [f(x_i + t_i v) - f(x_i)]$$

and $r_i \geq f(x_i)$ (since $(x_i, r_i) \in$ epi f); (12) is the result. We now turn to (ii).

Suppose first that f is regular at x. Let the function $g: X \times R \to R$ be defined via

$$g(x', r) = f(x') - r,$$

and note that g is regular at $(x, f(x))$, and that epi f is the set $\{g \leq 0\}$. Since $0 \notin \partial g(x, f(x)) = \partial f(x) \times \{-1\}$, Theorem 2.4.7 applies to yield the regularity of epi f at $(x, f(x))$.

To finish the proof we require the following result. Let f'_+ signify the following Dini derivate:

$$f'_+(x; v) = \liminf_{t \downarrow 0} \frac{f(x + tv) - f(x)}{t}.$$

Lemma

$$K_{\text{epi } f}(x, f(x)) = \text{epi } f'_+(x; \cdot).$$

The proof is an exercise along the lines of the proof of (i) above; we omit it.

2.4 Associated Geometric Concepts

Now suppose that epi f is regular at $(x, f(x))$. Then, in view of the lemma and (i) of the theorem, one has

$$\text{epi } f'_+(x; \cdot) = \text{epi } f°(x; \cdot),$$

which implies that $f'_+(x; v) = f°(x; v)$ for all v. It follows that $f'(x; v)$ exists and coincides with $f°(x; v)$; that is, that f is regular at x. □

Corollary

An element ζ of X^* belongs to $\partial f(x)$ iff $(\zeta, -1)$ belongs to $N_{\text{epi} f}(x, f(x))$.

Proof. We know that ζ belongs to $\partial f(x)$ iff, for any v, we have $f°(x; v) \geq \langle \zeta, v \rangle$; that is, iff for any v and for any $r \geq f°(x; v)$, we have

$$\langle (\zeta, -1), (v, r) \rangle \leq 0.$$

By the theorem, this is equivalent to the condition that this last inequality be valid for all elements (v, r) in $T_{\text{epi} f}(x, f(x))$; that is, that $(\zeta, -1)$ lie in $N_{\text{epi} f}(x, f(x))$. □

An Extended Definition of ∂f

The above corollary, which of course is analogous to the fact that $[f'(x), -1]$ is normal to the graph of the smooth function $f: R \to R$, provides a strong temptation: to define, for any f, locally Lipschitz or not, $\partial f(x)$ as the set of ζ for which $(\zeta, -1)$ lies in $N_{\text{epi} f}(x, f(x))$. (The corollary would then guarantee that this new definition is consistent with the previous one for the locally Lipschitz case.) It would be feasible then to define $\partial f(x)$ even for extended-valued functions f on X (i.e., those taking values in $R \cup \{\pm \infty\}$), as long as f is finite at x. In similar fashion, the concept of regularity could be extended to such f. We proceed to succumb to temptation:

2.4.10 Definition

Let $f: X \to R \cup \{\pm \infty\}$ be finite at a point x. We define $\partial f(x)$ to be the set of all ζ in X^* (if any) for which $(\zeta, -1) \in N_{\text{epi} f}(x, f(x))$. We say that f is regular at x provided epi f is regular at $(x, f(x))$.

It follows that $\partial f(x)$ is a weak*-closed subset of X^*, which may no longer be compact and, as the definition implies, may be empty (but not, of course, if f is locally Lipschitz). It turns out that $\partial f(x)$ is never empty if f attains a local minimum at x:

2.4.11 Proposition

Let $f: X \to R \cup \{\pm \infty\}$ be finite at x, and suppose that for all x' in a neighborhood of x, $f(x') \geq f(x)$. Then $0 \in \partial f(x)$.

Proof. We wish to prove that $(0, -1) \in N_{\text{epi} f}(x, f(x))$, or equivalently, that $\langle (0, -1), (v, r) \rangle \leq 0$ for all (v, r) in $T_{\text{epi} f}(x, f(x))$, or again that $r \geq 0$ for all such (v,r). But for any (v,r) in $T_{\text{epi} f}(x, f(x))$, for any sequence $t_i \downarrow 0$, there is a sequence $(v_i, r_i) \to (v, r)$ such that $(x, f(x)) + t_i(v_i, r_i) \in \text{epi } f$. But then

$$f(x) + t_i r_i \geq f(x + t_i v_i) \geq f(x)$$

for all i sufficiently large, and so $t_i r_i \geq 0$. It follows that one has $\lim r_i = r \geq 0$. □

Indicators

The *indicator* of a set C in X is the extended-valued function $\psi_C: X \to R \cup \{\infty\}$ defined as follows:

$$\psi_C(x) = \begin{cases} 0 & \text{if } x \in C \\ +\infty & \text{otherwise.} \end{cases}$$

Here is more evidence confirming the compatibility between the analytical and the geometrical concepts we have introduced.

2.4.12 Proposition

Let the point x belong to C. Then

$$\partial \psi_C(x) = N_C(x),$$

and ψ_C is regular at x iff C is regular at x.

Proof. ζ belongs to $\partial \psi_C(x)$ iff $(\zeta, -1) \in N_{\text{epi } \psi_C}(x, 0)$. But clearly

$$\text{epi } \psi_C = C \times [0, \infty),$$

so this is equivalent, by the corollary to Theorem 2.4.5, to

$$\zeta \in N_C(x), \quad -1 \in N_{[0, \infty)}(0).$$

The second of these is always true (Proposition 2.4.4); the formula follows. The regularity assertion follows easily from the definition as well. □

2.5 THE CASE IN WHICH X IS FINITE-DIMENSIONAL

We explore in this section the additional properties of generalized gradients, normals, and tangents that follow when $X = R^n$. As usual, we identify X^* with X in this case, so that $\partial f(x)$ is viewed as a subset of R^n. Our most important

2.5 The Case in Which X is Finite-Dimensional

result is the characterization given by Theorem 2.5.1. It facilitates greatly the calculation of ∂f in finite dimensions. We recall Rademacher's Theorem, which states that a function which is Lipschitz on an open subset of R^n is differentiable almost everywhere (a.e.) (in the sense of Lebesgue measure) on that subset. The set of points at which a given function f fails to be differentiable is denoted Ω_f.

2.5.1 Theorem

Let f be Lipschitz near x, and suppose S is any set of Lebesgue measure 0 in R^n. Then

(1) $$\partial f(x) = \mathrm{co}\{\lim \nabla f(x_i) : x_i \to x, x_i \notin S, x_i \notin \Omega_f\}.$$

(The meaning of Eq. (1) is the following: consider any sequence x_i converging to x while avoiding both S and points at which f is not differentiable, and such that the sequence $\nabla f(x_i)$ converges; then the convex hull of all such limit points is $\partial f(x)$.)

Proof. Let us note to begin with that there are "plenty" of sequences x_i which converge to x and avoid $S \cup \Omega_f$, since the latter has measure 0 near x. Further, because ∂f is locally bounded near x (Proposition 2.1.2(a)) and $\nabla f(x_i)$ belongs to $\partial f(x_i)$ for each i (Proposition 2.2.2), the sequence $\{\nabla f(x_i)\}$ admits a convergent subsequence by the Bolzano–Weierstrass Theorem. The limit of any such sequence must belong to $\partial f(x)$ by the closure property of ∂f proved in Proposition 2.1.5(b). It follows that the set

$$\{\lim \nabla f(x_i) : x_i \to x, x_i \notin S \cup \Omega_f\}$$

is contained in $\partial f(x)$ and is nonempty and bounded, and in fact compact, since it is rather obviously closed. Since $\partial f(x)$ is convex, we deduce that the left-hand side of (1) contains the right. Now, the convex hull of a compact set in R^n is compact, so to complete the proof we need only show that the support function of the left-hand side of (1) (i.e., $f^\circ(x; \cdot)$) never exceeds that of the right. This is what the following lemma does:

Lemma. For any $v \neq 0$ in R^n, for any $\varepsilon > 0$, we have

$$f^\circ(x; v) - \varepsilon \leq \limsup\{\nabla f(y) \cdot v : y \to x, y \notin S \cup \Omega_f\}.$$

To prove this, let the right-hand side be α. Then by definition, there is a $\delta > 0$ such that the conditions

$$y \in x + \delta B, y \notin S \cup \Omega_f$$

imply $\nabla f(y) \cdot v \leq \alpha + \varepsilon$. We also choose δ small enough so that $S \cup \Omega_f$ has measure 0 in $x + \delta B$. Now consider the line segments $L_y = \{y + tv : 0 < t < \delta/(2|v|)\}$. Since $S \cup \Omega_f$ has measure 0 in $x + \delta B$, it follows from Fubini's Theorem that for almost every y in $x + (\delta/2)B$, the line segment L_y meets $S \cup \Omega_f$ in a set of 0 one-dimensional measure. Let y be any point in $x + (\delta/2)B$ having this property, and let t lie in $(0, \delta/(2|v|))$. Then

$$f(y + tv) - f(y) = \int_0^t \nabla f(y + sv) \cdot v \, ds,$$

since Df exists a.e. on L_y. Since one has $|y + sv - x| < \delta$ for $0 < s < t$, it follows that $\nabla f(y + sv) \cdot v \leq \alpha + \varepsilon$, whence

$$f(y + tv) - f(y) \leq t(\alpha + \varepsilon).$$

Since this is true for all y within $\delta/2$ of x except those in a set of measure 0, and for all t in $(0, \delta/(2|v|))$, and since f is continuous, it is in fact true for all such y and t. We deduce

$$f^\circ(x; v) \leq \alpha + \varepsilon,$$

which completes the proof. □

Corollary

$$f^\circ(x; v) = \limsup_{y \to x} \{\nabla f(y) \cdot v : y \notin S \cup \Omega_f\}.$$

2.5.2 *Example*

Let us use the theorem to calculate $\partial f(0, 0)$ where $f: R^2 \to R$ is given by

$$f(x, y) = \max\{\min[x, -y], y - x\}.$$

Define

$$C_1 = \{(x, y) : y \leq 2x \text{ and } y \leq -x\}$$

$$C_2 = \{(x, y) : y \leq x/2 \text{ and } y \geq -x\}$$

$$C_3 = \{(x, y) : y \geq 2x \text{ or } y \geq x/2\}.$$

Then $C_1 \cup C_2 \cup C_3 = R^2$, and we have

$$f(x, y) = \begin{cases} x & \text{for } (x, y) \in C_1 \\ -y & \text{for } (x, y) \in C_2 \\ y - x & \text{for } (x, y) \in C_3. \end{cases}$$

2.5 The Case in Which X is Finite-Dimensional

Note that the boundaries of these three sets form a set S of measure 0, and that if (x, y) does not lie in S, then f is differentiable and $\nabla f(x, y)$ is one of the points $(1, 0)$, $(0, -1)$, or $(-1, 1)$. It follows from the theorem that $\partial f(0, 0)$ is the convex hull of these three points.

Note that $f(0, y) = \max[0, y]$, so that $\partial_y f(0, 0)$ is the interval $[0, 1]$. Similarly, $\partial_x f(0, 0) = [-1, 0]$. For this example therefore we have

$$\partial_x f(0,0) \times \partial_y f(0,0) \not\subset \partial f(0,0) \not\subset \partial_x f(0,0) \times \partial_y f(0,0).$$

We saw in Proposition 2.3.15 that the second of these potential inclusions holds if f is regular; here is a result in a similar vein.

2.5.3 Proposition

Let $f: R^n \times R^m \to R$ be Lipschitz on a convex neighborhood $U \times V$ of a point $x = (\alpha, \beta)$, and suppose that for each α' near α, the function $f(\alpha', \cdot)$ is convex on V. Then whenever (z, w) belongs to $\partial f(x)$ we also have $w \in \partial_2 f(\alpha, \beta)$.

Proof. In view of Theorem 2.5.1, it suffices to prove the implication for points (z, w) of the form

$$\lim_{i \to \infty} \nabla f(x_i), \quad \text{where } x_i = (\alpha_i, \beta_i) \to x, \ x_i \notin \Omega_f,$$

since it will follow for convex combinations of such points, and hence for $\partial f(x)$. Let $\nabla f(x_i) = [z_i, w_i]$. We have $w_i \in \partial_2 f(\alpha_i, \beta_i)$ by Proposition 2.2.2 so that (by Proposition 2.2.7) w_i is a subgradient of the convex function $f(\alpha_i, \cdot)$ on V. Thus for all v in R^m sufficiently small, and for all i sufficiently large, we have

$$f(\alpha_i, \beta_i + v) - f(\alpha_i, \beta_i) \geq v \cdot w_i.$$

Taking limits gives us

$$f(\alpha, \beta + v) - f(\alpha, \beta) \geq v \cdot w.$$

This implies $f_2^\circ(\alpha, \beta; v) \geq v \cdot w$ for all v, so w belongs to $\partial_2 f(\alpha, \beta)$. □

The Euclidean Distance Function

Let C be an arbitrary nonempty subset of R^n. We can also use Theorem 2.5.1 to characterize the generalized gradient of the distance function d_C relative to the usual Euclidean norm:

$$d_C(x) = \inf\{|x - c| : c \in C\},$$

where $|x - c| = \{\sum_{i=1}^n |x_i - c_i|^2\}^{1/2}$. (The symbol $|\cdot|$ is reserved for this norm on R^n.) Recall that d_C was shown to be (globally) Lipschitz of rank 1 in Proposition 2.4.1.

2.5.4 Proposition

Let $\nabla d_C(x)$ exist and be different from zero. Then $x \notin \text{cl } C$, there is a unique closest point c_0 in cl C to x, and $\nabla d_C(x) = (x - c_0)/|x - c_0|$.

Proof. If x lies in cl C, then for any v in R^n we calculate

$$v \cdot \nabla d_C(x) = \lim_{t \downarrow 0} \frac{d_C(x + tv) - d_C(x)}{t}$$

$$= \lim_{t \downarrow 0} \frac{d_C(x + tv)}{t} \geq 0,$$

whence $\nabla d_C(x) = 0$, contrary to assumption. Thus x does not lie in cl C, and admits at least one closest point c_0 in cl C (i.e., a point c_0 in cl C such that $d_C(x) = |x - c_0|$). We shall now show that $\nabla d_C(x) = (x - c_0)/|x - c_0|$, from which it follows that there are no other closest points.

For t in $(0, 1)$, the closest point in cl C to $x + t(c_0 - x)$ is still c_0, whence

$$d_C(x + t(c_0 - x)) = (1 - t)|x - c_0|.$$

Subtracting $d_C(x) = |x - c_0|$ from both sides, dividing by t, and taking the limit as $t \downarrow 0$, produces

$$d'_C(x; c_0 - x) = \nabla d_C(x) \cdot (c_0 - x) = -|c_0 - x|.$$

Now $|\nabla d_C(x)| \leq 1$, since d_C is Lipschitz of rank 1, so this last equation yields $\nabla d_C(x) = (x - c_0)/|x - c_0|$. (This is where the special nature of the Euclidean norm $|\cdot|$ is used.) This completes the proof. □

Let us define a nonzero vector v to be *perpendicular* to C at $x \in \text{cl } C$ (symbolically, $v \perp C$ at x), provided $v = x' - x$, where x' has unique closest point x in cl C. The following property of perpendiculars will prove useful:

2.5.5 Proposition

Let v be perpendicular to C at x, where $x \in \text{cl } C$. Then, for all c in cl C, one has

$$\langle v, c - x \rangle \leq \tfrac{1}{2}|x - c|^2.$$

Proof. We have by definition, for all c in cl C,

$$|x' - c| \geq |x' - x|,$$

2.5 The Case in Which X is Finite-Dimensional

where $v = x' - x$. This is equivalent to

$$\langle x' - c, x' - c \rangle \geq \langle x' - x, x' - x \rangle.$$

To obtain the conclusion of the proposition, it suffices to replace $x' - c$ by $v + (x - c)$ on the left in this last inequality, and to expand the inner product. □

2.5.6 Theorem
Let x belong to cl C. Then $\partial d_C(x)$ equals the convex hull of the origin and the set

$$\left\{ v = \lim \frac{v_i}{|v_i|} : v_i \perp C \text{ at } x_i \to x, v_i \to 0 \right\}.$$

Proof. The gradient of d_C, whenever it exists, is either 0 or a unit vector of the type described in Proposition 2.5.4 (i.e., a normalized perpendicular). We immediately derive from Theorem 2.5.1 the conclusion that $\partial d_C(x)$ is contained in the given set. The opposite inclusion requires first that we show that $0 \in \partial d_C(x)$. This follows from Proposition 2.3.2, since d_C attains a (global) minimum at x. Finally we need to show that any $v = \lim v_i/|v_i|$ as described belongs to $\partial d_C(x)$. Let $v_i = y_i - x_i$ as in the definition of perpendicular. By the mean-value theorem,

$$d_C(y_i) - d_C(x_i) \in \langle \partial d_C(x_i^*), y_i - x_i \rangle$$

for some point x_i^* in (x_i, y_i). But $d_C(x_i) = 0$ and $d_C(y_i) = |y_i - x_i|$, so that

$$1 \in \left\langle \partial d_C(x_i^*), \frac{(y_i - x_i)}{|y_i - x_i|} \right\rangle.$$

Since $|\partial d_C| \leq 1$, we deduce $v_i/|v_i| \in \partial d_C(x_i^*)$. Taking limits, and observing that x_i^* converges to x, we obtain $v \in \partial d_C(x)$ by Proposition 2.1.5(d). □

Corollary 1
If x lies on the boundary of cl C, then $\partial d_C(x)$ contains nonzero points.

Proof. If x lies on the boundary of cl C, then for every $\varepsilon > 0$, the set of points y which lie within ε of x but not in cl C is of positive measure. In consequence, any such set contains points y at which $\nabla d_C(y)$ exists, and we know the latter is of the form $v/|v|$ for some perpendicular v. Letting y converge to x (i.e., $\varepsilon \downarrow 0$) leads to a sequence of unit vectors, any of whose limit points must belong to $\partial d_C(x)$. □

Corollary 2
If x lies on the boundary of cl C, then $N_C(x)$ contains nonzero points.

Proof. Invoke Proposition 2.4.2 and Corollary 1. □

A Characterization of Normal Vectors

The alternate characterization of normal vectors to sets in R^n given below is useful in many particular calculations; it is a geometric analogue of Theorem 2.5.1.

2.5.7 Proposition

Let x belong to cl C. Then $N_C(x)$ is the closed convex cone generated by the origin and the set

$$\left\{ v = \lim \frac{v_i}{|v_i|} : v_i \perp C \text{ at } x_i \to x, v_i \to 0 \right\}.$$

Proof. Proposition 2.4.2 asserts that $N_C(x)$ is the closed convex cone generated by $\partial d_C(x)$. The result follows from Theorem 2.5.6. □

Interior of the Tangent Cone

2.5.8 Theorem

Let x belong to a closed subset C of R^n. Then one has $v \in \text{int } T_C(x)$ iff there is an $\varepsilon > 0$ such that

(2) $\quad d_C(y + tw) \leq d_C(y) \quad$ for all $y \in x + \varepsilon B, w \in v + \varepsilon B, t \in [0, \varepsilon)$.

Proof. Suppose first that v satisfies (2) for some $\varepsilon > 0$. We shall show that any w in $v + \varepsilon B$ lies in $T_C(x)$ (and hence that $v \in \text{int } T_C(x)$). To see this we shall use the characterization of T_C given by Theorem 2.4.5. Let x_i be a sequence in C converging to x, t_i a sequence decreasing to 0. Then, in light of (2), $d_C(x_i + t_i w) \leq 0$ for all i sufficiently large. But then $x_i + t_i w$ lies in C for all such i. This confirms $w \in T_C(x)$.

To prove the necessity of (2), let $v \in \text{int } T_C(x)$ be given. It follows that $v \cdot z < 0$ for every nonzero $z \in N_C(x)$, for if $v \cdot z = 0$, then every neighborhood of v would contain points w for which $w \cdot z > 0$, which cannot be since $T_C(x) \cdot z \leq 0$. Equivalently, there exists $k > 0$ such that

$$v \cdot z \leq -k|z| \quad \text{for all } z \in N_C(x).$$

In view of Proposition 2.4.2, this transmutes to

(3) $\quad\quad\quad v \cdot z \leq -k|z| \quad$ for all $z \in \partial d_C(x)$.

Recall that whenever y is such that $\nabla d_C(y)$ exists and is nonzero, it is a unit vector (by Proposition 2.5.4) and that $\partial d_C(x)$ contains all limits of such gradients as $y \to x$ (any such limit is of course a unit vector). In light of this, we derive from (3) the conclusion that for some $\delta > 0$,

(4) $$v \cdot \nabla d_C(y) \leq 0 \qquad \text{for all } y \in x + \delta B,$$

if $\nabla d_C(y)$ exists and is nonzero; of course the inequality is also valid if $\nabla d_C(y) = 0$.

Now choose $\varepsilon > 0$ such that $y + tw$ lies in $x + \delta B$ whenever $y \in x + \varepsilon B$, $w \in v + \varepsilon B$, $t \in [0, \varepsilon)$. Fix any such w, and note that for almost all y in $x + \varepsilon B$, the ray $t \to y + tw$ meets in a set of 0 one-dimensional measure the set where d_C is not differentiable (since this latter set has 0 measure in R^n). For any such y, and for any t in $[0, \varepsilon)$,

$$d_C(y + tw) = d_C(y) + \int_0^t \nabla d_C(y + sw) \cdot w \, ds.$$

Since the integrand is negative or zero by (4), we derive $d_C(y + tw) \leq d_C(y)$, for almost all y in $x + \varepsilon B$. Since d_C is continuous, this must in fact be the case for all y in $x + \varepsilon B$; that is, (2) holds. □

Corollary 1 (Rockafellar)

When C is a closed subset of R^n, the set of hypertangents to C at a point $x \in C$ coincides with int $T_C(x)$.

Proof. It follows from the theorem, by taking y in (2) to lie in C, that v is a hypertangent whenever v lies in int $T_C(x)$. The converse results from Theorem 2.4.8. □

Corollary 2 (Rockafellar)

Let C be a closed subset of R^n containing x, and suppose int $T_C(x) \neq \phi$. Then the multifunction N_C is closed at x; that is,

$$x_i \to x, \quad \zeta_i \in N_C(x_i), \quad \zeta_i \to \zeta \text{ implies } \zeta \in N_C(x).$$

Proof. Invoke Corollary 1 and the corollary to Theorem 2.4.8. □

2.6 GENERALIZED JACOBIANS

Let us now consider a vector-valued function $F: R^n \to R^m$, written in terms of component functions as $F(x) = [f^1(x), f^2(x), \ldots, f^m(x)]$. We assume that each f^i (and hence F) is Lipschitz near a given point x of interest. As before, Rademacher's Theorem asserts that F is differentiable (i.e., each f^i is differentiable) a.e. on any neighborhood of x in which F is Lipschitz. We shall continue to denote the set of points at which F fails to be differentiable by Ω_F.

We shall write $JF(y)$ for the usual $m \times n$ Jacobian matrix of partial derivatives whenever y is a point at which the necessary partial derivatives exist.

2.6.1 Definition

The *generalized Jacobian* of F at x, denoted $\partial F(x)$, is the convex hull of all $m \times n$ matrices Z obtained as the limit of a sequence of the form $JF(x_i)$, where $x_i \to x$ and $x_i \notin \Omega_F$.

Symbolically, then, one has

$$\partial F(x) = \text{co}\{\lim JF(x_i) : x_i \to x, x_i \notin \Omega_F\}.$$

Of course, there is no problem in taking convex combinations in the space $R^{m \times n}$ of $m \times n$ matrices. We shall endow this space with the norm

$$\|M\|_{m \times n} = \left\{ \sum_{i=1}^{m} |r_i|^2 : r_i \text{ is the } i\text{th row of } M \right\}^{1/2},$$

and we denote by $B_{m \times n}$ the open unit ball in $R^{m \times n}$. We proceed to summarize some properties of ∂F.

2.6.2 Proposition

(a) $\partial F(x)$ is a nonempty convex compact subset of $R^{m \times n}$.
(b) ∂F is closed at x; that is, if $x_i \to x$, $Z_i \in \partial F(x_i)$, $Z_i \to Z$, then $Z \in \partial F(x)$.
(c) ∂F is upper semicontinuous at x: for any $\varepsilon > 0$ there is $\delta > 0$ such that, for all y in $x + \delta B$,

$$\partial F(y) \subset \partial F(x) + \varepsilon B_{m \times n}.$$

(d) *If each component function f^i is Lipschitz of rank K_i at x, then F is Lipschitz at x of rank $K = |(K_1, K_2, \ldots, K_m)|$, and $\partial F(x) \subset K\overline{B}_{m \times n}$.*
(e) $\partial F(x) \subset \partial f^1(x) \times \partial f^2(x) \times \cdots \times \partial f^m(x)$, where the latter denotes the set of all matrices whose i row belongs to $\partial f^i(x)$ for each i. If $m = 1$, then $\partial F(x) = \partial f^1(x)$ (i.e., the generalized gradient and the generalized Jacobian coincide).

2.6.3 Remark

We must now confess that $\partial f(x)$ is not really a subset of R^n (when $f: R^n \to R$) as we have pretended up to now. To be consistent with the generalized Jacobian, ∂f should consist of $1 \times n$ matrices (i.e., row vectors). On the other hand, the usual convention that linear operators from R^n to R^m, such as $F(x) = Ax$, are represented by matrix multiplication by an $m \times n$ matrix on the left, forces us to view R^n as $n \times 1$ matrices (i.e., column vectors), and we get $\partial F(x) = \{A\}$ as we would wish. This distinction is irrelevant as long as we

2.6 Generalized Jacobians

remain in the case $m = 1$, but must be adhered to in interpreting, for example, certain chain rules to come.

Proof. Assertions (a) and (d) together follow easily, the former as in the first part of the proof of Theorem 2.5.1. It is clear that (c) subsumes (b). To prove (c), suppose, for a given ε, that no such δ existed. Then (since the right-hand side is convex), for each i there would necessarily be an element $JF(y_i)$ such that $y_i \in x + (1/i)B$ yet $JF(y_i) \notin \partial F(x) + \varepsilon B_{m \times n}$. We may suppose (by local boundedness of ∂F) that $JF(y_i)$ converges to an element Z. By definition, $Z \in \partial F(x)$, a contradiction.

Assertion (e) follows immediately from the definition of ∂F and Theorem 2.5.1. □

In comparing the statement of Theorem 2.5.1 and the definition of the generalized Jacobian, we may well ask: what happened to the set S (of the former)? In fact, although the generalized gradient is "blind" to sets of measure 0, as proven in Theorem 2.5.1, it remains unknown whether (for $m > 1$) this is true of the generalized Jacobian. Conceivably (although we doubt it), an altered definition in which the points x_i are also restricted to lie outside a null set S could lead to a different generalized Jacobian $\partial_S F(x)$. The possibility that the generalized Jacobian is nonintrinsic in this sense is unresolved.

In most applications, it is the *images* of vectors under the generalized Jacobian that enter into the picture. In this sense, we now show that the putative ambiguity of the generalized Jacobian is irrelevant.

2.6.4 Proposition

For any v in R^n, w in R^m,

$$\partial F(x)v = \partial_S F(x)v$$

$$\partial F(x)^*w = \partial_S F(x)^*w,$$

where * denotes transpose.

Proof. We shall prove the first of these, the other being analogous. Since both sets in question are compact and convex, and since ∂F includes $\partial_S F$, we need only show that for any point u in R^m, the value σ_1 of the support function of $\partial F(x)v$ evaluated at u does not exceed the corresponding value σ_2 for $\partial_S F(x)v$. One has

$$\sigma_1 = \limsup \{ \langle u, JF(y)v \rangle : y \to x, y \notin \Omega_F \}$$

$$= \limsup \{ \langle JF(y)^*u, v \rangle : y \to x, y \notin \Omega_F \}.$$

Let g be the function $g(y) = u \cdot F(y)$. Recall that ∇g, when it exists, is a row vector (see Remark 2.6.3). For $y \notin \Omega_F$, one has $\nabla g(y)^* = JF(y)^* u$, so we may write

$$\sigma_1 = \limsup\{\nabla g(y) \cdot v : y \to x, y \notin \Omega_F\}$$

$$= g^\circ(x; v) \quad \text{(by corollary, Theorem 2.5.1)}$$

$$= \limsup\{\nabla g(y) \cdot v : y \to x, y \notin S \cup \Omega_F\}$$

$$= \limsup\{\langle u, JF(y)v \rangle : y \to x, y \notin S \cup \Omega_F\},$$

which is precisely σ_2. □

The following is a straightforward extension of the vector mean-value theorem. Note that as a consequence of Proposition 2.6.4, the right-hand side below is blind to sets of measure 0 in calculating ∂F.

2.6.5 Proposition

Let F be Lipschitz on an open convex set U in R^n, and let x and y be points in U. Then one has

$$F(y) - F(x) \in \operatorname{co} \partial F([x, y])(y - x).$$

Proof. (The right-hand side above denotes the convex hull of all points of the form $Z(y - x)$, where $Z \in \partial F(u)$ for some point u in $[x, y]$. Since $[\operatorname{co} \partial F([x, y])](y - x) = \operatorname{co}[\partial F([x, y])(y - x)]$, there is no ambiguity.) Let us fix x. It suffices to prove this inclusion for points y having the property that the line segment $[x, y]$ meets in a set of 0 one-dimensional measure the set Ω_F, for, by a now familiar argument, almost all y have this property, and the general case will follow by a limiting argument based on the continuity of F and the upper semicontinuity of ∂F. For such y, however, we may write

$$F(y) - F(x) = \int_0^1 JF(x + t(y - x))(y - x)\, dt,$$

which directly expresses $F(y) - F(x)$ as a (continuous) convex combination of points from $\partial F([x, y])(y - x)$. This completes the proof. □

A Jacobian Chain Rule

2.6.6 Theorem

Let $f = g \circ F$, where $F: R^n \to R^m$ is Lipschitz near x and where $g: R^m \to R$ is Lipschitz near $F(x)$. Then f is Lipschitz near x and one has

(1) $$\partial f(x) \subset \operatorname{co}\{\partial g(F(x))\, \partial F(x)\}.$$

2.6 Generalized Jacobians

If in addition g is strictly differentiable at x, then equality holds (and co is superfluous).

Proof. It is easy to verify that f is Lipschitz near x. The right-hand side of (1) means the convex hull of all points of the form ζZ where $\zeta \in \partial g(F(x))$ and $Z \in \partial F(x)$. (Recall that ζ is $1 \times m$ and Z $m \times n$.) We shall establish the inclusion by showing that for any v in R^n, there is such an element $\zeta_0 Z_0$ for which $(\zeta_0 Z_0)v \geq f^\circ(x; v)$. Since the latter is the support function of $\partial f(x)$, (1) will result. (Note that the right-hand side of (1) is closed.)

For y near x, any expression of the form $[f(y + tv) - f(y)]/t$ can, by the mean-value theorem 2.3.7, be written as $\zeta(F(y + tv) - F(y))/t$, where ζ belongs to $\partial g(u)$ for some point u in $[F(y), F(y + tv)]$. In turn, $[F(y + tv) - F(y)]/t$ equals some element w of co $\partial F([y, y + tv])v$ (by Proposition 2.6.5), so there exists an element Z of $\partial F([y, y + tv])$ such that $\zeta w \leq \zeta(Zv)$. Gathering the threads, we have deduced

$$\text{(2)} \qquad \frac{f(y + tv) - f(y)}{t} \leq (\zeta Z)v,$$

where $\zeta \in \partial g(u)$, $Z \in \partial F([y, y + tv])$. Now choose sequences $y_i \to x$, $t_i \downarrow 0$, for which the corresponding terms on the left-hand side of (2) converge to $f^\circ(x; v)$. Note that u_i must converge to $F(x)$, and that the line segment $[y, y + tv]$ shrinks to x. By extracting subsequences, we may suppose that $\zeta_i \to \zeta_0$, which must belong to $\partial g(F(x))$, and that $Z_i \to Z_0$, which must belong to $\partial F(x)$ (by Proposition 2.6.2(b)). We derive from (2):

$$f^\circ(x; v) \leq \zeta_0 Z_0 v,$$

as required.

Now suppose that g is strictly differentiable at $F(x)$, and let $D_s g(F(x)) = \zeta$. We first prove a technical result.

Lemma. *For any $\varepsilon > 0$ so that F is Lipschitz on $x + \varepsilon B$, there exists $\delta > 0$ such that for all $y \notin \Omega_f \cup \Omega_F$, $y \in x + \delta B$, one has*

$$\nabla f(y) \in \zeta \partial F(y) + \varepsilon B.$$

To see this, pick δ in $(0, \varepsilon)$ so small that for all y in $x + \delta B$, F is Lipschitz near y, g is Lipschitz near $F(y)$, and finally so that $\partial g(F(y)) \subset \zeta + (\varepsilon/K)B$ (where K is the Lipschitz rank of F on $x + \varepsilon B$); this is possible because $\partial g(F(x)) = \{\zeta\}$. We claim first that $\nabla f(y) \in \partial g(F(y)) JF(y)$ if $y \in x +$

δB, $y \notin \Omega_f \cup \Omega_F$. To see this, take any vector v and observe

$$\nabla f(y)v = \lim_{t \downarrow 0} \frac{g(F(y + tv)) - g(F(y))}{t}$$

$$= \lim_{t \downarrow 0} \zeta_t \frac{F(y + tv) - F(y)}{t}$$

(where $\zeta_t \in \partial g(u_t)$ and u_t lies in $[F(y), F(y + tv)]$, by the mean-value theorem)

$$\in \partial g(F(y))JF(y)v,$$

by taking a convergent subsequence of ζ_t, and since F is differentiable at y. Since v is arbitrary and $\partial g(F(y))JF(y)$ is convex, this establishes that $\nabla f(y)$ belongs to $\partial g(F(y))JF(y)$.

Now for any $y \in x + \delta B$, $y \notin \Omega_f \cup \Omega_F$, one has

$$\nabla f(y) \in \partial g(F(y))JF(y)$$

$$\subset \left(\zeta + \frac{\varepsilon}{K}B\right)JF(y) \quad \text{(by choice of } \delta\text{)}$$

$$\subset \zeta JF(y) + \varepsilon B.$$

This establishes the lemma.

Consider now the quantity

$$q = \max \zeta \partial F(x)v,$$

where v is any vector in R^n. Then by definition

$$q = \limsup \{\zeta JF(y)v : y \to x, y \notin \Omega_F\}$$

$$\geq \limsup \{\zeta JF(y)v : y \to x, y \notin \Omega_f \cup \Omega_F\}$$

$$= \limsup \{\nabla f(y)v : y \to x, y \notin \Omega_f \cup \Omega_F\},$$

in view of the lemma. This last quantity is precisely $f°(x; v)$, by the corollary to Theorem 2.5.1. Thus the support functions of the two sides of (1) are equal, and the proof is complete. □

Corollary

With F as in the theorem, let $G: R^m \to R^k$ be Lipschitz near $F(x)$. Then, for any v in R^n, one has

$$\partial(G \circ F)(x)v \subset \text{co}\{\partial G(F(x))\partial F(x)v\}.$$

If G is continuously differentiable near $F(x)$, then equality holds (and co is superfluous).

Proof. Let α be any vector in R^k. Then

$$\alpha^* \partial(G \circ F)(x)v = \partial(\alpha^*[G \circ F])(x)v$$

(by the theorem applied to $F = G \circ F$, $g(y) = \alpha \cdot y$)

$$\subset \text{co}\{\partial(\alpha^* G)(F(x))\partial F(x)v\}$$

(by the theorem applied to $F = F$, $g = \alpha^* G$)

$$= \text{co}\{\alpha^* \partial G(F(x))\partial F(x)v\}$$

$$= \alpha^* \text{co}\{\partial G(F(x))F(x)v\}.$$

Since α is arbitrary, the result follows. We turn now to the last assertion.

When G is C^1, then it is clear from the definition that $\partial(G \circ F)(x) \supset DG(F(x))\partial F(x)$, since $\Omega_{G \circ F} \subset \Omega_F$. Consequently equality must hold in the corollary. □

2.7 GENERALIZED GRADIENTS OF INTEGRAL FUNCTIONALS

In Section 3, we studied generalized gradients of finite sums. We now extend that study to integrals, which will be taken over a positive measure space $(T, \mathcal{T}, \mu)$. Suppose that U is an open subset of a Banach space X, and that we are given a family of functions $f_t: U \to R$ satisfying the following conditions:

2.7.1 Hypotheses

(i) For each x in U, the map $t \to f_t(x)$ is measurable.
(ii) For some $k(\cdot) \in L^1(T, R)$ (the space of integrable functions from T to R), for all x and y in U and t in T, one has

$$|f_t(x) - f_t(y)| \leq k(t)\|x - y\|.$$

We shall consider the integral functional f on X given by

$$f(x) = \int_T f_t(x)\,\mu(dt)$$

whenever this integral is defined. Our goal is to assert the appealing formula

$$\tag{1} \partial f(x) = \partial \int_T f_t(x)\,\mu(dt) \subset \int_T \partial f_t(x)\,\mu(dt),$$

where ∂f_t denotes the generalized gradient of the (locally Lipschitz) function $f_t(\cdot)$. The interpretation of (1) is as follows: To every ζ in $\partial f(x)$ there corresponds a mapping $t \to \zeta_t$ from T to X^* with

$$\zeta_t \in \partial f_t(x) \qquad \mu\text{-a.e.,}$$

(that is, almost everywhere relative to the measure μ) and having the property that for every v in X, the function $t \to \langle \zeta_t, v \rangle$ belongs to $L^1(X, R)$ and one has

$$\langle \zeta, v \rangle = \int_T \langle \zeta_t, v \rangle\,\mu(dt).$$

Thus, every ζ in the right-hand side of (1) is an element of X^* that can be written $\zeta(\cdot) = \int_T \langle \zeta_t, \cdot \rangle\,\mu(dt)$ where $t \to \zeta_t$ is a (measurable) selection of $\partial f_t(x)$. The theorem below is valid in any of the three following cases:

(a) T is countable.
(b) X is separable.
(c) T is a separable metric space, μ is a regular measure, and the mapping $t \to \partial f_t(y)$ is upper semicontinuous (weak*) for each y in U.

2.7.2 Theorem
Suppose that f is defined at some point x in U. Then f is defined and Lipschitz in U. If at least one of (a), (b), or (c) is satisfied, then formula (1) holds. If in addition each $f_t(\cdot)$ is regular at x, then f is regular at x and equality holds in expression (1).

Proof. Let y be any point in U. The map $t \to f_t(y)$ is measurable by hypothesis, and one has

$$|f(y) - f(x)| \leq \int_T |f_t(y) - f_t(x)|\,\mu(dt)$$

$$\leq \int_T k(t)\|y - x\|\,\mu(dt) \quad \text{(by 2.7.1(ii))}$$

$$= K\|y - x\|,$$

where K is given by $\int_T k(t)\,\mu(dt)$. We conclude that f is defined and Lipschitz on U. Now let ζ belong to $\partial f(x)$, and let v be any element of X. Then, by

2.7 Generalized Gradients of Integral Functionals

definition,

$$f°(x; v) = \limsup_{\substack{y \to x \\ \lambda \downarrow 0}} \int_T \frac{f_t(y + \lambda v) - f_t(y)}{\lambda} \mu(dt).$$

Conditions (i) and (ii) of 2.7.1 allow us to invoke Fatou's Lemma to bring the lim sup under the integral and deduce

(2) $$\int_T f_t°(x; v) \mu(dt) \geq f°(x; v) \geq \langle \zeta, v \rangle,$$

where the second inequality results from the definition of $\partial f(x)$. Let us define $\hat{f}_t, \hat{f}$ as follows:

$$\hat{f}_t(v) = f_t°(x; v), \qquad \hat{f}(v) = \int_T \hat{f}_t(v) \mu(dt).$$

Then each $\hat{f}_t(\cdot)$ is convex (Proposition 2.1.1(a)) and thus, so is $\hat{f}$. If we observe that $\hat{f}_t(0) = \hat{f}(0) = 0$, we may rewrite (2) in the form

$$\hat{f}(v) - \hat{f}(0) \geq \langle \zeta, v \rangle,$$

and this holds for all v. In other words, ζ belongs to the subdifferential $\partial \hat{f}(0)$ of the integral functional $\hat{f}(v) = \int_T \hat{f}_t(v) \mu(dt)$. The subdifferentials of such convex functionals have been characterized, for example, by Ioffe and Levin (1972); the only requirement to be able to apply their results (which amount to saying that formula (1) is true in the convex case) to our situation is to verify that the map $t \to \hat{f}_t(v)$ is measurable for each v (in each of the possible cases (a), (b), or (c)). Let us postpone this to the lemma below and proceed now to show how the known result in the convex case yields the required conclusion.

Since (1) is true for $\hat{f}$ (where the role of x is played by zero) (by p. 8 of Ioffe and Levin (1972) if (a) or (b), by p. 13 if (c)), then there is a map $t \to \zeta_t$ with $\zeta_t \in \partial \hat{f}_t(0)$, μ-a.e., such that for every v in X,

$$\langle \zeta, v \rangle = \int_T \langle \zeta_t, v \rangle \mu(dt).$$

However, $\partial \hat{f}_t(0) = \partial f_t(x)$ by construction, so the result would follow.

An Alternate Proof

The possibility of reducing to the convex case is one of the useful features of generalized gradients. However, one might wish for a self-contained derivation of (1). We sketch one here in a brief digression, picking up the proof above at

(2). If Z denotes the set of all measurable maps $t \to \zeta_t$ with $\zeta_t \in \partial f_t(x)$ μ-a.e., then we have

$$\int_T f_t^\circ(x; v) \mu(dt) = \int_T \max \langle \partial f_t(x), v \rangle \mu(dt)$$

$$= \max_Z \int_T \langle \zeta_t, v \rangle \mu(dt)$$

(by a measurable selection theorem). This allows us to write (2) as follows:

$$\min_{v \in X} \max_Z \left\{ \int_T \langle \zeta_t, v \rangle \mu(dt) - \langle \zeta, v \rangle \right\} = 0.$$

Next we apply the "lop-sided" minimax theorem of Aubin (1978a) to deduce the existence of an element $\langle \zeta_t \rangle$ of Z such that

$$\min_{v \in X} \left\{ \int_T \langle \zeta_t, v \rangle \mu(dt) - \langle \zeta, v \rangle \right\} = 0.$$

It follows that for each v, the quantity in braces vanishes, and so once again we have expressed ζ in the required form.

Back to the Proof. As in our first proof, this one imposes measurability (selection) requirements. Here is the missing link we promised earlier:

Lemma. In either of cases (a), (b), or (c), the map $t \to \hat{f}_t(v) = f_t^\circ(x; v)$ is measurable for each t.

In case (a) this is automatic; so consider (b). Since $f_t(\cdot)$ is continuous, we may express $f_t^\circ(x; v)$ as the upper limit of

(3) $$\frac{f_t(y + \lambda v) - f_t(y)}{\lambda},$$

where $\lambda \downarrow 0$ taking rational values, and $y \to x$ taking values in a countable dense subset $\{x_i\}$ of X. But expression (3) defines a measurable function of t by hypothesis, so $f_t^\circ(x; v)$, as the "countable lim sup" of measurable functions of t, is measurable in t.

To complete the proof of the lemma, we consider now case (c). Consider the formula

$$f_t^\circ(x; v) = \max \{\langle \zeta, v \rangle : \zeta \in \partial f_t(x)\}.$$

Since now the multifunction $t \to \partial f_t(x)$ is upper semicontinuous, and weak*-

2.7 Generalized Gradients of Integral Functionals

compact for each t, it follows routinely that $f_t^\circ(x; v)$ is upper semicontinuous as a function of t, and therefore measurable. The lemma is proved.

The Final Steps. There remain the regularity and equality assertions of the theorem to verify. When f_t is regular at x for each t, the existence of $f'(x; v)$ for any v in X and the equality

(4) $$f'(x; v) = \int_T f_t'(x; v)\, \mu(dt)$$

follow from the dominated convergence theorem. The left-hand side of (4) is bounded above by $f^\circ(x; v)$, and the right-hand side coincides with $\int_T f_t^\circ(x; v)\, \mu(dt)$, which is no less than $f^\circ(x; v)$ by (2). It follows that the terms in (4) coincide with $f^\circ(x; v)$, and hence that f is regular.

Finally, let $\zeta = \int_T \langle \zeta_t, \cdot \rangle\, \mu(dt)$ belong to the right-hand side of relation (1). Then, since $\zeta_t \in \partial f_t(x)$ μ-a.e., one has

$$f^\circ(x; v) = \int_T f_t^\circ(x; v)\, \mu(dt)$$

$$\geq \int_T \langle \zeta_t, v \rangle\, \mu(dt)$$

$$= \langle \zeta, v \rangle,$$

which shows that ζ belongs to $\partial f(x)$, and completes the proof. □

A type of integral functional which occurs frequently in applications is that in which X is a space of functions on T and $f_t(x) = g(t, x(t))$. When the "evaluation map" $x \to x(t)$ is continuous (e.g., when $X = C[0, 1]$), Theorem 2.7.2 will apply. However, there are important instances when it is not, notably when X is an L^p space. We now treat this situation for $p = \infty$, and subsequently for $1 \leq p < \infty$.

Integrals on Subspaces of L^∞

We suppose now that $(T, \mathcal{T}, \mu)$ is a σ-finite positive measure space and Y a separable Banach space. $L^\infty(T, Y)$ denotes the space of measurable essentially bounded functions mapping T to Y, equipped with the usual supremum norm. We are also given a closed subspace X of $L^\infty(T, Y)$, and a family of functions $f_t: Y \to R$ ($t \in T$). We define a function f on X by the formula

$$f(x) = \int_T f_t(x(t))\, \mu(dt).$$

Suppose that f is defined (finitely) at a point x. We wish to characterize $\partial f(x)$. To this end, we suppose that there exist $\varepsilon > 0$ and a function k in $L^1(T, R)$ such that, for all $t \in T$, for all v_1, v_2 in $x(t) + \varepsilon B_Y$,

$$|f_t(v_1) - f_t(v_2)| \leq k(t)\|v_1 - v_2\|_Y.$$

We also suppose that the mapping $t \to f_t(v)$ is measurable for each v in Y.

2.7.3 Theorem

Under the hypotheses described above, f is Lipschitz in a neighborhood of x and one has

(5) $$\partial f(x) \subset \int_T \partial f_t(x(t))\, \mu(dt).$$

Further, if f_t is regular at $x(t)$ for each t, then f is regular at x and equality holds.

Proof. Let us note first that the right-hand side of expression (5) is interpreted analogously to that of (1) in Theorem 2.7.2. Thus it consists of those maps $\zeta(\cdot) = \int_T \langle \zeta_t, \cdot \rangle \mu(dt)$ such that $\zeta_t \in \partial f_t(x(t)) \subset Y^*$ μ-a.e., and such that for any v in X, $\int_T \langle \zeta_t, v(t) \rangle \mu(dt)$ is defined and equal to $\langle \zeta, v \rangle$.

The fact that f is defined and Lipschitz near x follows easily, as in Theorem 2.7.2. To prove (5), suppose that ζ belongs to $\partial f(x)$, and let v be any element of X.

Let $\hat{f}_t(\cdot)$ signify the function $f_t^\circ(x(t); \cdot)$. The measurability of $t \to \hat{f}_t(v)$ follows as in the lemma in Theorem 2.7.2, case (b), and of course $\hat{f}_t(\cdot)$ is continuous. It follows that the map $t \to \hat{f}_t(v(t))$ is measurable, and arguing precisely as in the proof of Theorem 2.7.2, we deduce

(6) $$\int_T f_t^\circ(x(t); v(t))\, \mu(dt) \geq f^\circ(x; v) \geq \langle \zeta, v \rangle.$$

This can be interpreted as saying that ζ belongs to the subdifferential at 0 of the integral functional on X defined by

$$\hat{f}(v) = \int_T \hat{f}_t(v(t))\, \mu(dt).$$

The requisites of Ioffe and Levin (1972, p. 22) are met and, as in Theorem 2.7.2, the truth of relation (5) for $\hat{f}, \hat{f}_t$ at 0 immediately gives (5) for f, f_t.

We remark that a direct argument avoiding recourse to convex analysis is available, one quite analogous to the one sketched in the proof of Theorem 2.7.2.

2.7 Generalized Gradients of Integral Functionals

Regularity. Now suppose that each f_t is regular at $x(t)$. Fix any v in X and set

$$\theta = \liminf_{\lambda \downarrow 0} \frac{f(x + \lambda v) - f(x)}{\lambda}.$$

Invoking Fatou's Lemma, we derive

$$f^\circ(x; v) \geqslant \theta \geqslant \int_T f'_t(x(t); v(t)) \mu(dt)$$

$$= \int_T f^\circ_t(x(t); v(t)) \mu(dt) \geqslant f^\circ(x; v) \quad \text{(by (6))}.$$

Consequently all are equal and f is regular at x.

Finally, if $\zeta = \int_T \langle \zeta_t, \cdot \rangle \mu(dt)$ belongs to the right-hand side of expression (5), then

$$\langle \zeta, v \rangle = \int_T \langle \zeta_t, v(t) \rangle \mu(dt)$$

$$\leqslant \int_T f^\circ_t(x(t); v(t)) \mu(dt)$$

$$= \int_T f'_t(x(t); v(t)) \mu(dt) = f'(x; v),$$

by the preceding calculation. Thus ζ belongs to $\partial f(x)$, and the proof is complete. □

2.7.4 Example (Variational Functionals)

A common type of functional in the calculus of variations is the following:

$$J(y) = \int_a^b L(t, y(t), \dot{y}(t)) \, dt,$$

where y is an absolutely continuous function from $[a, b]$ to R^n. We can calculate ∂J by means of Theorem 2.7.3 with the following cast of characters:

$(T, \mathcal{T}, \mu) = [a, b]$ with Lebesgue measure, $Y = R^n \times R^n$

$$X = \left\{ (s, v) \in L^\infty(T, Y) : \text{for some } c \text{ in } R^n, s(t) = c + \int_a^t v(\tau) \, d\tau \right\}$$

$f_t(s, v) = L(t, s, v)$.

Note that for any (s, v) in X, we have $f(s, v) = J(s)$. With $(\hat{s}, \hat{v})$ a given element of X (so that $\hat{v} = (d/dt)\hat{s}$), we assume that the integrand L is measurable in t, and that for some $\varepsilon > 0$ and some function $k(\cdot)$ in $L^1[a, b]$, one has

$$|L(t, s_1, v_1) - L(t, s_2, v_2)| \leq k(t)|(s_1 - s_2, v_1 - v_2)|$$

for all (s_i, v_i) in $(\hat{s}(t), \hat{v}(t)) + \varepsilon B$. It is easy to see that the hypotheses of the theorem are satisfied, so that if ζ belongs to $\partial f(\hat{s}, \hat{v})$, we deduce the existence of a measurable function $(q(t), p(t))$ such that

$$(q(t), p(t)) \in \partial L(t, \hat{s}(t), \hat{v}(t)) \quad \text{a.e.}$$

(where ∂L denotes generalized gradient with respect to (s, v)), and where, for any (s, v) in X, one has

$$\langle \zeta, (s, v) \rangle = \int_a^b \{q(t) \cdot s(t) + p(t) \cdot v(t)\} \, dt.$$

In particular, if $\zeta = 0$ (as when J attains a local minimum at $\hat{s}$), it then follows easily that $p(\cdot)$ is absolutely continuous and that $q = \dot{p}$ a.e.

In this case then we have

$$(\dot{p}, p) \in \partial L(t, \hat{s}, \dot{\hat{s}}) \quad \text{a.e.,}$$

which implies the classical Euler–Lagrange equation if L is C^1:

$$\frac{d}{dt} L_v(t, \hat{s}, \dot{\hat{s}}) = L_s(t, \hat{s}, \dot{\hat{s}}) \quad \text{a.e.}$$

Integral Functionals on L^p

We assume now that $(T, \mathcal{T}, \mu)$ is a positive complete measure space with $\mu(T) < \infty$, and that Y is a separable Banach space. Let X be a closed subspace of $L^p(T, Y)$ (for some p in $[1, \infty)$), the space of p-integrable functions from T to Y. We define a functional f on X via

$$f(x) = \int_T f_t(x(t)) \mu(dt),$$

where $f_t: Y \to R$ $(t \in T)$ is a given family of functions. We shall again suppose that for each y in Y the function $t \to f_t(y)$ is measurable, and that x is a point at which $f(x)$ is defined (finitely). We shall characterize $\partial f(x)$ under either of two additional hypotheses. Let q solve $1/p + 1/q = 1$ ($q = \infty$ if $p = 1$).

Hypothesis A. There is a function k in $L^q(T, R)$ such that, for all $t \in T$,

$$|f_t(y_1) - f_t(y_2)| \leq k(t)\|y_1 - y_2\|_Y \quad \text{for all } y_1, y_2 \text{ in } Y.$$

Hypothesis B. Each function f_t is Lipschitz (of some rank) near each point of Y, and for some constant c, for all $t \in T, y \in Y$, one has

$$\zeta \in \partial f_t(y) \quad \text{implies} \quad \|\zeta\|_{Y^*} \leq c\{1 + \|y\|_Y^{p-1}\}.$$

2.7.5 Theorem
Under the conditions described above, under either of Hypotheses A or B, f is uniformly Lipschitz on bounded subsets of X, and one has

$$(7) \qquad \partial f(x) \subset \int_T \partial f_t(x(t))\, \mu(dt).$$

Further, if each f_t is regular at $x(t)$ then f is regular at x and equality holds.

2.7.6 Remark
The interpretation of expression (7) is in every way analogous to expressions (1) and (5) of Theorems 2.7.2 and 2.7.3. When $X = L^p(T, Y)$, then any element ζ of $\partial f(x)$ must belong to $L^q(T, Y^*) = X^*$.

Proof (f Lipschitz). That f is defined and globally Lipschitz on X when Hypothesis A holds is immediate, for one has

$$|f(x_1) - f(x_2)| \leq \int_T |f_t(x_1(t)) - f_t(x_2(t))|\, \mu(dt)$$

$$\leq \int_T k(t)\|x_1(t) - x_2(t)\|_Y \mu(dt)$$

$$\leq K\|x_1 - x_2\|_X,$$

the last by Hölder's inequality, where $K = \|k\|_q$. We now prove under Hypothesis B that f is uniformly Lipschitz on bounded subsets of X. Let any $m > \|x\|_X$ be given, and let u be any element of X satisfying $\|u\|_p \leq m$. By the mean-value theorem we write

$$(8) \qquad f_t(u(t)) - f_t(x(t)) = \langle \zeta_t, u(t) - x(t)\rangle,$$

where $\zeta_t \in \partial f_t(x_t^*)$ and x_t^* lies in the interval $[u(t), x(t)]$. It follows from

Hypothesis B that

$$\|\zeta_t\|_{Y^*} \leq c\{1 + \|x_t^*\|^{p-1}\}$$
$$\leq c\{1 + \|u(t)\|^{p-1} + \|x(t)\|^{p-1}\}.$$

Let us call this last quantity $\theta(t)$; note $\theta \in L^q(T, R)$. If we substitute this into Eq. (8), take absolute values, and integrate, we obtain

$$\int_T |f_t(u(t)) - f_t(x(t))|\,\mu(dt) \leq \int_T \theta(t)\|u(t) - x(t)\|\,\mu(dt)$$
$$\leq \|\theta\|_q \|u - x\|_p \quad \text{(by Hölder's inequality)}.$$

It is easy to see that $\|\theta\|_q$ is bounded above by a number K depending only on c and m (but not on u!). We deduce then that $f(u)$ is not only finite, but satisfies

$$|f(u) - f(x)| \leq K\|u - x\|_p = K\|u - x\|_X.$$

The process by which we got this inequality can now be repeated for x replaced by any v satisfying $\|v\|_p \leq m$, so f is indeed Lipschitz on bounded subsets of X.

Deriving (7). In Theorem 2.7.2 we began by proving that for any v in X, for any ζ in $\partial f(x)$, one has

(9) $$\int_T f_t^\circ(x(t); v(t))\,\mu(dt) \geq f^\circ(x; v) \geq \langle \zeta, v \rangle.$$

Under Hypothesis A this is done just as in Theorem 2.7.2. Under Hypothesis B the process is essentially the same, except that the use of Fatou's Lemma must be justified by appealing to Eq. (8) (for $u = x + \lambda v$). Again we let $\hat{f}_t(v) = f_t^\circ(x(t); v)$, and we interpret (9) as saying that ζ belongs to the subgradient at 0 of the convex functional $\hat{f}(v) = \int_T \hat{f}_t(v(t))\,\mu(dt)$. This remains true if we restrict v to $X \cap L^\infty(T, Y)$, which is a closed subspace of $L^\infty(T, Y)$. Consequently we can apply Theorem 2.7.3 to $\hat{f}$, provided we can verify its hypotheses. Postponing that to the lemma below, we conclude that ζ is expressible in the form $\zeta(\cdot) = \int_T \langle \zeta_t, \cdot \rangle\,\mu(dt)$, where $\zeta_t \in \partial\hat{f}_t(0) = \partial f_t(x(t))$. This proves relation (7).

Lemma. The hypotheses of Theorem 2.7.3 are satisfied when $x = 0$, $f_t = \hat{f}_t$.

The required measurability of $t \to \hat{f}_t(v)$ follows exactly as in case (b) of the lemma in the proof of Theorem 2.7.2. Of course, $\int_T \hat{f}_t(0)\,\mu(dt) = \hat{f}(0)$ is defined, so we need only verify the Lipschitz hypothesis. Now, the set $\partial f_t(x(t))$

is bounded by an integrable function $r(t)$: if Hypothesis A holds, take $r = k$ (in view of Proposition 2.1.2(a)); if Hypothesis B holds, take $r = c\{1 + \|x(t)\|_Y^{p-1}\}$. Thus $\hat{f}_t(\cdot)$, as the support function of $\partial f_t(x(t))$, is globally Lipschitz with constant $r(t)$, and the lemma is proved.

Regularity. We now turn to verifying the last conclusion of the theorem. For any v in X, one derives

$$\liminf_{\lambda \downarrow 0} \frac{f(x + \lambda v) - f(x)}{\lambda} \geq \int_T f_t'(x(t); v(t)) \mu(dt)$$

$$= \int_T f_t^\circ(x(t)); v(t)) \mu(dt) \geq f^\circ(x; v).$$

It follows that $f'(x; v)$ exists and equals $f^\circ(x; v)$; that is, that f is regular at x. Given any ζ in the right-hand side of expression (7), one has

$$\langle \zeta, v \rangle = \int_T \langle \zeta_t, v(t) \rangle \mu(dt)$$

$$\leq \int_T f_t'(x(t); v(t)) \mu(dt) = f^\circ(x; v),$$

and so $\zeta \in \partial f(x)$. Thus relation (7) holds with equality. □

2.8 POINTWISE MAXIMA

A General Formula

Functionals which are explicitly expressed as pointwise maxima of some indexed family of functions will play an important role in later chapters. We have already studied the case in which the index set is finite, in Proposition 2.3.12. This section is devoted to the much more complex situation in which the family is an infinite one.

Let f_t be a family of functions on X parametrized by $t \in T$, where T is a topological space. Suppose that for some point x in X, each function f_t is Lipschitz near x. We shall find it convenient to define a new kind of partial generalized gradient, one which takes account of variations in parameters other than the primary one.

We denote by $\partial_{[T]} f_t(x)$ the set ($\overline{\text{co}}$ denotes weak*-closed convex hull)

$$\overline{\text{co}} \{\zeta \in X^* : \zeta_i \in \partial f_{t_i}(x_i),$$

$$x_i \to x, t_i \to t, t_i \in T, \zeta \text{ is a weak* cluster point of } \zeta_i\}.$$

2.8.1 Definition

The multifunction $(\tau, y) \to \partial f_\tau(y)$ is said to be (weak*) closed at (t, x) provided $\partial_{[T]} f_t(x) = \partial f_t(x)$.

(Note that this condition certainly holds if t is isolated in T (in view of Proposition 2.1.5(b)).)

We make the following hypotheses:
(i) T is a sequentially compact space.
(ii) For some neighborhood U of x, the map $t \to f_t(y)$ is upper semicontinuous for each y in U.
(iii) Each f_t, $t \in T$, is Lipschitz of given rank K on U, and $\{f_t(x) : t \in T\}$ is bounded.

We define a function $f : X \to R$ via

$$f(y) = \max\{f_t(y) : t \in T\},$$

and we observe that our hypotheses imply that f is defined and finite (with the maximum defining f attained) on U. It also follows readily that f is Lipschitz on U (of rank K), since each f_t is.

We denote by $M(y)$ the set $\{t \in T : f_t(y) = f(y)\}$. It is easy to see that $M(y)$ is nonempty and closed for each y in U. Finally, for any subset S of T, $P[S]$ signifies the collection of probability Radon measures supported on S.

2.8.2 Theorem

In addition to the hypotheses given above, suppose that either
(iv) *X is separable, or*
(iv)' *T is metrizable (which is true in particular if T is separable).*
Then one has

(1) $$\partial f(x) \subset \left\{ \int_T \partial_{[T]} f_t(x) \mu(dt) : \mu \in P[M(x)] \right\}.$$

Further, if the multifunction $(\tau, y) \to \partial f_\tau(y)$ is closed at (t, x) for each $t \in M(x)$, and if f_t is regular at x for each t in $M(x)$, then f is regular at x and equality holds in expression (1) (with $\partial_{[T]} f_t(x) = \partial f_t(x)$).

2.8.3 Remark

The interpretation of the set occurring on the right-hand side of expression (1) is completely analogous to that of (1) in Theorem 2.7.2. Specifically, an element ζ of that set is an element of X^* to which there corresponds a mapping $t \to \zeta_t \in \partial_{[T]} f_t(x)$ from T to X^* and an element μ of $P[M(x)]$ such that, for every v in X, $t \to \langle \zeta_t, v \rangle$ is μ-integrable, and

$$\langle \zeta, v \rangle = \int_T \langle \zeta_t, v \rangle \mu(dt).$$

Let us note two important special cases before proving the theorem.

Corollary 1
In addition to the basic hypotheses (i)–(iii) and (iv) or (iv)', assume that U is convex, and that $f_t(x)$ is continuous as a function of t and convex as a function of x. Then f is convex on U, and, for each x in U,

$$\partial f(x) = \left\{ \int_T \partial f_t(x) \mu(dt) : \mu \in P[M(x)] \right\}.$$

Proof. A pointwise maximum of convex functions is convex, so f is convex in this case. Because convex functions are regular (Proposition 2.3.6), the corollary follows immediately from the theorem once one establishes that the multifunction $(t, x) \to \partial f_t(x)$ is closed. To see this, let ζ_i belong to $\partial f_{t_i}(x_i)$, where $x_i \to x$, $t_i \to t$, and let ζ be a cluster point of $\{\zeta_i\}$. We wish to show that ζ belongs to $\partial f_t(x)$. Now for any v sufficiently small,

$$f_{t_i}(x_i + v) - f_{t_i}(x_i) \geq \langle \zeta_i, v \rangle.$$

We deduce (from hypothesis (iii))

$$f_{t_i}(x + v) - f_{t_i}(x) \geq \langle \zeta_i, v \rangle - 2K \|x_i - x\|.$$

There is a subsequence $\{\zeta_i'\}$ such that $\lim \langle \zeta_i', v \rangle = \langle \zeta, v \rangle$. Taking limits along this subsequence yields

$$f_t(x + v) - f_t(x) \geq \langle \zeta, v \rangle.$$

This implies

$$\zeta \in \partial f_t(x). \quad \square$$

Corollary 2
In addition to the basic hypotheses (i)–(iii) and (iv) or (iv)', assume that each f_t admits a strict derivative $D_s f_t(x)$ on U, and that $D_s f_t(x)$ is continuous as a function of (t, x). Then for each $x \in U$, f is regular at x and one has

$$\partial f(x) = \left\{ \int_T D_s f_t(x) \mu(dt) : \mu \in P[M(x)] \right\}.$$

This follows immediately from the theorem, since $\partial f_t(x) = \{D_s f_t(x)\}$ by Proposition 2.2.4.

2.8.4 Example (Supremum Norm)

Let us apply Corollary 1 to obtain a result originally due to Banach. Let X be the space of continuous functions x mapping the interval $[a, b]$ to R^n, and define

$$\|x\| = \max_{a \leqslant t \leqslant b} |x(t)|.$$

At which points x in X is this norm differentiable?

We appeal to Corollary 1 in the case:

$$T = [a, b], \quad f_t(x) = |x(t)|.$$

Since $f(x) = \|x\|$ is convex, f is regular by Proposition 2.3.6 and therefore differentiable iff $\partial f(x)$ reduces to a singleton. In view of the corollary, this happens only when $M(x)$ consists of just one point. We have proved: $\|x\|$ is differentiable at those x which admit a unique t maximizing $|x(t)|$.

Proof of Theorem 2.8.2

Step 1. Let us define the function $g: U \times X \to R$ as follows:

(2) $\qquad g(x; v) = \max\{\langle \zeta, v \rangle : \zeta \in \partial_{[T]} f_t(x), t \in M(x)\}.$

The use of max is justified by the fact that $\partial_{[T]} f_t$ is weak*-compact (by hypothesis (iii)) and $M(x)$ is closed. We wish to establish, for any v in X,

Lemma. $f^\circ(x; v) \leqslant g(x; v).$

To see this, let $y_i \to x$ and $\lambda_i \downarrow 0$ be sequences such that the terms

$$\Delta_i = \frac{f(y_i + \lambda_i v) - f(y_i)}{\lambda_i}$$

converge to $f^\circ(x; v)$. Pick any $t_i \in M(y_i + \lambda_i v)$. Then it follows that

(3) $\qquad \Delta_i \leqslant \dfrac{f_{t_i}(y_i + \lambda_i v) - f_{t_i}(y_i)}{\lambda_i}.$

By the mean-value theorem, the expression on the right-hand side of (3) may be expressed $\langle \zeta_i, v \rangle$, where ζ_i belongs to $\partial f_{t_i}(y_i^*)$ for some y_i^* between y_i and $y_i + \lambda_i v_i$. By extracting a subsequence (without relabeling) we may suppose

2.8 Pointwise Maxima

that $t_i \to t \in T$, and that $\{\zeta_i\}$ admits a weak* cluster point with $\lim_{i\to\infty}\langle\zeta_i, v\rangle = \langle\zeta, v\rangle$. Note that y_i^* has no choice but to converge to x, so that $\zeta \in \partial_{[T]}f_t(x)$ by definition. We derive from all this, in view of (3), the following:

$$f^\circ(x; v) = \lim_{i\to\infty} \Delta_i \leq \langle\zeta, v\rangle.$$

Now the lemma will follow if $t \in M(x)$, for then $\langle\zeta, v\rangle \leq g(x; v)$ by definition. To see that this is the case, observe that for any τ in T, one has

$$f_{t_i}(y_i + \lambda_i v) \geq f_\tau(y_i + \lambda_i v),$$

(since $t_i \in M(y_i + \lambda_i v)$), whence (by hypothesis (iii)):

$$f_{t_i}(x) \geq f_\tau(x) - 2K\|y_i + \lambda_i v - x\|.$$

From hypothesis (ii) we now conclude, upon taking limits:

$$f_t(x) \geq f_\tau(x).$$

Since τ is arbitrary, it follows that $t \in M(x)$, and the lemma is proved.

Step 2. Now let ζ be an element of $\partial f(x)$. We deduce from the lemma that, for all v in X,

(4) $$g(x; v) \geq \langle\zeta, v\rangle.$$

Since $g(x; 0) = 0$, this means that ζ belongs to the subdifferential of the convex function $g(x; \cdot)$ at 0, which has been characterized in convex analysis, for example by Ioffe and Levin (1972, p. 33 if (iv) holds, or p. 34 if (iv)' holds). It follows that ζ can be expressed in the form $\int_T \langle\zeta_t, \cdot\rangle \mu(dt)$ for some μ in $P[M(x)]$ and measurable mapping $t \to \zeta_t$ assuming values in $\partial_{[T]}f_t(x)$ μ-a.e.. This establishes expression (1).

It is possible to proceed directly rather than appeal to convex analysis. We sketch the alternate approach before proceeding with the proof of the theorem. Let us call the right-hand side of expression (1) Z. Note that Z is the weak*-closed convex hull of all points in $\partial_{[T]}f_t(x)$ for t in $M(x)$. Then (4) and the lemma assert

$$\min_{v \in X} \max_{w \in Z} \{\langle w, v\rangle - \langle\zeta, v\rangle\} = 0.$$

We apply Aubin's lop-sided Minimax Theorem to deduce the existence of a point w in Z, satisfying, for all v in X,

$$\langle w - \zeta, v\rangle \geq 0.$$

It follows that $\zeta = w$, and we obtain expression (1) as before.

Step 3. We now assume the additional hypotheses. Let

$$\alpha = \liminf_{\lambda \downarrow 0} \frac{f(x + \lambda v) - f(x)}{\lambda}.$$

To conclude that f is regular at x, it suffices to show that $f^\circ(x; v) \leq \alpha$. Pick any t in $M(x)$. One has

$$\frac{f(x + \lambda v) - f(x)}{\lambda} \geq \frac{f_t(x + \lambda v) - f_t(x)}{\lambda},$$

and taking the lower limit of each side yields $\alpha \geq f_t'(x; v) = f_t^\circ(x; v)$, the last equality resulting from the fact that f_t is regular at x. We conclude

(5) $$\alpha \geq \max\{f_t^\circ(x; v) : t \in M(x)\}.$$

By Definition 2.8.1, we have $\partial_{[T]} f_t(x) = \partial f_t(x)$, so the function $g(x; v)$ defined at the beginning of the proof reduces to $\max\{\langle \zeta, v \rangle : \zeta \in \partial f_t(x), t \in M(x)\}$, which is precisely the right-hand side of (5). The lemma of Step 1, together with (5), thus combine to imply $f^\circ(x; v) \leq \alpha$. As noted, this proves that f is regular at x.

In order to show that expression (1) holds with equality, let $\zeta = \int_T \langle \zeta_t, \cdot \rangle \mu(dt)$ be any element of the right-hand side of (1). Then, for any v in X,

$$\langle \zeta, v \rangle = \int_T \langle \zeta_t, v \rangle \mu(dt)$$

$$\leq \int_T f_t^\circ(x; v) \mu(dt) = \int_T f_t'(x; v) \mu(dt)$$

$$= \lim_{\lambda \downarrow 0} \frac{\int f_t(x + \lambda v) \mu(dt) - \int f_t(x) \mu(dt)}{\lambda}$$

(by dominated convergence)

$$\leq \limsup_{\lambda \downarrow 0} \frac{\int f(x + \lambda v) \mu(dt) - \int f(x) \mu(dt)}{\lambda}$$

(since $f_t(x) = f(x)$ for $t \in M(x)$ and μ is supported on $M(x)$)

$$= \limsup_{\lambda \downarrow 0} \frac{f(x + \lambda v) - f(x)}{\lambda}$$

$$= f'(x; v) = f^\circ(x; v) \quad \text{(as just proven above)}.$$

We have shown $\langle \zeta, v \rangle \leq f^\circ(x; v)$. Thus $\zeta \in \partial f(x)$, and the proof of the theorem is complete. □

When there is Lipschitz behavior of f_t in t (as well as in x), it has been demonstrated by Hiriart-Urruty (1978) that a different sort of estimate for ∂f is possible. We use Theorem 2.8.2 to extend his result to the case in which X is infinite-dimensional.

We now write $\phi(t, x)$ for $f_t(x)$, and we make a single hypothesis which subsumes all the hypotheses (i)–(iii) and (iv)':

(v) T is a compact subset of a finite-dimensional Banach space Y, and for some $\varepsilon > 0$, for all t_1, t_2 in $T + 2\varepsilon B_Y$, and for all x_1, x_2 in U, one has

$$|\phi(t_1, x_1) - \phi(t_2, x_2)| \leq K(\|t_1 - t_2\|_Y + \|x_1 - x_2\|_X).$$

As usual, d_T denotes the distance function of T.

Corollary 3

If z belongs to $\partial f(x)$, then $(0, z)$ belongs to the set

(6) $$\overline{\mathrm{co}}\{\partial \phi(t, x) - \hat{K} \, \partial d_T(t) \times \{0\} : t \in M(x)\},$$

where $\hat{K}$ is any number greater than K.

Proof. We are going to place ourselves in the context of Theorem 2.8.2, but with altered data. We define

$$\hat{T} = T + \varepsilon B_Y, \qquad \hat{U} = \varepsilon B_Y \times U \subset Y \times X$$

$$\hat{f}_t(y, x) = \phi(t + y, x) - \hat{K} d_T(t + y)$$

$$\hat{f}(y, x) = \max\{\hat{f}_t(y, x) : t \in \hat{T}\}.$$

The mapping $(t, y, x) \to \partial \hat{f}_t(y, x)$ is closed because so are $\partial \phi$ and ∂d_T, so that $\partial_{[\hat{T}]} \hat{f}_t$ and $\partial \hat{f}_t$ coincide (Definition 2.8.1). The basic hypotheses of the theorem are satisfied. We conclude that $\partial \hat{f}(0, x)$ is contained in the set

(7) $$\overline{\mathrm{co}}\{\partial \phi(t, x) - \hat{K} \, \partial d_T(t) \times \{0\} : t \in \hat{M}(0, x)\},$$

where we have replaced $\partial \hat{f}_t(0, x)$ by the (possibly larger) set $\partial \phi(t, x) - \hat{K} \partial d_T(t) \times \{0\}$ (Proposition 2.3.3).

Lemma. For all (y, x) in $\hat{U}$, $\hat{M}(y, x) = M(x) - y$ and $\hat{f}(y, x) = f(x)$.

Note that the second assertion follows from the first by substitution. Once the lemma is proven, the resulting formula $\partial \hat{f}(0, x) = \{0\} \times \partial f(x)$, together with the fact that expression (6) and (7) coincide, immediately yield the corollary. To prove the lemma's first assertion, note that t lies in $\hat{M}(y, x)$ iff $\tau = t + y$ maximizes $\phi(\cdot, x) - \hat{K} d_T(\cdot)$ over $\hat{T} + y$. By Proposition 2.4.3 (with $S = \hat{T} + y$, $C = T$), this is equivalent to the statement that τ lies in T and maximizes $\phi(\cdot, x)$ over T; that is, that τ belongs to $M(x)$. □

2.8.5 Remark

When T is a subset of a finite-dimensional Banach space Y, as in Corollary 3 above, and when $M(x)$ admits a locally bounded selection $m(x)$, we may replace $M(x)$ in the preceding results by the set $\tilde{M}(x) = $ all limit points of $m(y)$ as y approaches x. (This observation is also due to Hiriart–Urruty.) To see this, simply apply the theorem to the modified problem in which T is replaced by $(\tilde{M}(x) + \varepsilon B_Y) \cap T$. For arbitrarily small $\varepsilon > 0$, this has no effect on f locally.

When X is Finite-Dimensional

When $X = R^n$ we can exploit the characterization of ∂f given by Theorem 2.5.1 to obtain a result along the lines of Theorem 2.8.2, but requiring far less structure on the data. As far as T is concerned, we now let it be simply a given abstract set. We define

$$f(x) = \sup\{f_t(x) : t \in T\}$$

as before, but our hypotheses will no longer imply that this supremum is attained. Now we posit only hypothesis (iii), namely,

$$|f_t(x_1) - f_t(x_2)| \leq K\|x_1 - x_2\|$$

for $t \in T$, x_1, x_2 in U. We assume that f is finite at some point in U, from which it follows easily that f is finite and Lipschitz on all of U.

2.8.6 Theorem

Let x be any point in U, and let S be a subset of measure 0 in U. Then $\partial f(x)$ is contained in the set C defined by

$$\mathrm{co}\left\{ \lim_{i \to \infty} \nabla f_{t_i}(x_i) : x_i \to x,\, x_i \notin S,\, t_i \in T,\, f_{t_i}(x) \to f(x) \right\}.$$

Proof. The definition of C expresses it as the convex hull of all points z of the form $\lim_{i \to \infty} \nabla f_{t_i}(x_i)$, where $x_i \notin S$, $\nabla f_{t_i}(x_i)$ exists, $x_i \to x$, and t_i is a maximizing sequence for $f_t(x)$ in T. It is easy to see that C is nonempty and

2.8 Pointwise Maxima

compact. Let v be any point in R^n, and define

$$m = \max\{\langle \zeta, v \rangle : \zeta \in C\}.$$

It suffices to prove that for any $\varepsilon > 0$, one has

(8) $$f^\circ(x; v) \leq m + \varepsilon,$$

since $f^\circ(x; \cdot)$ is the support function of $\partial f(x)$. We may assume $|v| = 1$.

The definition of C implies the existence of some δ in $(0, 1)$ such that whenever $y \in x + 2\delta B$ and $t \in T$ satisfy

$$y \notin S, \quad \nabla f_t(y) \text{ exists}, \quad f_t(x) \geq f(x) - \delta,$$

then one has

(9) $$\nabla f_t(y) \cdot v \leq m + \varepsilon.$$

Let λ be any point in $(0, \delta)$ and t any point in T satisfying $f_t(x) \geq f(x) - \delta$. Let Ω be the set of points in U at which $Df_t(\cdot)$ fails to exist. Now let y be any point in $x + \delta B$ such that the line segment $[y, y + \delta v]$ meets $S \cup \Omega$ in a set of 0 one-dimensional measure. (Note that almost all y in $x + \delta B$ have this property, since $S \cup \Omega$ has measure zero.) With the help of (9) we obtain

(10) $$f_t(y + \lambda v) - f_t(y) = \int_0^\lambda \nabla f_t(y + \tau v) \cdot v \, d\tau \leq \lambda(m + \varepsilon).$$

Since f_t is continuous, this inequality must in fact hold for all y in $x + \delta B$.

Pick r in $(0, \delta)$ such that $r^2 + 4rK < \delta$.

Lemma. For all y in $x + rB$ and λ in $(0, r)$, one has

$$f(y + \lambda v) - f(y) \leq \lambda(m + \varepsilon) + \lambda^2.$$

Clearly this will imply (8), and thus complete the proof.

To prove the lemma, given y and λ as stated, choose any t such that $f_t(y + \lambda v) \geq f(y + \lambda v) - \lambda^2$. Then

$$f_t(x) \geq f_t(y + \lambda v) - K\|x - y - \lambda v\|$$

$$\geq f(y + \lambda v) - \lambda^2 - K\|x - y - \lambda v\| \quad \text{(by choice of } t\text{)}$$

$$\geq f(x) - \lambda^2 - 2K\|x - y - \lambda v\|$$

$$\geq f(x) - r^2 - 4rK \geq f(x) - \delta.$$

This shows that (10) is applicable to this t and λ, whence

$$f(y + \lambda v) - f(y) \leq f_t(y + \lambda v) - f_t(y) + \lambda^2$$
$$\leq \lambda(m + \varepsilon) + \lambda^2,$$

which proves the lemma. □

2.8.7 Example (Greatest Eigenvalue)

Let $A(x)$ be an $n \times n$ matrix-valued function of a variable x in R^m. A function which arises naturally as a criterion in many engineering design problems, and which is known to be nondifferentiable in general, is

$$f(x) = \text{greatest eigenvalue of } A(x).$$

We shall assume that $A(\cdot)$ is C^1, and that for each x, $A(x)$ is positive-definite and symmetric. We wish to calculate $\partial f(x)$.

Some notation is in order. We know that there exists for each x a unitary matrix U (depending on x) such that

$$U^*A(x)U = D,$$

where D is the diagonal matrix of eigenvalues $\lambda_1, \lambda_2, \ldots, \lambda_n$ of $A(x)$. We shall assume without loss of generality that (for the particular x of interest) $\lambda_1 \geq \lambda_2 \geq \cdots \geq \lambda_n$. For any integer j between 1 and n, let S_j denote the unit vectors w of R^n whose components beyond the jth are zero. (Thus, for example, S_1 consists of the two vectors $[\pm 1, 0, \ldots, 0]$.) Finally, let r be the multiplicity of the eigenvalue λ_1, and let $A'_i(x)$ denote the matrix $A(x)$ differentiated with respect to x_i.

2.8.8 Proposition

The function f is Lipschitz and regular at x, and one has

$$\partial f(x) = \text{co}\{[\langle U^*A'_1 Uw, w\rangle, \ldots, \langle U^*A'_m Uw, w\rangle] : w \in S_r\}.$$

f is differentiable at x if λ_1 has multiplicity 1, and in that case one has

$$\frac{\partial}{\partial x_i} f(x) = \text{the } (1,1)\text{-entry in } U^*\left[\frac{\partial}{\partial x_i} A(x)\right] U.$$

Proof. From the characterization

(11) $$f(x) = \max\{\langle A(x)w, v\rangle : w, v \in \bar{B}\}$$

(where B is the unit ball in R^n) combined with Theorem 2.8.2, it ensues that f

is Lipschitz and regular at x. From Corollary 2 we have

$$(12) \quad \partial f(x) = \text{co}\left\{\frac{d}{dx}\langle A(x)w, v\rangle : (w, v) \in M(x)\right\}$$

$$= \text{co}\{[\langle A'_1(x)w, v\rangle, \ldots, \langle A'_m(x)w, v\rangle] : (w, v) \in M(x)\},$$

where $M(x)$ is the set of maximizing (w, v) in Eq. (11). Let U be the unitary matrix defined earlier. Since $U\bar{B} = \bar{B}$, one has

$$f(x) = \max\{\langle A(x)U\tilde{w}, U\tilde{v}\rangle : \tilde{w}, \tilde{v} \in \bar{B}\}$$

$$= \max\{\langle U^*A(x)U\tilde{w}, \tilde{v}\rangle : \tilde{w}, \tilde{v} \in \bar{B}\}$$

$$= \max\{\langle D\tilde{w}, \tilde{v}\rangle : \tilde{w}, \tilde{v} \in \bar{B}\}.$$

It is a simple exercise to verify that the vectors $(\tilde{w}, \tilde{v})$ at which this last maximum is attained are precisely those of the form $(\tilde{w}, \tilde{w})$, where $\tilde{w}$ belongs to S_r. It follows that the elements (w, v) of $M(x)$ are those of the form $(U\tilde{w}, U\tilde{w})$, where $\tilde{w}$ belongs to S_r. Substitution into Eq. (12) now yields the formula as stated in the proposition. The last assertion is an immediate consequence of this formula. □

2.9 EXTENDED CALCULUS

In Section 2.4 we defined $\partial f(x)$ for any extended-valued function f provided only that it be finite at x. With few exceptions, however, the calculus we have developed has been for locally Lipschitz functions; it is natural to ask to what extent it would be possible to treat more general functions. Recently, significant progress on this issue has been made by Rockafellar, Hiriart–Urruty, and Aubin.

We must expect somewhat sparser results and greater technical detail in such an endeavor. A major contribution of Rockafellar's has been to suitably extend the definition of the generalized directional derivative f°, and to identify a subclass of extended-valued functions (those that are "directionally Lipschitz") which is general enough for most applications yet well-behaved enough to lead to several useful extensions of the Lipschitz calculus. These results have been obtained in the broader context of locally convex spaces rather than in Banach spaces, with new limiting concepts. For our purposes, however, it shall suffice to remain in a Banach space X. The results and proofs of this section are due to Rockafellar.

First on the agenda then, is the study of what becomes of f° in the non-Lipschitz case. In view of Theorem 2.4.9 we would expect f° to be such that the epigraph of $f^\circ(x; \cdot)$ is the set $T_{\text{epi}\,f}(x, f(x))$ (assuming for the moment that the latter is an epigraph). But what is the direct characterization of $f°$? The answer below involves some complicated limits for which some additional notation is in order. Following Rockafellar, the expression

$$(y, \alpha) \downarrow_f x$$

shall mean that $(y, \alpha) \in \text{epi } f$, $y \to x$, $\alpha \to f(x)$.

2.9.1 Theorem (Rockafellar)

Let $f: X \to R \cup \{\infty\}$ be an extended real-valued function and x a point where f is finite. Then the tangent cone $T_{\text{epi} f}(x, f(x))$ is the epigraph of the function $f°(x; \cdot): X \to R \cup \{\pm \infty\}$ defined as follows:

$$f°(x; v) = \lim_{\varepsilon \downarrow 0} \limsup_{\substack{(y, \alpha) \downarrow_f x \\ t \downarrow 0}} \inf_{w \in v + \varepsilon B} \frac{f(y + tw) - \alpha}{t}.$$

Proof. We must show that a point (v, r) of $X \times R$ belongs to $T_{\text{epi} f}(x, f(x))$ iff $f°(x; v) \leq r$.

Necessity. Let $(v, r) \in T_{\text{epi} f}(x, f(x))$, and let any $\varepsilon > 0$ be given. It suffices to show

(1) $$\limsup_{\substack{(y, \alpha) \downarrow_f x \\ t \downarrow 0}} \inf_{w \in v + \varepsilon B} \frac{f(y + tw) - \alpha}{t} \leq r.$$

To see this, let $(y_i, \alpha_i) \downarrow_f x$ and $t_i \downarrow 0$. Since $(v, r) \in T_{\text{epi} f}(x, f(x))$, there exists (by Theorem 2.4.5) a sequence (v_i, r_i) converging to (v, r) such that $(y_i, \alpha_i) + t_i(v_i, r_i) \in \text{epi } f$, that is, such that

$$\alpha_i + t_i r_i \geq f(y_i + t_i v_i).$$

Consequently, for i sufficiently large,

$$\inf_{w \in v + \varepsilon B} \frac{f(y_i + t_i w) - \alpha_i}{t_i} \leq \frac{f(y_i + t_i v_i) - \alpha_i}{t_i} \leq r_i,$$

and (1) follows.

Sufficiency. Suppose $f°(x; v) \leq r$ and let (y_i, α_i) be any sequence in epi f converging to $(x, f(x))$ and t_i any sequence decreasing to 0. To prove that (v, r) belongs to $T_{\text{epi} f}(x, f(x))$, it suffices to produce a sequence (v_i, r_i) converging to (v, r) such that

$$(y_i, \alpha_i) + t_i(v_i, r_i) \in \text{epi } f \quad \text{infinitely often;}$$

that is,

(2) $$\alpha_i + t_i r_i \geq f(y_i + t_i v_i) \quad \text{infinitely often}$$

(by an evident "subsequence version" of the characterization given in Theorem 2.4.5). For each positive integer n, there is an $\varepsilon_n < 1/n$ such that

$$\limsup_{\substack{(y,\alpha)\downarrow_f x \\ t\downarrow 0}} \inf_{w \in v + \varepsilon_n B} \frac{f(y + tw) - \alpha}{t} \leq r + \frac{1}{n}$$

(since $f°(x; v) \leq r$). Consequently, since $(y_i, \alpha_i) \downarrow_f x$, there must be an index $i = i(n) > n$ such that

$$\inf_{w \in v + \varepsilon_n B} \frac{f(y_i + t_i w) - \alpha_i}{t_i} \leq r + \frac{2}{n},$$

and therefore a point v_i in $v + \varepsilon_n B$ such that

(3) $$\frac{f(y_i + t_i v_i) - \alpha_i}{t_i} \leq r + \frac{3}{n}.$$

Let us define, for indices i in the subsequence $i(1), i(2), \ldots,$

(4) $$r_i = \max\left\{r, \frac{f(y_i + t_i v_i) - \alpha_i}{t_i}\right\}.$$

Note that the inequality in (2) is satisfied, so we shall be through provided we verify that r_i converges to r. But this is evident from (3) and (4), and so the proof is complete. $\square$

It is not hard to see that if f is lower semicontinuous at x, then $f°(x; v)$ is given by the slightly simpler expression

$$\lim_{\varepsilon \downarrow 0} \limsup_{\substack{y \downarrow_f x \\ t\downarrow 0}} \inf_{w \in v + \varepsilon B} \frac{f(y + tw) - f(y)}{t},$$

where $y \downarrow_f x$ signifies that y and $f(y)$ converge to x and $f(x)$, respectively. Note also (in all cases) that the limit over $\varepsilon > 0$ is equivalent to a supremum over $\varepsilon > 0$. As a consequence of the theorem, we see that the extended $f°$ plays the same role vis-à-vis ∂f as it did in the Lipschitz case:

Corollary
One has $\partial f(x) = \emptyset$ iff $f°(x; 0) = -\infty$. Otherwise, one has

$$\partial f(x) = \{\zeta \in X^* : f°(x; v) \geq \langle \zeta, v \rangle \text{ for all } v \in X\},$$

and

$$f°(x; v) = \sup\{\langle \zeta, v \rangle : \zeta \in \partial f(x)\}.$$

Proof. ζ belongs to $\partial f(x)$, by definition, iff $\langle (\zeta, -1), (v, r) \rangle \leq 0$ for all $(v, r) \in T_{\text{epi } f}(x, f(x)) = \text{epi } f°(x; \cdot)$ (by the theorem). Equivalently, $\zeta \in \partial f(x)$ iff $\langle \zeta, v \rangle \leq r$ for all v in X and r in R such that $r \geq f°(x; v)$. If $f°(x; v) = -\infty$ for some v, there can be no such ζ; that is, $\partial f(x) = \emptyset$. On the other hand, we have $f°(x; \lambda v) = \lambda f°(x; v)$ for all v and $\lambda > 0$, since epi $f°(x; \cdot)$ is a cone (we use the convention $\lambda(-\infty) = -\infty$). Thus $f°(x; \cdot)$ is $-\infty$ somewhere iff it is $-\infty$ at 0, and the first assertion follows.

Assuming then that $f°(x; \cdot) > -\infty$, we have $\zeta \in \partial f(x)$ iff $\langle \zeta, v \rangle \leq f°(x; v)$ for all v, as we have seen above. This is the second assertion, which is equivalent to the statement that $f°(x; \cdot)$ is the support function of $\partial f(x)$; that is, the third assertion (see Proposition 2.1.4). □

In view of Theorem 2.4.9, the expression for $f°$ given in the theorem must reduce, when f is locally Lipschitz, to the one now so familiar from the preceding sections. This is in fact quite easy to show but, again following Rockafellar, we pursue the more ambitious goal of identifying a useful intermediate class of functions f for which this is "essentially" the case.

2.9.2 Definition

f is *directionally Lipschitz at* x *with respect to* $v \in X$ if (f is finite at x and) the quantity

$$f^+(x; v) := \limsup_{\substack{(y, \alpha) \downarrow_f x, w \to v \\ t \downarrow 0}} \frac{f(y + tw) - \alpha}{t}$$

is not $+\infty$. We say f is *directionally Lipschitz at* x if f is directionally Lipschitz at x with respect to at least one v in X.

We can relate this concept to a geometrical one introduced in Section 2.4.

2.9.3 Proposition

f is directionally Lipschitz at x with respect to v iff for some $\beta \in R, (v, \beta)$ is hypertangent to epi f at $(x, f(x))$.

Proof. Suppose first that (v, β) is hypertangent to epi f at $(x, f(x))$. Then, for some $\varepsilon > 0$, one has by definition

(5) $\quad [(x, f(x)) + \varepsilon B] \cap C + t[(v, \beta) + \varepsilon B] \subset C \quad \text{for } t \in (0, \varepsilon),$

where $B = B_{X \times R}$ and $C = \text{epi } f$. Consequently, whenever $(y, \alpha) \in C$ is close to $(x, f(x))$, and w is close to v, one has $(y, \alpha) + t(w, \beta) \in C$; that is,

$$\alpha + t\beta \geq f(y + tw).$$

It follows that the expression $f^+(x; v)$ in Definition 2.9.2 is bounded above by β, so that f is directionally Lipschitz at x with respect to v.

Now let us assume this last conclusion, and let β be any number greater than $f^+(x; v)$. Then, for all $(y, \alpha) \in \text{epi } f$ near $(x, f(x))$, for all w near v and $t > 0$ near 0, one has

$$\frac{f(y + tw) - \alpha}{t} < \beta.$$

This clearly implies (5) for some ε; that is, that (v, β) is hypertangent to epi f at $(x, f(x))$. □

The proof actually showed:

Corollary

If $f^+(x; v)$ is less than $+\infty$, then it equals

$$\inf\{\beta : (v, \beta) \text{ is hypertangent to epi } f \text{ at } (x, f(x))\}.$$

Here are some criteria assuring that f is directionally Lipschitz.

2.9.4 Theorem

Let f be extended real-valued, and let x be a point at which f is finite. If any of the following hold, then f is directionally Lipschitz at x:

(i) *f is Lipschitz near x.*
(ii) *$f = \psi_C$, where C admits a hypertangent at x.*
(iii) *f is convex, and bounded in a neighborhood of some point (not necessarily x).*
(iv) *f is nondecreasing with respect to the partial ordering induced by some closed convex cone K with nonempty interior.*
(v) *$X = R^n$, f is lower semicontinuous on a neighborhood of x, and $\partial f(x)$ is nonempty and does not include an entire line.*

Proof. (i) has already been noted, and (ii) is immediate from Proposition 2.9.3. Let us turn to (iii).

In view of Proposition 2.9.3, it suffices to prove that $C = \text{epi } f$ admits a hypertangent at $(x, f(x))$. If f is bounded above by $\beta - 1$ on a neighborhood of $x + \lambda v$ for some $\lambda > 0$, it follows that (v, α) satisfies $(x, f(x)) + \lambda(v, \alpha) \in \text{int } C$, where $\alpha = (\beta - f(x))/\lambda$. Thus the following lemma is all we need:

Lemma. *If a set C is convex and contains, along with x, a neighborhood of $x + \lambda v$ for some vector v and $\lambda > 0$, then v is hypertangent to C at x.*

To see this, suppose that $x + \lambda v + \varepsilon B \subset C$ for $\varepsilon > 0$. Choose $\delta > 0$ such that $(x + \delta B) + \lambda(v + \delta B) \subset C$. Then, for any t in $(0, \lambda)$, $y \in x + \delta B$, $w \in v + \delta B$, one has

$$y + tw = \left(1 - \frac{t}{\lambda}\right) y + \frac{t}{\lambda}(y + \lambda w).$$

If also $y \in C$, then this last expression belongs to C by convexity. That is, we have shown $(x + \delta B) \cap C + t(v + \delta B) \subset C$ for $t \in (0, \lambda)$, which establishes that v is hypertangent to C at x.

Next on the agenda is the proof of (iv). The condition is that $f(x_1) \leq f(x_2)$ whenever $x_1 \leq_K x_2$; that is, $x_2 - x_1 \in K$. Suppose $-v \in \text{int } K$, so that $-v + \varepsilon B \subset K$ for some $\varepsilon > 0$. Then for all points w such that $-w \in -v + \varepsilon B$, and for all $t \geq 0$, one has $-tw \in K$, so that $y + tw \leq_K y$ for all y. Consequently,

$$\frac{f(y + tw) - f(y)}{t} \leq 0$$

for y near x, w near v, and we derive the result that $f^+(x; v)$ is finite, so that f is directionally Lipschitz at x (with respect to v).

Finally we turn to (v). The statement that $\partial f(x)$ does not include a line is equivalent, in light of the corollary to Theorem 2.9.1, to the condition that the set $D = \{v : f°(x; v) < \infty\}$ is not contained in any subspace of dimension less than n, which is equivalent to saying that D has nonempty interior; suppose $v \in \text{int } D$. Now $f°(x; \cdot)$ is locally bounded in int D, because it is convex and X is finite-dimensional. It follows that for some $\beta \in R$, $(v, \beta) \in \text{int epi } f°(x; \cdot)$ = int $T_{\text{epi } f}(x, f(x))$. But when $X = R^n$, the existence of hypertangents to a closed set at a point is equivalent to the tangent cone at the point having nonempty interior (Corollary 1, Theorem 2.5.8). It now follows from Proposition 2.9.3 that f is directionally Lipschitz at x. (Note that the lower semicontinuity hypothesis of (v) could be weakened to: epi f is locally closed near $(x, f(x))$.) □

We denote by $D_f(x)$ the set of vectors v, if any, with respect to which f is directionally Lipschitz at x.

2.9.5 Theorem

Let f be finite at x, and suppose $D_f(x) \neq \emptyset$. Then

$$D_f(x) = \text{int}\{v : f°(x; v) < \infty\}.$$

Furthermore, $f°(x; \cdot)$ is continuous at each $v \in D_f(x)$ and agrees there with $f^+(x; \cdot)$.

Proof. We shall derive this as a consequence of Theorem 2.4.8 (for $C = \text{epi } f$). Let K denote the set of all hypertangents to C at $(x, f(x))$; K is nonempty by assumption. Then $D_f(x) = \pi_X K$ (projection of K on X) by Proposition 2.9.3, and $K = \text{int } T_C(x, f(x))$ by Theorem 2.4.8. So in order to prove the first assertion of the theorem, it suffices to establish

(6) $$\pi_X \text{int } T_C(x, f(x)) = \text{int}\{v : f^\circ(x; v) < \infty\},$$

which we now proceed to do.

If v lies in the left-hand side of Eq. (6), then $(v, \beta) \in \text{int } T_C(x, f(x))$ for some β. By Theorem 2.9.1 then, $(v, \beta) \in \text{int epi } f^\circ(x; \cdot)$. Consequently, $f^\circ(x; w) < \infty$ for all w near v, and v lies in the right-hand side of Eq. (6).

If now v lies in the right-hand side of Eq. (6), then for some $\varepsilon > 0$ and β in R, $f^\circ(x; w) < \beta - 1$ for all w in $v + \varepsilon B$. It follows that (w, α) lies in epi $f^\circ(x; \cdot) = T_C(x, f(x))$ for all w near v and α near β; that is, $(v, \beta) \in \text{int } T_C(x, f(x))$, so $v \in \pi_X \text{int } T_C(x, f(x))$.

A convex function such as $f^\circ(x; \cdot)$ is always continuous on the interior of the set $\{v : f^\circ(x; v) < \infty\}$, provided it is bounded above in a neighborhood of one point; this is the case precisely because $D_f(x) \neq \varnothing$.

Finally let us examine the expressions

$$f^\circ(x; v) = \inf\{\beta : (v, \beta) \in T_C(x, f(x))\}$$

$$f^+(x; v) = \inf\{\beta : (v, \beta) \in K\},$$

where v belongs to $D_f(x) = \pi_X K$. The first equality is a result of the fact that $T_C(x, f(x))$ is the epigraph of $f^\circ(x; \cdot)$ (and $f^\circ(x; v) < \infty$); the second is the corollary to Proposition 2.9.3. Since, as we have seen, $K = \text{int } T_C(x, f(x))$, it follows that $f^+(x; v) = f^\circ(x; v)$, and the theorem is proved. □

In view of Corollary 1 to Theorem 2.5.8, the proof above can be modified to yield the following:

Corollary

Let X be finite-dimensional, and suppose epi f is locally closed near $(x, f(x))$. Then one has

$$D_f(x) = \text{int}\{v : f^\circ(x; v) < \infty\},$$

and so f is directionally Lipschitz at x precisely when the set $\{v : f^\circ(x; v) < \infty\}$ has nonempty interior.

The Asymptotic Generalized Gradient

A convenient measure of the degree to which a given function f fails to be Lipschitz near a point x (at which f is finite) is provided by the *asymptotic generalized gradient* of f at x, denoted $\partial^\infty f(x)$, which is defined to be the set

$$\{\zeta \in X^* : (\zeta, 0) \in N_{\text{epi } f}(x, f(x))\}$$

(compare with Definition 2.4.10). Note that $\partial^\infty f(x)$ always contains 0. The first two assertions below follow from the definition and the theory of recession cones (see Rockafellar, 1979b, p. 350); the third is a consequence of Theorem 2.5.6, Corollary 2.

2.9.6 Proposition

$\partial^\infty f(x)$ is a closed convex cone. If $\partial f(x)$ is nonempty, then one has

$$N_{\text{epi } f}(x, f(x)) = \bigcup_{\lambda > 0} \lambda [\partial f(x), -1] \cup [\partial^\infty f(x), 0],$$

and in this case $\partial^\infty f(x)$ reduces to $\{0\}$ iff $\partial f(x)$ is bounded. If X is finite-dimensional, and if epi f is locally closed near $(x, f(x))$ (e.g., if f is lower semicontinuous) then $\partial f(x) \cup (\partial^\infty f(x) \setminus \{0\}) \neq \emptyset$.

When f is Lipschitz near x, then of course $\partial f(x)$ is nonempty and bounded (and so $\partial^\infty f(x) = \{0\}$). The converse holds for a large class of functions:

2.9.7 Proposition

Suppose that X is finite-dimensional, that f is finite at x, and that epi f is locally closed near $(x, f(x))$. Then the following are equivalent:
(a) $\partial f(x)$ is nonempty and bounded.
(b) f is Lipschitz near x.
(c) $\partial^\infty f(x) = \{0\}$.

Proof. We shall show that (a) implies (b) implies (c) implies (a). If (a) holds, then f is directionally Lipschitz at x by Theorem 2.9.4. It then follows from the corollary to Theorem 2.9.1 together with Theorem 2.9.5 that $f^+(x; 0)$ is finite. This is readily seen to imply (b). We have already observed that (b) implies (c), and the fact that (c) implies (a) is immediate from Proposition 2.9.6. □

Sum of Two Functions

We are now in a position to prove the extended formula for the generalized gradient of a sum of two functions. Recall that regularity in the extended setting was defined in Definition 2.4.10.

2.9.8 Theorem (Rockafellar)

Let f_1 and f_2 be finite at x, and let f_2 be directionally Lipschitz at x. Suppose that

(7) $\qquad \{v : f_1^\circ(x; v) < \infty\} \cap \text{int}\{v : f_2^\circ(x; v) < \infty\} \neq \emptyset.$

Then

(8) $$\partial(f_1 + f_2)(x) \subset \partial f_1(x) + \partial f_2(x),$$

where the set on the right is weak*-closed. If in addition f_1 and f_2 are regular at x, then equality holds, and if $\partial f_1(x)$ and $\partial f_2(x)$ are nonempty, then $f_1 + f_2$ is regular at x.

Proof. As in the proof of Proposition 2.3.3 in the locally Lipschitz case, we begin by proving:

Lemma.

$$(f_1 + f_2)^\circ(x; v) \leq f_1^\circ(x; v) + f_2^\circ(x; v).$$

Let $f_0 = f_1 + f_2$, $l_i(v) = f_i^\circ(x; v)$, for $i = 0, 1, 2$. We wish to prove

(9) $$l_0(v) \leq l_1(v) + l_2(v),$$

where the convention $\infty - \infty = \infty$ is in force, and we start with the case in which v lies in the left-hand side of (1).

Let $\beta > l_2(v) = f_2^\circ(x; v)$. By Theorem 2.9.5, we have $f_2^\circ(x; v) = f_2^+(x; v) < \beta$. Hence, for some $\delta > 0$, one has

(10) $$\frac{f_2(y + tw) - \alpha_2}{t} < \beta$$

whenever $t \in (0, \delta)$, $y \in x + \delta B$, $w \in v + \delta B$, $\alpha_2 \geq f_2(y)$, $|\alpha_2 - f(y)| < \delta$. Recall that by definition

$$l_0(y) = \lim_{\varepsilon \downarrow 0} \limsup_{\substack{(y, \alpha) \downarrow_f x \\ t \downarrow 0}} \inf_{w \in v + \varepsilon B} \frac{f_0(y + tw) - \alpha}{t}.$$

We can write this difference quotient as

(11) $$\frac{f_1(y + tw) - \alpha_1}{t} + \frac{f_2(y + tw) - \alpha_2}{t},$$

where $\alpha_1 + \alpha_2 = \alpha$, and the lim sup can be taken equivalently as $t \downarrow 0$, $y \to x$, $\alpha_i \to f_i(x)$ with $\alpha_i \geq f_i(x)$. Invoking (10) in (11), we obtain

$$l_0(y) \leq \lim_{\varepsilon \downarrow 0} \limsup_{\substack{(y, \alpha) \downarrow_f x \\ t \downarrow 0}} \inf_{w \in v + \varepsilon B} \left\{ \frac{f_1(y + tw) - \alpha_1}{t} + \beta \right\}$$

$$= f_1^\circ(x; v) + \beta = l_1(v) + \beta.$$

Since this is true for all $\beta > l_2(v)$, we derive (9), which we still need to establish for general v. If either $l_1(v)$ or $l_2(v) = +\infty$, then (9) is trivial, so suppose that v belongs to $D_1 \cap D_2$, where D_i is the convex set $\{w: l_i(w) < \infty\}$. By (7) there is a point $\tilde{v}$ in $D_1 \cap \text{int } D_2$; by convexity, we have $v_\varepsilon := (1 - \varepsilon)v + \varepsilon\tilde{v} \in D_1 \cap \text{int } D_2$ for ε in $(0, 1)$. By the case already treated, then, we know

$$l_0(v_\varepsilon) \leq l_1(v_\varepsilon) + l_2(v_\varepsilon).$$

Because the functions l_i are convex and lower semicontinuous, we have $\lim_{\varepsilon \to 0} l_i(v_\varepsilon) = l_i(v)$, so (9) results, and hence the lemma is proved.

Let us now prove (8). If either $l_1(0)$ or $l_2(0)$ is $-\infty$, then, as the proof of the lemma actually showed, $l_0(0) = -\infty$ also. Then (corollary, Theorem 2.9.1) both sides of (8) are empty and there is nothing to prove. Assume therefore that $l_1(0) = l_2(0) = 0$ (the only other possibility, since l_i is sublinear). Then one has

(12) $$\partial f_0(x) = \{z : l_0(v) \geq \langle z, v \rangle \text{ for all } v\}$$

$$\subset \{z : (l_1 + l_2)(v) \geq \langle z, v \rangle \text{ for all } v\}$$

$$= \partial(l_1 + l_2)(0),$$

where the final equality is valid because $l_1 + l_2$ is a convex function which is 0 at 0. The assumption (7) is precisely what is needed to make the following formula from convex analysis valid:

$$\partial(l_1 + l_2)(0) = \partial l_1(0) + \partial l_2(0) \quad (= \partial f_1(x) + \partial f_2(x)).$$

This together with (12) yields (8). As the subdifferential of $l_1 + l_2$ at 0, $\partial f_1(x) + \partial f_2(x)$ is automatically weak*-closed.

The final component to be proved concerns the equality in expression (8). If either $l_1(0)$ or $l_2(0)$ is $-\infty$, then, as we have seen, $l_0(0)$ is also, and both sides of (8) are empty (and hence, equal). Assuming therefore that this is not the case, and that f_1, f_2 are regular at x, we have (as an easy consequence of regularity) that, for any v,

$$l_i(v) = \liminf_{\substack{w \to v \\ t \downarrow 0}} \frac{f_i(x + tw) - f_i(x)}{t} \quad (i = 1, 2).$$

Taking lower limits in the expression

$$\frac{f_1(x + tw) - f_1(x)}{t} + \frac{f_2(x + tw) - f_2(x)}{t} = \frac{f_0(x + tw) - f_0(x)}{t},$$

we derive (since the first two terms tend separately to $l_1(v), l_2(v) > -\infty$) the inequality

(13) $$l_1(v) + l_2(v) \leq \liminf_{\substack{w \to v \\ t \downarrow 0}} [f_0(x + tw) - f_0(x)] \leq l_0(v).$$

In view of the lemma, we have therefore

$$l_1(v) + l_2(v) = l_0(v) \quad \text{for all } v.$$

This permits the reversal of the inclusion in (12), so (8) holds with equality. Further, $l_0(\cdot)$ is now seen to be the support function of $\partial f_0(x)$, so $l_0(v) = f_0^\circ(x; v)$ and (13) implies that f_0 is regular at x. □

Corollary 1

Suppose that f_1 is finite at x and f_2 is Lipschitz near x. Then one has $\partial(f_1 + f_2)(x) \subset \partial f_1(x) + \partial f_2(x)$, and there is equality if f_1 and f_2 are also regular at x.

Proof. The theorem may be applied directly, since 0 belongs to the set in (7). □

Corollary 2

Let C_1 and C_2 be subsets of X and let $x \in C_1 \cap C_2$. Suppose that

$$T_{C_1}(x) \cap \operatorname{int} T_{C_2}(x) \neq \emptyset,$$

and that C_2 admits at least one hypertangent vector v at x. Then one has

(14) $$T_{C_1 \cap C_2}(x) \supset T_{C_1}(x) \cap T_{C_2}(x)$$

(15) $$N_{C_1 \cap C_2}(x) \subset N_{C_1}(x) + N_{C_2}(x),$$

where the set on the right in expression (15) is weak*-closed. Equality holds in (14) and (15) if C_1 and C_2 are regular at x, in which case $C_1 \cap C_2$ is also regular at x.

Proof. Apply the theorem to $f_1 = \psi_{C_1}$, $f_2 = \psi_{C_2}$. The existence of v as above is equivalent to f_2 being directionally Lipschitz (by Proposition 2.9.3). □

Corollary 3

Suppose $X = R^n$. Then the hypothesis in Theorem 2.9.8 that f_2 be directionally Lipschitz can be replaced by the assumption that f_2 is lower semicontinuous in a neighborhood of x. Likewise, the existence of v in Corollary 2 can be replaced by the assumption that C_2 is locally closed near x.

Proof. Invoke the corollary to Theorem 2.9.5 in the first case, Corollary 1 to Theorem 2.5.8 in the second. □

Composition with a Differentiable Map

The following is Rockafellar's extended analogue of Theorem 2.3.10:

2.9.9 Theorem
Let $f = g \circ F$, where F is a strictly differentiable mapping from X to another Banach space Y and g is an extended-valued function on Y. Suppose that g is finite and directionally Lipschitz at $F(x)$ with

$$(\text{range } D_s F(x)) \cap \text{int}\{v : g^\circ(F(x); v) < \infty\} \neq \emptyset.$$

Then one has

(16) $$\partial f(x) \subset D_s F(x)^* \circ \partial g(F(x)),$$

and equality holds if g is regular at x.

Proof. Let $y = F(x)$, and define h on $X \times Y$ by

$$h(x', y') = \begin{cases} f(x') & \text{if } y' = F(x') \\ +\infty & \text{otherwise.} \end{cases}$$

Note that $h = f_1 + f_2$, where f_1 is the indicator of the graph of F and where $f_2(x', y') = g(y')$. We let A stand for $D_s F(x)$.

Lemma.

$$h^\circ(x, y; v, w) = \begin{cases} f^\circ(x; v) & \text{if } w = A(v) \\ +\infty & \text{otherwise.} \end{cases}$$

To see this, recall that h° above is given by

$$\lim_{\substack{\varepsilon \to 0 \\ (x', y', \alpha) \downarrow_h (x, y) \\ t \downarrow 0}} \limsup_{(v', w') \in (v, w) + \varepsilon B} \inf \frac{h(x' + tv', y' + tw') - \alpha}{t},$$

where the condition $\alpha \geq h(x', y')$ implicit above is equivalent to $\alpha \geq f(x')$ and $y' = F(x')$, and where

$$h(x' + tv', y' + tw') = \begin{cases} f(x' + tv') & \text{if } w' = \dfrac{F(x' + tv') - F(x')}{t} \\ +\infty & \text{otherwise.} \end{cases}$$

2.9 Extended Calculus

Since F is strictly differentiable at x, we have

$$\limsup_{\substack{x'\to x,\, v'\to v \\ t\downarrow 0}} \frac{F(x'+tv') - F(x')}{t} = A(v),$$

as an easy consequence of Proposition 2.2.1. It follows that the expression for $h°$ above is $+\infty$ if $w \neq A(v)$, while otherwise it is

$$\lim_{\varepsilon\to 0} \limsup_{\substack{(x',\alpha)\downarrow_f x \\ t\downarrow 0}} \inf_{v'\in v+\varepsilon B} \frac{f(x'+tv') - \alpha}{t} = f°(x; v),$$

as stated in the lemma. We now calculate as follows:

$$\partial h(x, y) = \{(\zeta, \phi) : h°(x, y; v, w) \geq \langle \zeta, v\rangle + \langle \phi, w\rangle \text{ for all } v, w\}$$

$$= \{(\zeta, \phi) : f°(x, v) \geq \langle \zeta, v\rangle + \langle \phi, A(v)\rangle \text{ for all } v\}$$

$$= \{(\zeta, \phi) : f°(x; v) \geq \langle \zeta + A^*(\phi), v\rangle \text{ for all } v\}$$

$$= \{(\zeta, \phi) : \zeta + A^*(\phi) \in \partial f(x)\},$$

and consequently,

(17) $$\partial f(x) = \{\zeta : (\zeta, 0) \in \partial h(x, y)\}.$$

The next step is to apply Theorem 2.9.8 to the representation $h = f_1 + f_2$ noted above. The strict differentiability implies for the set $G = \text{graph } F$ that

$$T_G(x, y) = K_G(x, y) = \text{graph } A.$$

Thus, f_1 is regular at (x, y) and

$$f_1°(x, y; v, w) = \begin{cases} 0 & \text{if } w = A(v) \\ +\infty & \text{otherwise,} \end{cases}$$

(18) $$\partial f_1(x, y) = N_G(x, y) = \text{polar of graph } A$$

$$= \{(\zeta, \phi) : \zeta = -A^*(\phi)\}.$$

Clearly f_2 inherits the directionally Lipschitz property from g, and one has

$$f_2°(x, y; v, w) = g°(y; w),$$

(19) $$\partial f_2(x, y) = \{(0, \phi) : \phi \in \partial g(y)\}.$$

In particular, the set

$$\{(v,w): f_1^\circ(x,y;v,w) < \infty\} \cap \text{int}\{(v,w): f_2^\circ(x,y;v,w) < \infty\}$$
$$= \{(v, A(v)): A(v) \in \text{int}\{w: g^\circ(y;w) < \infty\}\}$$

is nonempty by assumption. Therefore the hypotheses of Theorem 2.9.8 are satisfied, and we have

$$\partial h(x,y) \subset \partial f_1(x,y) + \partial f_2(x,y).$$

Combining this with (17) (18) (19) gives (16). The remaining assertions about equality are also direct translations from Theorem 2.9.8. □

Corollary 1

Let $x \in C := F^{-1}(\Omega)$, where $\Omega \subset Y$ and where F is strictly differentiable at x. Suppose that Ω admits a hypertangent at $F(x)$, and that

$$(\text{range } D_s F(x)) \cap \text{int } T_\Omega(F(x)) \neq \varnothing.$$

Then one has

$$T_C(x) \supset D_s F(x)^{-1}[T_\Omega(F(x))]$$
$$N_C(x) \subset D_s F(x)^* \circ N_\Omega(F(x)),$$

and equality holds if Ω is also regular at $F(x)$.

Proof. Apply Theorem 2.9.9 to $g = \psi_\Omega$. □

Corollary 2

Suppose $Y = R^n$. Then the hypothesis in Theorem 2.9.9 that g is directionally Lipschitz can be replaced with the condition that g is lower semicontinuous on a neighborhood of $F(x)$. The existence of a hypertangent in Corollary 1 can be replaced with the condition that Ω is closed in a neighborhood of $F(x)$.

Proof. Invoke the corollary to Theorem 2.9.5 in the first case, Corollary 1 to Theorem 2.5.8 in the second. □

Sets Defined by Inequalities

2.9.10 Theorem

Let $C = \{x': f(x') \leq 0\}$, and let x be a point satisfying $f(x) = 0$. Suppose that f is directionally Lipschitz at x with $0 \notin \partial f(x) \neq \varnothing$, and let $D = \{v: f^\circ(x;v) <$

$\infty\}$. *Then C admits a hypertangent at x and one has*

(20) $$T_C(x) \supset \{v : f^\circ(x; v) \leq 0\}$$

(21) $$\text{int } T_C(x) \supset \{v \in \text{int } D : f^\circ(x; v) < 0\} \neq \varnothing$$

(22) $$N_C(x) \subset \bigcup_{\lambda > 0} \lambda \, \partial f(x) \cup \partial^\infty f(x),$$

where the set on the right of (22) is weak*-closed. If in addition f is regular at x, then equality holds in each of the above and C is regular at x.

Proof. The result will be derived by invoking Corollary 2 of Theorem 2.9.8, for the sets $C_1 = \{(z, \mu) : \mu = 0\}$ and $C_2 = \text{epi } f$ at the point $(x, 0)$ (note that $C \times \{0\} = C_1 \cap C_2$). One has

(23) $$T_{C_1}(x, 0) = \{(z, \mu) : \mu = 0\}, \qquad T_{C_2}(x, 0) = \text{epi } f^\circ(x; \cdot),$$

and by polarity

(24) $$N_{C_1}(x, 0) = \{(\zeta, \mu) : \zeta = 0\}$$

$N_{C_2}(x, 0) = $ the cone in the statement of Proposition 2.9.6.

From Theorem 2.9.5, the function $f^\circ(x; \cdot)$ is continuous on int D, so one also has

(25) $$\text{int epi } f^\circ(x; \cdot) = \{(v, \beta) : v \in \text{int } D, f^\circ(x; v) < \beta\},$$

and this set is nonempty. Therefore by Eq. (23),

$$T_{C_1}(x, 0) \cap \text{int } T_{C_2}(x, 0) = \{(v, 0) : v \in \text{int } D, f^\circ(x; v) < 0\}.$$

If this latter set were empty, then the set (25) would be contained in the half-space $\{(v, \beta) : \beta \geq 0\}$, and the same would then be true of its closure which includes epi $f^\circ(x; \cdot)$. This would imply $f^\circ(x; v) \geq 0$ for all v in contradiction with the hypothesis $0 \notin \partial f(x)$. The hypotheses of Corollary 2, Theorem 2.9.8, are therefore satisfied. (C_2 admits a hypertangent at $(x, f(x))$ since f_2 is directionally Lipschitz at x.) The conclusions of the theorem follow directly from those of the corollary together with Eqs. (23), (24), and (25). □

Chapter Three

Differential Inclusions

Hamiltonians were then all the rage—together with elliptic functions.

L. C. YOUNG, *Calculus of Variations and Optimal Control*

Nobody wants to read anyone else's formulas.

Finman's Law

There are many situations, such as in optimal control theory and the study of differential inequalities, in which one is dealing with a class of functions whose derivatives satisfy given constraints. A fundamental example is the case in which one is considering the solutions $x(\cdot)$ of the ordinary differential equation

$$\dot{x}(t) = f(t, x(t)).$$

We can capture this and other situations by considering instead the *differential inclusion*

$$\dot{x}(t) \in F(t, x(t)),$$

where F is a mapping whose values are sets in R^n. There is a fairly recent and still growing body of theory that extends to such differential inclusions many results from differential equations, such as those pertaining to the existence and nature of solutions, stability, and invariance. In this chapter, however, our attention will be focused for the most part upon (dynamic) optimization problems in which the differential inclusion is one of the constraints; that is, we shall deal with the optimal control of differential inclusions. The central character is the problem P_D (whose family tree was discussed in Sections 1.3 and 1.4), although its formal appearance under that name is postponed until the stage is set.

3.1 MULTIFUNCTIONS AND TRAJECTORIES

A *multifunction* $\Gamma: R^m \to R^n$ is a mapping from R^m to the subsets of R^n; thus, for each x in R^m, $\Gamma(x)$ is a (possibly empty) set in R^n. If S is a subset of R^m, we say that Γ is closed, compact, convex, or nonempty on S, provided that for each x in S the set $\Gamma(x)$ has that particular property.

A multifunction $\Gamma: S \to R^n$ is said to be *measurable* provided that for every open subset C of R^n, the set

$$\{x \in S : \Gamma(x) \cap C \neq \varnothing\}$$

is Lebesgue measurable. We obtain an equivalent definition by requiring this condition for every closed subset C.

3.1.1 Theorem (Measurable Selection)
Let Γ be measurable, closed, and nonempty on S. Then there exists a measurable function $\gamma: S \to R^n$ such that $\gamma(x)$ belongs to $\Gamma(x)$ for all x in S.

Proof (Castaing and Valadier, 1977). We begin by noting that for any ζ in R^n the function $s \to d_{\Gamma(s)}(\zeta)$ is measurable on S (where $d_{\Gamma(s)}$ is as usual the Euclidean distance function), since

$$\{s \in S : d_{\Gamma(s)}(\zeta) \leq \alpha\} = \{s \in S : \Gamma(s) \cap [\zeta + \alpha \bar{B}] \neq \varnothing\}.$$

Now let $\{\zeta_i\}$ be a countable dense subset of R^n, and define a function $\gamma_0: S \to R^n$ as follows:

$$\gamma_0(s) = \text{the first } \zeta_i \text{ such that } d_{\Gamma(s)}(\zeta_i) \leq 1.$$

Lemma. The functions $s \to \gamma_0(s)$ and $s \to d_{\Gamma(s)}(\gamma_0(s))$ are measurable.

To see this, observe that γ_0 assumes countably many values, and that, for each i,

$$\{s : \gamma_0(s) = \zeta_i\} = \bigcap_j \{s : d_{\Gamma(s)}(\zeta_j) > 1\} \cap \{s : d_{\Gamma(s)}(\zeta_i) \leq 1\},$$

where the intersection is over $j = 1, \ldots, i - 1$. This implies that γ_0 is measurable. To complete the proof, we need only note

$$\{s: d_{\Gamma(s)}(\gamma_0(s)) > \alpha\} = \bigcup_j \left[\{s: \gamma_0(s) = \zeta_j\} \cap \{s: d_{\Gamma(s)}(\zeta_j) > \alpha\}\right],$$

where the union is over the positive integers j.

We pursue the process begun above by defining for each integer i a function γ_{i+1} such that $\gamma_{i+1}(s)$ is the first ζ_j for which both the following hold:

$$|\zeta_j - \gamma_i(s)| \leq \tfrac{2}{3} d_{\Gamma(s)}(\gamma_i(s)), \quad d_{\Gamma(s)}(\zeta_j) \leq \tfrac{2}{3} d_{\Gamma(s)}(\gamma_i(s)).$$

It follows much as above that each γ_i is measurable. Furthermore, we deduce the inequalities

$$d_{\Gamma(s)}(\gamma_{i+1}(s)) \leq \left(\tfrac{2}{3}\right)^i d_{\Gamma(s)}(\gamma_0(s)) \leq \left(\tfrac{2}{3}\right)^i,$$

together with $|\gamma_{i+1}(s) - \gamma_i(s)| \leq \left(\tfrac{2}{3}\right)^{i+1}$. It follows that $\{\gamma_i(s)\}$ is a Cauchy sequence converging to a value $\gamma(s)$ for each s, and that γ is a measurable selection for Γ. □

The proof of the following result is left as an exercise.

3.1.2 Proposition

Let Γ be closed and measurable and let $g: R^m \times R^n \to R$ be such that
(a) for each x in S the function $y \to g(x, y)$ is continuous, and
(b) for each y in R^n the function $x \to g(x, y)$ is measurable on S.
Then the multifunction $G: R^m \to R^n$ defined by

$$G(x) = \{y \in \Gamma(x): g(x, y) = 0\}$$

is measurable and closed on S.

Aumann's Theorem

Consider now the case in which $S = [a, b]$, an interval in R. We say that Γ is *integrably bounded* provided there is an integrable function $\phi(t)$ such that for all t in $[a, b]$, for all γ in $\Gamma(t)$, $|\gamma| \leq \phi(t)$. In the following, co Γ denotes the multifunction whose value at t is the convex hull of $\Gamma(t)$. The *integral* of Γ (over $[a, b]$) is the set

$$\int \Gamma := \left\{\int_a^b \gamma(t): \gamma(\cdot) \text{ is a measurable selection for } \Gamma\right\}.$$

In Chapter 7, Section 7.2, we shall apply the theory of Chapter 2 to prove:

3.1.3 Theorem (Aumann (1965))
If Γ is measurable, closed, nonempty, and integrably bounded on $[a, b]$, then

$$\int \Gamma = \int \operatorname{co} \Gamma.$$

Lipschitz Multifunctions and Tubes

Γ is said to be Lipschitz on S (of rank k) provided that for all x_1, x_2 in S and for all γ_1 in $\Gamma(x_1)$ there exists γ_2 in $\Gamma(x_2)$ such that

$$|\gamma_1 - \gamma_2| \leq k|x_1 - x_2|,$$

where $|\cdot|$ denotes (as usual) the Euclidean norm. (Of course this reduces to the definition of Section 2.1 if Γ is actually a function.)

We shall be dealing repeatedly in this chapter with a multifunction $F(t, x)$ from $R \times R^n$ to R^n whose argument (t, x) lies in some prescribed subset Ω of $R \times R^n$, and such that the behaviors of F in t and x differ. It is a convenient convention to set $F(t, x) = \emptyset$ whenever (t, x) lies outside the given domain Ω of definition.

Let S, Ω_t be the sets defined by

$$S = \{t : (t, x) \in \Omega \text{ for some } x \text{ in } R^n\}$$

$$\Omega_t = \{x : (t, x) \in \Omega\}.$$

Ω is called a *tube* provided the set S is an interval ($[a, b]$, say) and provided there exist a continuous function $w(t)$ and a continuous positive function ε on $[a, b]$ such that $\Omega_t = w(t) + \varepsilon(t)B$ for t in $[a, b]$. We call such a tube Ω a tube on $[a, b]$. If x is a given continuous function on $[a, b]$, the "ε-tube about x," denoted $T(x; \varepsilon)$, is the tube on $[a, b]$ obtained by setting

$$\Omega = \{(t, x') : a \leq t \leq b, x' \in x(t) + \varepsilon B\}.$$

A function x on $[a, b]$ for which $(t, x(t))$ lies in Ω for all t in $[a, b]$ is simply said to lie in Ω. A tube Ω has the property that whenever a continuous function x lies in Ω, then Ω contains $T(x; \varepsilon)$ for some positive ε.

3.1.4 Definition
Let Ω be a tube on $[a, b]$. F is said to be *measurably Lipschitz* on Ω provided:
(i) For each x in R^n, the multifunction $t \to F(t, x)$ is measurable on $[a, b]$.
(ii) There is an integrable function $k(t)$ on $[a, b]$ such that for each t in $[a, b]$, the multifunction $x \to F(t, x)$ is nonempty and Lipschitz of rank $k(t)$ on Ω_t.

To the multifunction F we associate the function $\rho: \Omega \times R^n \to [0, \infty]$ defined by

$$\rho(t, x, v) := \inf\{|v - y| : y \in F(t, x)\}.$$

(By the usual convention, the infimum over the empty set is $+\infty$.) When $F(t, x)$ is closed and nonempty, then v belongs to $F(t, x)$ iff $\rho(t, x, v) = 0$. The proof of the following is straightforward (see Clarke, 1973, Prop. 1).

3.1.5 Proposition

If F is measurably Lipschitz, then
(a) *For each x and v in R^n, the function $t \to \rho(t, x, v)$ is measurable.*
(b) *For any (t, x_1) and (t, x_2) in Ω, and for any v_1 and v_2 in R^n, one has*

$$|\rho(t, x_1, v_1) - \rho(t, x_2, v_2)| \leq k(t)|x_1 - x_2| + |v_1 - v_2|.$$

It can be shown that if F is closed and nonempty on Ω, properties (a) and (b) are actually equivalent to the measurably Lipschitz condition. This is one way to derive the result that the multifunction $t \to F(t, x(t))$ is measurable (see Clarke, 1973, Lemma 3.8), a fact we shall use later.

Arcs and Trajectories

From now on we work with a given interval $[a, b]$, and a given tube Ω on $[a, b]$. The term *arc* will refer to any absolutely continuous function $x: [a, b] \to R^n$. An arc, then, is a function $x(\cdot)$ having a derivative at t (denoted $\dot{x}(t)$) for almost all t in $[a, b]$, and which is the integral of its derivative. (This is the customary class within which one studies differential equations.) We denote by $\|x\|$ the supremum norm:

$$\|x\| := \max\{|x(t)| : a \leq t \leq b\}.$$

A *trajectory* (for F, or for the differential inclusion) is an arc x such that for almost all t in $[a, b]$, $\dot{x}(t)$ belongs to the set $F(t, x(t))$. Symbolically

$$\dot{x}(t) \in F(t, x(t)) \text{ a.e.}$$

In the theorem below, we assume that F is measurably Lipschitz on Ω (see Definition 3.1.4), and we set $K = \exp\{\int_a^b k(t)\, dt\}$. If x is an arc lying in Ω, we define

$$\rho_F(x) = \int_a^b \rho(t, x(t), \dot{x}(t))\, dt.$$

(Proposition 3.1.5 implies that the function $t \to \rho(t, x(t), \dot{x}(t))$ is measurable, so the integral is well defined, possibly $+\infty$.)

3.1.6 Theorem

If x is an arc for which $T(x; \varepsilon)$ is contained in Ω, and if $\rho_F(x) < \varepsilon/K$, then there exists a trajectory y for F lying in $T(x; \varepsilon)$ with $y(a) = x(a)$ and

$$\|x - y\| \leq \int_a^b |\dot{x}(t) - \dot{y}(t)| dt \leq K\rho_F(x)$$

Proof. It follows from Propositions 3.1.2 and 3.1.5 that the multifunction

$$t \to \{v \in F(t, x(t)) : \rho(t, x(t), \dot{x}(t)) = |v - \dot{x}(t)|\}$$

is measurable, and hence admits a measurable (integrable) selection v_0 by Theorem 3.1.1. Set $x_1(t) = x(a) + \int_a^t v_0(\tau) d\tau$. We easily deduce

$$\|x_1 - x\| \leq \int_a^b |\dot{x}_1 - \dot{x}| dt = \rho_F(x) < \varepsilon$$

(since $K \geq 1$), so that x_1 lies in $T(x; \varepsilon)$. We now choose $v_1(t) \in F(t, x_1(t))$ a.e. such that

$$|v_1 - \dot{x}_1| = \rho(t, x_1, \dot{x}_1) \quad \text{a.e.}$$

By Proposition 3.1.5 we calculate

$$\rho(t, x_1, \dot{x}_1) \leq \rho(t, x, \dot{x}) + k(t)|x_1 - x| + |\dot{x}_1 - \dot{x}|$$

$$\leq \rho(t, x, \dot{x}) + |\dot{x}_1 - \dot{x}| + k(t)\rho_F(x).$$

It follows that v_1 is integrable. Set $x_2(t) = x(a) + \int_a^t v_1(\tau) d\tau$. Then a.e.

$$|\dot{x}_2 - \dot{x}_1| = |v_1 - \dot{x}_1| = \rho(t, x_1, \dot{x}_1)$$

$$\leq \rho(t, x, \dot{x}_1) + k(t)|x - x_1| = k(t)|x - x_1|.$$

We derive from this the inequality

$$|x_2(t) - x_1(t)| \leq \rho_F(x) \int_a^t k(\tau) d\tau.$$

Furthermore, x_2 lies in $T(x; \varepsilon)$ since

$$\|x_2 - x\| \leq \|x_2 - x_1\| + \|x_1 - x\| \leq \rho_F(x) \int_a^b k(\tau) d\tau + \|x_1 - x\|$$

$$\leq \rho_F(x) \int_a^b k(\tau) d\tau + \rho_F(x) = \rho_F(x)\left[1 + \int_a^b k(\tau) d\tau\right]$$

$$\leq \rho_F(x) \exp\left[\int_a^b k(\tau) d\tau\right] = \rho_F(x) K < \varepsilon.$$

Now let

$$S_i(t) = \int_a^t k(t_1) \int_a^{t_1} k(t_2) \cdots \int_a^{t_{i-1}} k(t_i) \, dt_i \cdots dt_2 \, dt_1$$

($i \geq 1$, $t_0 = t$) and recall the elementary formula

$$S_i(t) = \frac{\left[\int_a^t k(\tau) \, d\tau\right]^i}{i!}.$$

We continue the sequence x_i as above, getting at each step

$$\dot{x}_{i+1}(t) \in F(t, x_i(t)) \quad \text{a.e.}$$

$$|\dot{x}_{i+1}(t) - \dot{x}_i(t)| \leq k(t)|x_i(t) - x_{i-1}(t)|.$$

By induction we obtain

(1) $\quad |\dot{x}_{i+1}(t) - \dot{x}_i(t)| \leq \rho_F(x) k(t) \dfrac{\left[\int_a^t k(\tau) \, d\tau\right]^{i-1}}{(i-1)!} \quad (i = 1, 2, \ldots)$

(2) $\quad |x_{i+1}(t) - x_i(t)| \leq \rho_F(x) \dfrac{\left[\int_a^t k(\tau) \, d\tau\right]^i}{i!} \quad (i = 0, 1, 2, \ldots).$

At each step, x_i lies in $T(x; \varepsilon)$:

$$|x_i(t) - x(t)| \leq \rho_F(x) + \rho_F(x) S_1(b) + \rho_F(x) S_2(b) + \cdots$$

$$= \rho_F(x) \left[1 + \int_a^b k(\tau) \, d\tau + \left\{\int_a^b k(\tau) \, d\tau\right\}^2 / 2! \right.$$

$$\left. + \left\{\int_a^b k(\tau) \, d\tau\right\}^3 / 3! + \cdots \right]$$

$$= \rho_F(x) \exp\left\{\int_a^b k(\tau) \, d\tau\right\} = K\rho_F(x) < \varepsilon.$$

From (1) we deduce that $\{\dot{x}_i\}$ is Cauchy in $L^1[a, b]$, and so converges to a function $v(\cdot)$ in $L^1[a, b]$, a subsequence converging a.e. From (2) we deduce that $\{x_i\}$ converges uniformly to a continuous function y (with $y(a) = x(a)$). It follows that

$$v(t) \in F(t, y(t)) \quad \text{a.e.}$$

From $x_{i+1}(t) = x(a) + \int_a^t \dot{x}_i(\tau)\,d\tau$ we derive $y(t) = x(a) + \int_a^t v(\tau)\,d\tau$, whence y is an arc and a trajectory for F. The required estimate for y follows easily from (1). □

Relaxed Trajectories

A *relaxed trajectory* is an arc x on $[a, b]$ which is a trajectory for co F; that is, such that

$$\dot{x}(t) \in \mathrm{co}\, F(t, x(t)) \quad \text{a.e.}$$

Let us now suppose, in addition to the hypotheses of Theorem 3.1.6, that F is *integrably bounded* on Ω (i.e., there is an integrable function $\phi(\cdot)$ such that $|v| \leq \phi(t)$ for all v in $F(t, x)$, for all (t, x) in Ω).

Corollary

Let y be any relaxed trajectory lying in Ω. Then, for any $\delta > 0$, there is a trajectory x for F with $y(a) = x(a)$ such that $\|y - x\| < \delta$.

Proof. We assume without loss of generality that $a = 0$, $b = 1$. Let $\lambda > 0$ be such that $T(y; \lambda)$ is contained in Ω. Set $\varepsilon = \lambda/2$, and choose any positive α such that

$$\alpha < \min\left\{\frac{\varepsilon}{K\ln(K)}, \frac{\delta}{1 + K\ln(K)}, \varepsilon\right\}.$$

Now choose a positive integer m such that for any subinterval I of $[0, 1]$ with measure no greater than $1/m$,

$$\int_I \phi(t)\,dt < \frac{\alpha}{2}.$$

Let I_j be the interval $[(j-1)/m, j/m)$, $i = 1, 2, \ldots, m$. Because $F(t, y(t))$ is measurable and integrably bounded, we may apply Aumann's Theorem 3.1.3 to deduce the existence of integrable functions f_j such that $f_j(t) \in F(t, y(t))$ for almost all t in I_j, and

$$\int_{I_j} f_j(t)\,dt = \int_{I_j} \dot{y}(t)\,dt \quad (j = 1, 2, \ldots, m).$$

Let f be the function which is equal to f_j on I_j, and define the arc x_0 by

$$x_0(t) = y(0) + \int_0^t f(\tau)\,d\tau.$$

We then have

$$|x_0(t) - y(t)| = \left|\int_0^t [f(\tau) - \dot{y}(\tau)] \, d\tau\right| \leq \int_{I_j} |f(\tau) - \dot{y}(\tau)| \, d\tau \quad \text{(for some } j\text{)}$$

$$\leq \int_{I_j} 2\phi(\tau) \, d\tau < \alpha.$$

It follows that $\|x_0 - y\| \leq \alpha < \varepsilon$ and hence that $T(x_0; \varepsilon)$ is contained in Ω. We also derive, since $\dot{x}_0(t) \in F(t, y(t))$ a.e.,

$$\rho(t, x_0(t), \dot{x}_0(t)) \leq k(t)|x_0(t) - y(t)| \leq k(t)\alpha \quad \text{a.e.,}$$

whence

$$\rho_F(x_0) \leq \alpha \int_0^1 k(t) \, dt < \frac{\varepsilon}{K}.$$

From Theorem 3.1.6 we conclude that a trajectory x exists such that $x(0) = x_0(0)$ and $\|x - x_0\| \leq K\rho_F(x_0)$. Then

$$\|x - y\| \leq \|x - x_0\| + \|x_0 - y\| \leq K\rho_F(x_0) + \alpha \leq K\alpha \int_0^1 k(t) \, dt + \alpha$$

$$= \alpha\left[1 + K\int_0^1 k(t) \, dt\right] < \delta. \quad \square$$

Compactness of Trajectories

The technical result that we pause to prove here will be used many times. Let Γ be a multifunction defined on a tube Ω on $[a, b]$. We suppose that Γ is integrably bounded by ϕ on Ω, and that Γ is nonempty, compact, and convex on Ω. We suppose further that there is a multifunction X from $[a, b]$ to R^n and a positive-valued function $\varepsilon(t)$ with the following properties:

(i) For all t in $[a, b]$, $X(t) + \varepsilon(t)B \subset \Omega_t$.
(ii) For every t in $[a, b]$, for every x in $X(t) + \varepsilon(t)B$, the multifunction $x' \to \Gamma(t, x')$ is upper semicontinuous at x (as defined in Section 2.1).
(iii) For every (t, x) in the interior of Ω, the multifunction $t' \to \Gamma(t', x)$ is measurable.

3.1.7 Theorem

Let $\{x_j\}$ be a sequence of arcs on $[a, b]$ satisfying:

(i) $x_j(t) \in X(t)$ and $|\dot{x}_j(t)| \leq \phi(t)$ for almost all t in $[a, b]$.
(ii) $\dot{x}_j(t) \in \Gamma(t, x_j(t) + y_j(t)) + r_j(t)B$ for $t \in A_j$, where $\{y_j\}$, $\{r_j\}$ are sequences of measurable functions on $[a, b]$ which converge uniformly to 0,

3.1 Multifunctions and Trajectories

and where $\{A_j\}$ is a sequence of measurable subsets of $[a, b]$ such that measure $(A_j) \to (b - a)$.

(iii) The sequence $\{x_j(a)\}$ is bounded.

Then there is a subsequence of $\{x_j\}$ which converges uniformly to an arc x which is a trajectory for Γ.

Proof. The set $\{x_j\}$ is bounded and equicontinuous, so there is a subsequence (we do not relabel) converging uniformly to a limit x. We then invoke the Dunford-Pettis criterion to extract a subsequence of $\{\dot{x}_j\}$ converging weakly to a limit v in $L^1[a, b]$. From $x_j(t) = x_j(a) + \int_a^t \dot{x}_j$ we deduce $x(t) = x(a) + \int_a^t v$, so x is an arc and $\dot{x} = v$ a.e.

Now let $h(t, s, p)$ be the function $\max\{\langle p, \gamma \rangle : \gamma \in \Gamma(t, s)\}$. Fix any p in R^n and any measurable set V in $[a, b]$. It is easy to prove the integrability of $t \to h(t, x_j(t), p)$. Letting χ_j be the characteristic function of A_j, one has

$$0 \leq \limsup_{j \to \infty} \int_{V \cap A_j} \left[h(t, x_j + y_j, p) - \langle p, \dot{x}_j \rangle + \|r_j\|\|p\| \right] dt$$

(since the integrand is nonnegative on the set A_j)

$$= \limsup_{j \to \infty} \int_{V \cap A_j} \left[h(t, x_j, p) - \langle p, \dot{x}_j \rangle \right] dt$$

$$\leq \int_V \limsup_{j \to \infty} \chi_j h(t, x_j, p) \, dt + \limsup_{j \to \infty} \int_{V \cap A_j} \langle -p, \dot{x}_j \rangle \, dt$$

$$\leq \int_V h(t, x, p) \, dt + \limsup_{j \to \infty} \int_V \langle -p, \dot{x}_j \rangle \, dt + \limsup_{j \to \infty} \int_{V \cap A_j^c} \langle p, \dot{x}_j \rangle \, dt$$

$$= \int_V \left[h(t, x, p) - \langle p, \dot{x} \rangle \right] dt.$$

Since V is arbitrary, it follows that $h(t, x(t), p) \geq \langle p, \dot{x}(t) \rangle$ a.e. Since h is continuous in p, this inequality can be obtained for all p, except for t in a fixed set of measure 0. It follows (see Proposition 2.1.4(a)) that $\dot{x}(t)$ belongs to $\Gamma(t, x(t))$ a.e., and the proof is complete. □

A related result which requires more measure theory, and which we shall use on several occasions, is the following one due to Vinter and Pappas (1982, Lemma 4.5), to whom we refer for the proof.

3.1.8 Proposition

Let Γ be a compact convex multifunction from $[a, b] \times R^n$ to R^n having closed graph, and whose images are contained in a given bounded set. Let v_i be a sequence of positive measures converging weak* to ν_0, and let x_i be a sequence of arcs converging uniformly to x, such that for each i, there is a measurable function γ_i such that $\gamma_i(t)$ belongs to $\Gamma(t, x_i(t))$ ν_i-a.e. Then there exists a measurable function γ_0, which is ν_0 integrable, such that $\gamma_0(t)$ belongs to $\Gamma(t, x(t))$ ν_0-a.e. and such that some subsequence of the measures $\gamma_i \, d\nu_i$ converges weak* to $\gamma_0 \, d\nu_0$.

3.2 A CONTROL PROBLEM

The first control problem involving differential inclusions that we shall study is the following:

3.2.1 Problem

To minimize $f(x(b))$ over the arcs x which are trajectories for F and which satisfy the initial point condition $x(a) \in C_0$ and the state constraint

$$g(t, x(t)) \leq 0, \quad a \leq t \leq b.$$

In the above, F is a multifunction defined on a tube Ω on $[a, b]$ as in Definition 3.1.4 (typically in applications Ω is a tube about a given arc of interest). Note that the requirement $\dot{x} \in F(t, x)$ implies that x must lie in Ω. The set C_0 is a given subset of R^n, and the functions $f: R^n \to R$ and $g: \Omega \to R$ are also given. An arc x is termed *admissible* if it satisfies the constraints of the problem.

3.2.2 Basic Hypotheses

(i) F is measurably Lipschitz, integrably bounded, closed, and convex on the tube Ω.
(ii) f is Lipschitz on Ω_b of rank K_f.
(iii) g is upper semicontinuous, and for each t in $[a, b]$ the function $g(t, \cdot)$ is Lipschitz on Ω_t of rank K_g (not depending on t).

3.2.3 Proposition (Existence of a Solution)

Along with the basic hypotheses, suppose that C_0 is compact, and that for almost every t the set $\{x: g(t, x) \leq 0\}$ is contained in Ω_t. Then, if there is at least one admissible arc, Problem 3.2.1 admits a solution.

Proof. Let $X(t) = \{x: g(t, x) \leq 0\}$. A straightforward application of Theorem 3.1.7 shows that any minimizing sequence $\{x_j\}$ admits a subsequence converging to a solution. □

3.2 A Control Problem

The Hamiltonian

The *Hamiltonian* for Problem 3.2.1 is the function $H: \Omega \times R^n \to R$ defined by

$$H(t, x, p) = \max\{\langle p, v \rangle : v \in F(t, x)\}.$$

In the following, k is the Lipschitz function associated with F (see Definition 3.1.4) and ϕ the bound. The notation ∂H refers to the generalized gradient of H with respect to the (x, p) variables, for t fixed.

3.2.4 Proposition

H is finite on $\Omega \times R^n$. If (t, x) is a point in int Ω, and if p is any point in R^n, then
(a) $t' \to H(t', x, p)$ is measurable near t.
(b) $x' \to H(t, x', p)$ is Lipschitz near x of rank $|p|k(t)$.
(c) $p' \to H(t, x, p')$ is Lipschitz near p of rank $\phi(t)$.
(d) If (α, β) belongs to $\partial H(t, x, p)$, then β belongs to $\partial_p H(t, x, p)$ (so that $H(t, x, p) = \langle \beta, p \rangle$); the latter is a subset of $F(t, x)$.
(e) If (α, β) belongs to $\partial H(t, x, p)$, then for any nonnegative scalar λ, $(\lambda\alpha, \beta)$ belongs to $\partial H(t, x, \lambda p)$; any point $(0, \beta)$, where β belongs to $F(t, x)$, lies in $\partial H(t, x, 0)$.

Proof. We leave (a)–(c) as exercises; consider (d). The first assertion follows from Proposition 2.5.3, while the second is an immediate consequence of Theorem 2.8.2. Finally we turn to (e). To prove the first assertion, it suffices in view of Theorem 2.5.1 to prove that if $[\alpha, \beta] = \nabla H(x, p)$, then $[\lambda\alpha, \beta] = \nabla H(x, \lambda p)$. But this is immediate from the identity $H(x, \lambda p) = \lambda H(x, p)$. As for the last assertion, we observe

$$H^\circ(t, x, 0; v, w) \geq \limsup_{\lambda \downarrow 0} \left[\frac{H(t, x, \lambda w)}{\lambda} \right]$$

$$= H(t, x, w) \geq \langle \beta, w \rangle = \langle (0, \beta), (v, w) \rangle$$

and so the result follows from Proposition 2.1.5(a). □

Hamiltonian Multipliers

We denote by $\partial_x^> g(t, x)$ the set

$$\text{co}\left\{ \gamma = \lim_{i \to \infty} \gamma_i : \gamma_i \in \partial_x g(t_i, x_i), t_i \to t, x_i \to x, g(t_i, x_i) > 0 \right\}.$$

Note that (under Hypothesis 3.2.2(iii)) $\partial_x^> g(t, x)$ is empty if $g(t, x) < 0$.

3.2.5 Definition

A *multiplier* corresponding to an admissible arc x is a five-tuple $[p, \gamma, \mu, \zeta, \lambda]$, where p is an arc and γ is a measurable function on $[a, b]$, μ is a nonnegative

Radon measure on $[a, b]$, ζ is a point in R^n, and λ is a nonnegative scalar. The multiplier satisfies:

(i) $\zeta \in \partial f(x(b))$.
(ii) μ is supported on the set
$$S := \{t \in [a, b] : \partial_x^> g(t, x(t)) \neq \emptyset\}.$$

(iii) $\gamma(t) \in \partial_x^> g(t, x(t))$ μ-a.e.
(iv) For almost all t in $[a, b]$, one has
$$\begin{bmatrix} -\dot{p}(t) \\ \dot{x}(t) \end{bmatrix} \in \partial H\left(t, x(t), p(t) + \int_{[a, t)} \gamma(s) \mu(ds)\right).$$

(v) For some $r > 0$, one has
$$p(a) \in r \partial d_{C_0}(x(a)),$$
where d_{C_0} is the Euclidean distance function for C_0.
(vi) $\lambda \zeta + p(b) + \int_{[a, b]} \gamma(s) \mu(ds) = 0$.

3.2.6 Theorem

If the arc x solves Problem 3.2.1, then there exists a multiplier $[p, \gamma, \mu, \zeta, \lambda]$ for x such that $\lambda + \|\mu\| = 1$.

3.2.7 Remarks

(i) If x is any admissible arc, we always obtain a multiplier $[p, \gamma, \mu, \zeta, \lambda]$ for x by taking $p = 0$, $\mu = 0$, $\lambda = 0$ (for any γ and ζ). This stems from the fact that $(0, \dot{x})$ always belongs to $\partial H(t, x, 0)$, and 0 always belongs to $\partial d_{C_0}(x(a))$. The purpose of the condition $\lambda + \|\mu\| = 1$ is to avoid this triviality ($\|\mu\|$ denotes the measure norm).

(ii) If $g(t, x(t)) < 0$ for all t (i.e., if the state constraint is *inactive* at x), then $\mu = 0$ and $\lambda = 1$, and condition (iv) reduces to $(-\dot{p}, \dot{x}) \in \partial H(t, x, p)$, which is reminiscent of a classical Hamiltonian system of equations. If in addition f is smooth, then the "transversality conditions" (v) and (vi) can be written in the more familiar form
$$p(a) \in N_{C_0}(x(a)), \quad p(b) = -\nabla f(x(b)),$$
where N_{C_0} refers to the (generalized) cone of normals (see Sections 1.2 and 2.4).

(iii) The set S of condition (ii) is contained in the set $\{t : g(t, x(t)) = 0\}$. Note that only the values of γ on S are relevant. If 0 belongs to $\partial_x^> g(t, x(t))$ for some t, then we obtain a multiplier by taking μ as the unit mass on

$\{t\}$, $\gamma(t) = 0$, and $p = 0$, $\lambda = 0$ (together with any ζ in $\partial f(x(b))$). Thus the theorem will generally yield useful information only when $0 \notin \partial_x^> g(t, x(t))$. The case in which the state constraints are given in the form $x(t) \in X(t)$ is handled by setting

$$g(t, x) := d_{X(t)}(x).$$

Although one always has $0 \in \partial_x g(t, x)$, it is the case for a large class of $X(t)$ (in particular, if $X(t)$ is a convex body) that one has $0 \notin \partial_x^> g(t, x)$.

(iv) The case of a family of constraints $g_\alpha(t, x) \leq 0$, $\alpha \in A$, can usually be handled by defining $g(t, x) = \sup_{\alpha \in A} g_\alpha(t, x)$. If a constraint $\theta(t, x) \leq 0$ is to hold only for t in a closed subset C of $[a, b]$, one can define $g(t, x) = \theta(t, x)$ for $t \in C$, $g(t, x)$ being a large negative number otherwise.

Proof of Theorem 3.2.6

Step 1. Let Ω_δ denote the closure of the tube $T(x; \delta/2)$, where $\delta > 0$ and $T(x; \delta)$ is contained in Ω. For any $\alpha > 0$, define a new multifunction F_α via $F_\alpha(t, x') = F(t, x') + \alpha \bar{B}$. Presently α will be shrinking to zero. For the moment, note that F_α inherits the various properties hypothesized for F. We also define two functions of arcs:

$$G(y) := \max_{a \leq t \leq b} g(t, y(t))$$

$$\phi_\varepsilon(y) := \max\{G(y), f(y(b)) - f(x(b)) + \varepsilon^2\},$$

where ε is any positive number. Finally, A_α signifies the set of all arcs y satisfying

$$(t, y(t)) \in \Omega_\delta, \quad a \leq t \leq b$$

$$y(a) \in C_0, \quad \dot{y} \in F_\alpha(t, y) \quad \text{a.e.}$$

It follows easily from Theorem 3.1.7 that the metric Δ defined by

$$\Delta(y, z) := \int_a^b |\dot{y}(t) - \dot{z}(t)|\, dt + |y(a) - z(a)|$$

turns A_α into a complete metric space. Note that x remains optimal if C_0 is replaced by $C_0 \cap (x + \bar{B})$, and that this does not affect the multipliers. Thus there is no loss of generality in assuming C_0 compact.

Lemma 1. Given $\varepsilon > 0$, then for all $\alpha > 0$ sufficiently small and for all y in A_α, one has $\phi_\varepsilon(y) > 0$.

If this were false, there would exist a sequence $\alpha_i \to 0$ and arcs y_i in A_{α_i} with $\phi_\varepsilon(y_i) \leq 0$. In view of Theorem 3.1.7, we may further suppose that y_i converges uniformly to an element y of A_0. It follows that $\phi_\varepsilon(y) \leq 0$, so that y satisfies the state constraint $G(y) \leq 0$, and also the inequality

$$f(y(b)) \leq f(x(b)) - \varepsilon^2.$$

This contradicts the fact that x solves Problem 3.2.1, and so the lemma is proved.

Step 2. We choose α as in the statement of Lemma 1 with $\alpha < \varepsilon$, and we note that $\phi_\varepsilon(x) = \varepsilon^2$. By Lemma 1 it follows that

$$\phi_\varepsilon(x) \leq \inf_{A_\alpha} \phi_\varepsilon + \varepsilon^2.$$

By Theorem 7.5.1 therefore there is an element z of A_α minimizing $\phi_\varepsilon(y) + \varepsilon\Delta(y, z)$ over y in A_α, and such that

$$\Delta(x, z) \leq \varepsilon, \qquad \phi_\varepsilon(z) \leq \varepsilon^2.$$

For ε small enough this implies that z lies in the tube $T(x; \delta/4)$; we assume that ε is so chosen. We let ρ_α correspond to F_α as ρ does to F (see Proposition 3.1.5). Recall the constants K_f, K_g defined in Hypotheses 3.2.2, and K in Theorem 3.1.6.

Lemma 2. For some $\eta > 0$ the arc z minimizes

$$\phi_\varepsilon(y) + \varepsilon\Delta(y, z) + (K_1 + \varepsilon K_2)d_{C_0}(y(a)) + (K_3 + \varepsilon K_4)\int_a^b \rho_\alpha(t, y, \dot{y})\, dt$$

over all arcs y satisfying $\|y - z\| < \eta$, where

$$K_1 = \max\{K_f, K_g\}[K\ln(K) + 1]$$

$$K_2 = \frac{(b - a + 1)K_1}{\max\{K_f, K_g\}}$$

$$K_3 = K\max\{K_f, K_g\}$$

$$K_4 = (b - a + 1)K.$$

Suppose the assertion to be false. Then there is a sequence y_i of arcs converging uniformly to z for which the expression in the lemma is less than its value at z. It follows that $\int_a^b \rho_\alpha(t, y_i, \dot{y}_i)\, dt$ goes to zero. Let c_i in C_0 be such that

3.2 A Control Problem

$d_{C_0}(y_i(a)) = |y_i(a) - c_i|$; again we have $|y_i(a) - c_i| \to 0$. Now let $\tilde{y}_i$ be the arc defined by $\tilde{y}_i(t) = y_i(t) + c_i - y_i(a)$. Then we derive, with the help of Proposition 3.1.5, the bound

$$\int_a^b \rho_\alpha(t, \tilde{y}_i, \dot{\tilde{y}}_i)\, dt \leq \int_a^b \rho_\alpha(t, y_i, \dot{y}_i)\, dt + |y_i(a) - c_i|\ln(K).$$

We are now able to apply Theorem 3.1.6 (with $x = \tilde{y}_i$, $F = F_\alpha$) for i sufficiently large to deduce the existence of a trajectory z_i for F_α such that $z_i(a) = \tilde{y}_i(a) = c_i \in C_0$ and such that

$$\int_a^b |\dot{z}_i - \dot{\tilde{y}}_i|\, dt \leq K \int_a^b \rho_\alpha(t, \tilde{y}_i, \dot{\tilde{y}}_i)\, dt.$$

The preceding leads to

$$\|z_i - y_i\| \leq \|z_i - \tilde{y}_i\| + \|\tilde{y}_i - y_i\|$$

$$\leq \int_a^b |\dot{z}_i - \dot{\tilde{y}}_i|\, dt + d_{C_0}(y_i(a))$$

$$\leq K\left\{ d_{C_0}(y_i(a))\ln(K) + \int_a^b \rho_\alpha(t, y_i, \dot{y}_i)\, dt \right\} + d_{C_0}(y_i(a)).$$

We now deduce

$$\phi_\varepsilon(z_i) + \varepsilon\Delta(z_i, z) \leq \phi_\varepsilon(y_i) + \varepsilon\Delta(y_i, z)$$
$$+ \left[\max\{K_f, K_g\} + \varepsilon + \varepsilon(b-a)\right]\|z_i - y_i\|$$
$$\leq \phi_\varepsilon(y_i) + \varepsilon\Delta(y_i, z) + [K_1 + \varepsilon K_2]d_{C_0}(y_i(a))$$
$$+ [K_3 + \varepsilon K_4]\int_a^b \rho_\alpha(t, y_i, \dot{y}_i)\, dt$$
$$< \phi_\varepsilon(z) \quad \text{(by assumption)}.$$

The upshot is that the element z_i of A_α assigns a lower value to $\phi_\varepsilon(\cdot) + \varepsilon\Delta(\cdot, z)$ than does z. This contradiction of the optimality of z completes the proof of Lemma 2.

Step 3. We are now ready to calculate some generalized gradients, namely those of the four functionals summed in the statement of Lemma 2. Let us set, for any arc y,

$$f_1(y) = \phi_\varepsilon(z + y)$$
$$f_2(y) = \varepsilon\Delta(z + y, z)$$
$$f_3(y) = (K_1 + \varepsilon K_2)d_{C_0}(z(a) + y(a))$$
$$f_4(y) = (K_3 + \varepsilon K_4)\int_a^b \rho_\alpha(t, z + y, \dot{z} + \dot{y})\, dt.$$

The assertion of the lemma is that $y = 0$ minimizes $f_1 + f_2 + f_3 + f_4$ over the arcs y for which $\|\dot{y}\|_\infty$ is sufficiently small. The four functionals are Lipschitz near 0 with respect to this norm, so the stationarity condition

$$0 \in \partial\{f_1 + f_2 + f_3 + f_4\}(0)$$

implies

(1) $$0 \in \partial f_1(0) + \partial f_2(0) + \partial f_3(0) + \partial f_4(0).$$

If h is any Lipschitz function on R^n, any element ζ of the generalized gradient of the mapping $y \to h(y(a))$ at y_0 is represented by an element ζ_0 of $\partial h(y_0(a))$, so that $\zeta(y) = \langle \zeta_0, y(a) \rangle$ for all y (this follows from Theorem 2.3.10). It follows that any element ζ of $\partial f_3(0)$ is of the form $\zeta(y) = \langle \zeta_0, y(a) \rangle$ where ζ_0 belongs to $\partial(K_1 + \varepsilon K_2)d_{C_0}(z(a))$.

Applying Example 2.7.4 tells us that any element ζ of $\partial f_4(0)$ is such that, for some function (q, s) with

$$(q, s)(t) \in \partial(K_3 + \varepsilon K_4)\rho_\alpha(t, z, \dot{z}) \quad \text{a.e.,}$$

one has, for all y,

$$\zeta(y) = \int_a^b [\langle q, y \rangle + \langle s, \dot{y} \rangle] \, dt.$$

Similarly, any element ζ of $\partial f_2(0)$ corresponds to a function r with $r(t) \in \varepsilon \bar{B}$ and a point r_0 in $\varepsilon \bar{B}$ such that

$$\zeta(y) = \langle r_0, y(a) \rangle + \int_a^b \langle r, y \rangle \, dt.$$

This leaves f_1 to consider. If ζ lies in the generalized gradient at 0 of the map $y \to f(z(b) + y(b))$, then as above $\zeta(y) = \langle \zeta_1, y(b) \rangle$ for some element ζ_1 of $\partial f(z(b))$. If ζ is an element of $\partial G(z)$, and if $G(z) > 0$, it is expressible (in view of Theorem 2.8.2) in the form

$$\zeta(y) = \int_{[a, b]} \langle \gamma(t), y(t) \rangle \mu(dt),$$

where μ is a probability Radon measure on $[a, b]$ and $\gamma(t)$ belongs to $\partial_x^> g(t, z(t))$ μ-a.e. Now if $G(z) < \phi_\varepsilon(z) = f_1(0)$, then $\phi_\varepsilon(y) = f(y(b)) - f(x(b)) + \varepsilon^2$ for y near z, so that $\partial f_1(0)$ is representable by the ζ_1 above. Otherwise we have $G(z) \geq \phi_\varepsilon(z) > 0$ (see Lemma 1), so that (by Theorem 2.8.2) any element ζ of $\partial f_1(0)$ has the form

$$\zeta(y) = \lambda \langle \zeta_1, y(b) \rangle + (1 - \lambda) \int_{[a, b]} \langle \gamma, y \rangle \mu(dt),$$

where $0 \leq \lambda \leq 1$ and ζ_1, γ, μ are as described above. We now summarize the interpretation of (1):

Lemma 3. *There exist a probability measure μ, a μ-selection γ for $\partial_x^> g(t, z(t))$, a scalar λ in $[0, 1]$, elements ζ_0 and ζ_1 of $(K_1 + \varepsilon K_2) \partial d_{C_0}(z(a))$ and $\partial f(z(b))$, respectively, a selection (q, s) of $(K_3 + \varepsilon K_4) \partial \rho_\alpha(t, z, \dot{z})$, a selection r of $\varepsilon \bar{B}$, and a point r_0 in $\varepsilon \bar{B}$ such that, for every arc y with bounded derivative, one has*

$$\langle \zeta_0 + r_0, y(a) \rangle + \lambda \langle \zeta_1, y(b) \rangle + (1 - \lambda) \int_{[a, b]} \langle \gamma, y \rangle \mu(dt)$$

$$+ \int_a^b [\langle q + r, y \rangle + \langle s, \dot{y} \rangle] \, dt = 0.$$

We are now at a very familiar point in variational arguments, one for which the classical lemma of Dubois–Reymond was proven. The identity of the lemma, by a standard argument (see Hestenes, 1966, p. 50) implies that

$$s(t) = \zeta_0 + r_0 + \int_a^t (q + r) \, d\tau + (1 - \lambda) \int_{[a, t)} \gamma(\tau) \mu(d\tau)$$

and that

$$s(b) + (1 - \lambda) \gamma(b) \mu\{b\} = -\lambda \zeta_1.$$

Let us define the arc p via

$$p(t) = \zeta_0 + r_0 + \int_a^t (q + r) \, d\tau.$$

Then p satisfies

(2) $\quad \left[\dot{p} - r, p + (1 - \lambda) \int_{[a, t)} \gamma \, d\mu \right] \in (K_3 + \varepsilon K_4) \partial \rho_\alpha(t, z, \dot{z}) \quad$ a.e.

(3) $\quad p(a) \in (K_1 + \varepsilon K_2) \partial d_{C_0}(z(a)) + \varepsilon \bar{B}$

(4) $\quad p(b) + (1 - \lambda) \int_{[a, b]} \gamma \, d\mu \in -\lambda \partial f(z(b)).$

Step 4. We now wish to bring in the Hamiltonian H.

Lemma 4. *The relations $(q, p) \in \partial K \rho_\alpha(t, y, v)$ and $\rho_\alpha(t, y, v) = 0$ imply $(-q, v) \in \partial H(t, y, p) + \alpha \bar{B}$.*

In the proof of the lemma, we suppress t, which is fixed. Note that $H_\alpha(y, p) = H(y, p) + \alpha|p|$, so that it suffices to prove the inclusion

$$(-q, v) \in \partial H_\alpha(y, p).$$

When $\nabla \rho_\alpha(y', v')$ exists at a point v' in $F_\alpha(y')$, it follows that it equals 0, since ρ_α attains a minimum there. In light of Theorem 2.5.1, it follows that (q, p) can be expressed as $\sum_i \lambda_i (Q_i, P_i)$, where the λ_i are positive and sum to $\gamma \leq 1$, and where each (Q_i, P_i) is of the form $\lim_j K \nabla \rho_\alpha(y_j, v_j)$, where $(y_j, v_j) \to (y, v)$ and $v_j \notin F_\alpha(y_j)$. For such points, $\nabla_v \rho_\alpha(y_j, v_j)$ is the unique unit vector normal to $F_\alpha(y_j)$ at the point in $F_\alpha(y_j)$ closest to v_j (by Proposition 2.5.4), and it follows that each P_i equals the unique vector P of length K that generates $N_{F_\alpha(y)}(v)$. Note that $p = \gamma P$, and that we may assume $\gamma > 0$ (for otherwise $p = q = 0$ and the assertion of the lemma follows from Proposition 3.2.4(e)).

We now claim that it suffices to prove that each $(Q_i, P_i) = (Q_i, P)$ satisfies $(-Q_i, v) \in \partial H_\alpha(y, P)$. For if this holds, then (because $\{\lambda_i/\gamma\}$ is a convex combination), one deduces

$$\left(-\sum_i \frac{\lambda_i Q_i}{\gamma}, v\right) \in \partial H_\alpha(y, P),$$

which implies (by Proposition 3.4.2(e))

$$\left(-\sum_i \lambda_i Q_i, v\right) \in \partial H_\alpha(y, \gamma P),$$

which is the desired conclusion.

We have reduced the proof of the lemma to proving that $(-Q_i, v)$ belongs to $\partial H_\alpha(y, P)$, where $(Q_i, P) = \lim K \nabla \rho_\alpha(y_j, v_j)$ as described above. To show this, it is clearly enough to show that any point of the form $(\tilde{q}, \tilde{p}) = K \nabla \rho_\alpha(\tilde{y}, \tilde{v})$, where $\tilde{v} \notin F_\alpha(\tilde{y})$, is such that $(-\tilde{q}, \tilde{v})$ belongs to $\partial H_\alpha(\tilde{y}, \tilde{p}) + [0, \rho_\alpha(\tilde{y}, \tilde{v})\bar{B}]$. In turn, this follows from the inequality

(5) $\qquad H_\alpha^\circ(\tilde{y}, \tilde{p}; u, w) \geq (-\tilde{q}, \tilde{v}) \cdot (u, w) - |w|\rho_\alpha(\tilde{y}, \tilde{v}),$

where (u, w) is an arbitrary element of $R^n \times R^n$. We note two facts:

(i)
$$H_\alpha(s, r) = \sup\{r \cdot \beta - K\rho_\alpha(s, \beta) : \beta \in R^n\}$$

whenever $|r| \leq K$ (this is a consequence of Proposition 2.4.3).

(ii)
$$H_\alpha(s, r) = r \cdot \beta - K\rho_\alpha(s, \beta) \quad \text{if } r = K \nabla_\beta \rho_\alpha(s, \beta)$$

(from (i), since the sup is attained at β).

3.2 A Control Problem

For any positive λ, set $K_\lambda = |\tilde{p} + \lambda w|$. We observe

$$H_\alpha(\tilde{y} + \lambda u, \tilde{p} + \lambda w) - H_\alpha(\tilde{y}, \tilde{p}) \geq (\tilde{p} + \lambda w) \cdot \tilde{v}$$

$$- K_\lambda \rho_\alpha(\tilde{y} + \lambda u, \tilde{v}) - \tilde{p} \cdot \tilde{v} + K \rho_\alpha(\tilde{y}, \tilde{v})$$

(using (i) and (ii))

$$= \lambda w \cdot \tilde{v} - \left[K \rho_\alpha(\tilde{y} + \lambda u, \tilde{v}) - K \rho_\alpha(\tilde{y}, \tilde{v}) \right] + (K - K_\lambda) \rho_\alpha(\tilde{y} + \lambda u, \tilde{v}).$$

Dividing across by λ and using the inequality $K - K_\lambda \geq -\lambda |w|$ (since $|\tilde{p}| = K$) yields (5) and completes the proof of Lemma 4.

We now invoke the lemma to rewrite (2):

$$(6) \qquad (-\dot{p}, \dot{z}) \in \partial H\!\left(t, z, p + (1 - \lambda) \int_{[a,t)} \zeta \, d\mu \right) + 2\varepsilon B,$$

where we have used the bounds $|r| \leq \varepsilon$, $\alpha < \varepsilon$. It is time to recall that to this point ε has been any (sufficiently small) positive parameter. The final phase of the proof will consist of going to the limit in (3), (4), and (6) along appropriate subsequences. Observe that $p, \gamma, \mu, \lambda, z$ all depend on ε, but that K_1, K_2, K_3, K_4 do not. Note that z converges uniformly to x as $\varepsilon \to 0$ (since $\Delta(x, z) \leq \varepsilon$). We may select a sequence of ε's converging to 0 such that $\lambda \to \lambda_0 \in [0, 1]$.

By an argument involving the Radon–Nikodym Theorem it can be shown that for a further subsequence, the (vector) measures η defined by $d\eta = (1 - \lambda) \gamma \, d\mu$ converge weak* to a measure η_0 of the form $d\eta_0 = \gamma_0 \, d\mu_0$, where the nonnegative measure μ_0 is the weak* limit of $(1 - \lambda)\mu$ and where γ_0 is a measurable selection of $\partial_x^> g(t, x(t))$ (μ_0-a.e.). (In consequence, $\lambda_0 + \|\mu_0\| = 1$.) The result just quoted is Proposition 3.1.8, and uses the fact that $\partial_x^> g$ is uniformly bounded, compact, and convex, and of closed graph. It follows that μ_0 is supported on the set S of Definition 3.2.5(ii). If this set is empty, then $\lambda = 1$ for ε small, so that μ_0 is "not really there" in the multiplier.

Let us now define a multifunction Γ and a measurable function y_ε as follows.

$$\Gamma(t, x, p) = \left\{ (-v, u) : (u, v) \in \partial H\!\left(t, x, p + \int_{[a,t)} \gamma_0 \, d\mu_0 \right) \right\}$$

$$y_\varepsilon(t) = (1 - \lambda_\varepsilon) \int_{[a,t)} \gamma \, d\mu - \int_{[a,t)} \gamma_0 \, d\mu_0.$$

Note that for each ε one has

$$(\dot{z}_\varepsilon, \dot{p}_\varepsilon) \in \Gamma(t, z_\varepsilon, p_\varepsilon + y_\varepsilon) + 2\varepsilon B$$

and that y_ε converges uniformly to 0 as $\varepsilon \to 0$. We invoke Theorem 3.1.7 to deduce the uniform convergence (along a subsequence) of $(z_\varepsilon, p_\varepsilon)$ to (x, p). We have produced a multiplier in the sense of Definition 3.2.5, thereby completing the proof of the theorem. □

In view of relation (3) above, the proof actually showed:

Corollary 1

The multiplier $[p, \gamma, \mu, \zeta, \lambda]$ whose existence is asserted in the theorem may be assumed to satisfy condition 3.2.5(v) with any $r > K_1$, where

$$K_1 := \max\{K_f, K_g\}[K\ln(K) + 1].$$

The proof adapts with obvious changes to the case in which the minimization is that of a function $f(x(a), x(b))$ of both $x(a)$ and $x(b)$. It can also be modified to treat the case in which x is merely assumed to minimize the quantity

$$\max\{f(y(b)) - f(x(b)), G(y)\},$$

which is a weaker assumption than supposing that x minimizes $f(y(b))$ subject to $G(y) \leq 0$. The only difference in the proof in this case arises in Lemma 1, where only $\phi_\varepsilon(y) \geq 0$ can be asserted. In calculating $\partial \phi_\varepsilon$ subsequently, therefore, $\partial G(z)$ may contribute even if $G(z) = 0$, so that instead of $\partial_x^> g(t, x)$, the following (possibly larger) set arises:

$$\bar{\partial}_x g(t, x) = \text{co}\{\lim \zeta_i : \zeta_i \in \partial_x g(t_i, x_i), (t_i, x_i) \to (t, x)\}.$$

We summarize:

Corollary 2

Let x minimize

$$\max\{f(y(a), y(b)) - f(x(a), x(b)), G(y)\}$$

over the trajectories y for F with $y(a) \in C_0$. Then there exists $[p, \gamma, \mu, \zeta, \lambda]$ with $\lambda + \|\mu\| = 1$ such that the conditions of Definition 3.2.5 hold with the following changes: $\partial_x^> g$ is replaced by $\bar{\partial}_x g$, $\zeta = (\zeta_0, \zeta_1)$ belongs to $\partial f(x(a), x(b))$, ζ_1 replaces ζ in (vi), and (v) is replaced by

$$p(a) \in \lambda \zeta_0 + r \partial d_{C_0}(x(a)).$$

3.2.8 Remark

As will be shown in Section 3.6, when F and g have no explicit dependence on t, we may add to the other conditions the constancy of the Hamiltonian $H(x(t), p(t) + \int_{[a,t)} \gamma(s)\mu(ds))$.

3.3 A PROBLEM IN RESOURCE ECONOMICS

The economic theory of resource extraction dates back to Hotelling's classic paper of 1931. The dynamics of that paper are very simple: if $y(t)$ measures the level of a (nonrenewable) resource at time t, then $\dot{y} = -q$, where q, the extraction rate, is a choice variable. As the models have become more complicated in order to reflect added features of the extraction process, the theory of optimal control has been increasingly useful in the analysis of optimal extraction policies. In this section we present a model which is new in emphasizing certain economic issues (as explained below). While the economic structure of the problem is quite natural, the resulting mathematical problem is one that is not amenable to the standard techniques. We shall solve the problem by the theory of the previous section, and in so doing illustrate some of the issues that arise in applying the theory to specific problems.

The Problem

In our model, the extraction process itself requires capital (machines) z and labour v in equal quantities (by normalizing units). ("One person, one shovel.") This can be expressed as follows:

(1) $$\dot{y} = -\min(z, v)r(y),$$

where the factor $r(y)$ is added to reflect the fact that, for a given effort (i.e., z and v), extraction rates vary with the level of the remaining stock y. Further we suppose that a component w of labour can be used to build more machines:

(2) $$\dot{z} = \gamma w.$$

The total labour force $L(t)$ is exogeneously determined, hence the constraint

(3) $$(v, w) \in U(t),$$

where $U(t)$ is the set determined by

$$0 \leq v, \quad 0 \leq w, \quad v + w \leq L(t).$$

In the language of optimal control theory, v and w are controls and y, z the resulting states, completely determined by Eqs. (1) and (2) and the initial conditions

(4) $$y(0) = y_0 > 0, \quad z(0) = z_0 > 0.$$

We suppose that the owners of the resource are in a competitive market and are concerned with maximizing net discounted revenue (or variable profit):

(5) $$\int_0^T e^{-\delta t}\{\pi q - \alpha v - \beta w\}\, dt = \int_0^T e^{-\delta t}\{\pi \min(z, v)r(y) - \alpha v - \beta w\}\, dt,$$

where T is the planning horizon, δ the discount rate, π the resource price, and α and β are the labour costs in resource production and capital investment, respectively.

The central issue for the planner is to decide on a division of labour between that used for immediate production (v) and that used to build more capital stock (w). Note that capital has no alternative use in this model. Suppose that the initial capital z_0 was very large ($> L(t)$). Then there can be no purpose in increasing z, so that our dynamics (1), (2) collapse essentially to the classic $\dot{y} = -q$ (except that we allow the varying factor $r(y)$). Viewed in this light, our model is also one that brings in capital formation as an issue.

In Campbell (1980), and in Clark, Clarke, and Munro (1979) (in the context of renewable resources) the capitalization issue is treated by allowing instantaneous jumps in z. This serves to simplify the issue by making an increase in capital an independent decision which does not reduce in any way the inputs available for current production.

The issue of capitalization (and investment) is treated in some of the modern models of resource extraction (see, for example, Dasgupta and Heal (1979) and the references therein). They incorporate production functions as follows:

$$(6) \qquad g = G(z, q, L, t),$$

where z is capital, L labour, t time, and q the rate of extraction as before, and where g is the quantity of the output which in turn can be used to produce z. Typically, the role of labour is deemphasized (e.g., by "factoring out" L), and the central question becomes that of choosing (1) the extraction rate q, and (2) the investment portion of production used to augment capital, in such a way as to maximize social welfare (or utility from consumption).

The nature of these production functions imposes hypotheses of substitutability and independence of inputs. The first refers to the fact that a decrease in one input (e.g., z) can be compensated by an increase in another (e.g., q). (This is a consequence of the customary assumption that G is smooth and that G_z and G_q are positive.) By "independence" we mean that q, for example, can be chosen without regard to z: extraction possibilities do not depend upon the capital level. In short, the process of extraction again is left out as it was in the Hotelling model. The model we study here suppresses these hypotheses and focuses attention on situations in which inputs are required in fixed proportions, and in which production is constrained by the current stock level of capital and by the segment of labour committed to it. Note that Eq. (6) has elaborated the production process by *adding* inputs to production while retaining q as an independent input. In contrast, our model defines q as a function of other inputs, thus emphasizing the mechanics of the extraction process itself. Puu (1977) has treated a model with some similar features, but in which there is only one input (capital), and in which production and capital

3.3 A Problem in Resource Economics

formation are independent of one another. Thus the issues we raise here are only indirectly broached. However, our conclusions are consistent with his.

The problem of maximizing the net revenue (5) subject to the dynamics (1), (2), the control constraints (3), the initial conditions (4), and of course the constraint $y \geq 0$, is of a standard form, but in fact is not one to which the usual methods apply. This is due to the nondifferentiability of the function $\min(v, z)$, which makes it infeasible to even write down the usual equations of the maximum principle, for example (see Chapter 1). Any formulation of constrained fixed-proportions production will incorporate "corners" in the model, and this undoubtedly accounts for the fact that this economically very natural form of production has received little attention.

As we shall see, however, the problem is readily framed in the form of Problem 3.2.1.

We suppose that r and L are continuously differentiable and nondecreasing, and we adopt the mild hypothesis that for some positive $\hat{r}$, $r(y) \leq \hat{r}y$. This ensures that the solution y of Eq. (1) remains positive, and obviates the necessity of considering the implicit constraint $y \geq 0$.

Reformulating the Problem

The main step in reformulating the preceding problem so as to give it the semblance of Problem 3.2.1 consists of "absorbing" the integral cost functional into the multifunction F. We set $n = 3$ and $[a, b] = [0, T]$, and points x will stand for triples (y, z, s) in R^3. Define

$$f(y, z, s) = s, \qquad C_0 = \{(y_0, z_0, 0)\}.$$

Now let M be so large that, for all relevant values of y, z, v, w, t, one has $e^{-\delta t}[\alpha v + \beta w - \pi \min(v, z)r(y)] < M$. We set

$$F(t, x) = F(t, y, z, s)$$

$$= \{[\min(v, z)r(y), \gamma w, \theta] : (v, w) \in U(t),$$

$$e^{-\delta t}[\alpha v + \beta w - \pi \min(v, z)r(y)] \leq \theta \leq M\},$$

and we leave the reader to verify that f, C_0, F satisfy Hypotheses 3.2.2. (Note that the state constraint is inactive, since there is an a priori positive lower bound on the admissible arcs y, while z is necessarily nondecreasing.)

It is a consequence of the measurable selection theorem that (y, z) satisfies Eqs. (1), (2), and (4) for some measurable (v, w) taking values in U, along with $Mt \geq s(t) \geq \int_0^t e^{-\delta \tau}[\alpha v + \beta w - \pi \min(v, z)r(y)] \, d\tau$, iff (y, z, s) is a trajectory for F emanating from C_0. It follows that the resource problem and this version of Problem 3.2.1 are equivalent (the optimal arcs are the same).

The Hamiltonian

We plan to apply Theorem 3.2.6, in which the central figure is the Hamiltonian (Proposition 3.2.4). If the p of Proposition 3.2.4 is identified with (p_1, p_2, p_3) in R^3, we observe that the Hamiltonian inclusion 3.2.5(iv) implies $\dot{p}_3 = 0$ (since F and hence H is independent of s). Note that γ and μ are absent from the multiplier (so that $\lambda = 1$). Condition 3.2.5(vi) gives $p_3(T) = -1$ (since $\zeta = [0, 0, 1]$), along with $p_1(T) = p_2(T) = 0$. It follows that p_3 is identically -1. Let us define

$$h(t, y, z, p, q) := H(t, y, z, s, p, q, -1).$$

Then it follows from Proposition 3.2.4(e) that the Hamiltonian inclusion implies (setting $p_1 = p$, $p_2 = q$):

(7) $\qquad (-\dot{p}, -\dot{q}, \dot{y}, \dot{z}) \in \partial h(t, y, z, p, q)$ a.e.

The analysis will reduce to "solving" this differential inclusion with the boundary conditions $y(0) = y_0$, $z(0) = z_0$, $p(T) = q(T) = 0$. We first need to calculate h, which is the maximum over (v, w) in $U(t)$ of the expression

$$-p \min(v, z)r + \gamma q w + e^{-\delta t}\{\pi \min(v, z)r - \alpha v - \beta w\}$$

(where $r = r(y)$). Note that this expression is piecewise linear in (v, w), so that the maximum can only occur at extreme points of U or at "corners" (nondifferentiable points) of the function; that is, at the points $(0, 0)$, $(L, 0)$, $(0, L)$, and (when $z \leqslant L(t)$) the points $(z, 0)$, $(z, L - z)$.

Let us set

$$\phi = (\pi e^{-\delta t} - p)r - \alpha e^{-\delta t}, \qquad \psi = \gamma q - \beta e^{-\delta t}$$

as a convenient shorthand. When $z \leqslant L$, the points above yield the values 0, $\phi z + \alpha(z - L)e^{-\delta t}$, ψL, ϕz, $\phi z + \psi(L - z)$, and h is the greatest of these (note that the fourth is greater than the second, so we may eliminate the latter). When $z > L$, h is the greatest among 0, ϕL, ψL. We summarize:

3.3.1 Proposition

We have in all cases

$$h = \max\{0, \phi \min(z, L), \psi L, \psi L + (\phi - \psi)\min(z, L)\}.$$

The Hamiltonian Flow

The proposition leads us to identify four cases defined by appropriate values of ϕ and ψ (see Figure 3.1).

3.3 A Problem in Resource Economics

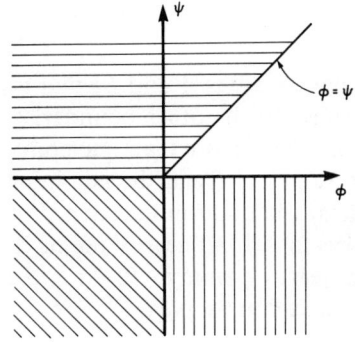

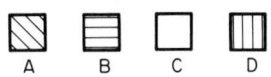

Figure 3.1 The four regions of definition for the Hamiltonian.

Case A: $\phi < 0, \psi < 0 : h = 0.$ Here, the generalized gradient ∂h is just the derivative of h in (y, z, p, q); that is, $\partial h = (0, 0, 0, 0)$. Hence (7) yields:

(8a) $$(-\dot{p}, -\dot{q}, \dot{y}, \dot{z}) = (0, 0, 0, 0),$$

and we also derive

(9a) $$\dot{\phi} = -\delta(\pi r - \alpha)e^{-\delta t}, \qquad \dot{\psi} = \delta\beta e^{-\delta t}.$$

Case B: $\psi > 0, \psi > \phi : h = \psi L.$

(8b) $$(-\dot{p}, -\dot{q}, \dot{y}, \dot{z}) = (0, 0, 0, \gamma L)$$

(9b) $$\dot{\phi} = -\delta(\pi r - \alpha)e^{-\delta t}, \qquad \dot{\psi} = \delta\beta e^{-\delta t}.$$

Case C: $\phi > \psi > 0 : h = \psi + (\phi - \psi)\min(z, L).$ When $z < L$,

(8c) $$(-\dot{p}, -\dot{q}, \dot{y}, \dot{z}) = \left(\frac{z(\phi + \alpha e^{-\delta t})r'}{r}, \phi - \psi, -zr, \gamma(L - z)\right)$$

(9c) $$\dot{\phi} = -\delta(\pi r - \alpha)e^{-\delta t}, \qquad \dot{\psi} = \gamma(\psi - \phi) + \delta\beta e^{-\delta t}.$$

Case D: $\psi < 0, \phi > 0 : h = \phi \min(z, L).$ When $z < L$,

(8d) $$(-\dot{p}, -\dot{q}, \dot{y}, \dot{z}) = \left(\frac{z(\phi + \alpha e^{-\delta t})r'}{r}, \phi, -zr, 0\right)$$

(9d) $$\dot{\phi} = -\delta(\pi r - \alpha)e^{-\delta t}, \qquad \dot{\psi} = -\gamma\phi + \delta\beta e^{-\delta t}.$$

Boundary Points

We shall denote AB the boundary between the open regions A and B; that is, AB is the set $\phi < 0$, $\psi = 0$. Similar notation applies to the other boundaries. When (ϕ, ψ) lies on one of these boundaries, ∂h contains more than one point. In fact, ∂h is the convex set determined by the values of ∂h in the adjoining regions A, B, C, or D at points infinitesimally near the point in question. For example, suppose $\phi < 0$, $\psi = 0$; that is, (ϕ, ψ) lies in AB. At nearby points in A, ∂h is the right-hand side of Eq. (8a), and at nearby points in B, ∂h is the right-hand side of Eq. (8b). Thus at the point in question we have

$$\partial h = \{(0,0,0,s) : 0 \leqslant s \leqslant \gamma L\}.$$

Phase Plane Analysis

The rest of the analysis merely consists of analyzing the solution curves of the system of differential equations. In some problems the analysis is too difficult to be carried out all the way. In such cases, we may nonetheless hope to combine partial information from the necessary conditions with trial and error and intuition in order to formulate a conjecture as to the solution, and then proceed to verify this conjecture rigorously (this was the procedure in Clark, Clarke, and Munro (1979)). We begin with a simple observation.

3.3.2 Proposition

If at some point t we have $\pi r(y(t)) - \alpha < 0$, it is optimal beyond that point to cease production and investment ($v = w = 0$).

Proof. It suffices to note that the revenue integrand

$$e^{-\delta t}\{\pi \min(v, z)r(y) - \alpha v - \beta w\}$$

is necessarily nonpositive beyond t, and equal to 0 when $v = w = 0$. $\square$

To avoid the trivial case in which "nothing happens," we henceforth assume $\pi r(y_0) - \alpha > 0$.

We may also suppose that if $\pi r(y) - \alpha$ becomes 0, then $v = w = 0$ afterwards.

3.3.3 Proposition

At time $t = T$, (ϕ, ψ) lies in either D or AD. At no time does (ϕ, ψ) lie in A, B, AB, or BC.

Proof. The condition $p(T) = q(T) = 0$ becomes, in terms of ϕ and ψ,

(10) $\quad \phi(T) = (\pi r(y(T)) - \alpha)e^{-\delta T}, \quad \psi(T) = -\beta e^{-\delta T}.$

3.3 A Problem in Resource Economics

In view of the preceding remarks, $\phi(T) \geq 0$ and $\psi(T) < 0$, so that at T, (ϕ, ψ) lies in either D or AD. Suppose that (ϕ, ψ) lies in B for some t. Note that Eq. (9b) applies near t, and so $\dot\phi \leq 0$, $\dot\psi > 0$. But then (ϕ, ψ) must remain in B thereafter, contrary to the requirement that it terminate in D or AD. Similarly, (ϕ, ψ) could never be in A, AB, or BC. □

The preceding result has an immediate and interesting consequence:

Corollary
Along the optimal path, we have $z(t) < L(t)$.

Proof. Note that $z(0) < L(0)$ by assumption, so that by continuity $z(t) < L(t)$ at least for t near 0. For any such t, only Eq. (8c) or Eq. (8d) (or a convex combination) can hold, so that in any case $\dot z \leq \gamma(L - z)$. If we define $\theta(t) = L(t) - z(t)$, this last inequality implies $\dot\theta = -\gamma\theta + \varepsilon$, where $\varepsilon(t)$ is a nonnegative function. Solving this differential equation gives

$$\theta(t) = \theta_0 e^{-\gamma t} + e^{-\gamma t}\int_0^t \varepsilon(s)e^{\gamma s}\,ds,$$

where $\theta_0 = \theta(0) > 0$; it follows that $\theta(t) > 0$ for all t. □

Intervals of time during which (ϕ, ψ) lies on a boundary correspond to what are known as *singular* intervals in optimal control theory. As we shall see, this can happen on AD. We now examine the case of CD. We call c a *constant value* of the function $r(\cdot)$ if, for all y throughout some interval, we have $r(y) = c$. The function $r(\cdot)$ is nondecreasing and in consequence admits at most countably many constant values c_i. In what follows, we rule out by assumption the coincidence that among these c_i is the number $(\alpha + \delta\beta/\gamma)/\pi$. Note that this assumption is superfluous if r is strictly decreasing.

3.3.4 Proposition
There is no interval of time throughout which (ϕ, ψ) lies in CD.

Proof. Suppose the contrary. Then $\psi = 0$ gives $q\gamma = \beta e^{-\delta t}$ and differentiation yields $-\dot q = \delta\beta e^{-\delta t}/\gamma$. From Eqs. (8c) and (8d), we deduce the existence of λ in $[0, 1]$ such that

$$-\dot q = \lambda\phi + (1 - \lambda)(\phi - \psi) = \phi.$$

Thus $\delta\beta e^{-\delta t} = \gamma\phi$, and differentiation yields $\gamma\dot\phi = -\delta^2\beta e^{-\delta t}$. Substituting $\dot\phi = -\delta(\pi r - \alpha)e^{-\delta t}$ leads to $r(y) = (\alpha + \delta\beta/\gamma)/\pi$. But $\dot y = -zr < 0$ so $(\alpha + \delta\beta/\gamma)\pi$ is a constant value of r, contrary to assumption. □

3.3.5 Remark
The multiple boundary $ABCD$ can also be ruled out, as we shall see.

Time Reversal

To complete the analysis, we proceed by a method not uncommon in control theory. We reverse time via the transformation $s = T - t$, $\phi(s) = \phi(T - s)$, and so on. For convenience we define, for $0 \leq s \leq T$,

$$k(s) = (\pi r_s - \alpha)e^{\delta(s-T)},$$

where $r_s = r(y(T - s)) = r(y(t))$. Note that $k \geq 0$, and that $s \to r_s$ is nondecreasing. Then the reversed functions ϕ, ψ satisfy, in D,

(11d) $$\dot{\phi}(s) = \delta k(s)$$

(12d) $$\dot{\psi}(s) = \gamma\phi(s) - \delta\beta e^{\delta(s-T)},$$

and in C,

(11c) $$\dot{\phi}(s) = \delta k(s)$$

(12c) $$\dot{\psi}(s) = \gamma(\phi(s) - \psi(s)) - \delta\beta e^{\delta(s-T)}.$$

If (ϕ, ψ) lies in AD on some interval, then, in view of Eqs. (9a) and (9d), we have during that interval:

(13) $$\dot{\phi}(s) = \delta k(s) = 0$$

(14) $$\dot{\psi}(s) = -\delta\beta e^{\delta(s-T)},$$

and also $\dot{z} = 0$.

Our transversality conditions now become initial conditions at $s = 0$:

(15) $$\phi(0) = k(0), \quad \psi(0) = -\beta e^{-\delta T}.$$

It follows that for all s, one has

(16) $$\phi(s) = k(0) + \delta \int_0^s k(\tau)\, d\tau.$$

The function (ϕ, ψ) begins in AD if $k(0) = 0$, or in D if $k(0) > 0$. If the former, D must be entered at some point, otherwise Eq. (13) would imply $k(T) = \pi r(y(0)) - \alpha = 0$, contrary to assumption. Once the switch from AD to D occurs (at $s = s_0 \geq 0$, say), then $k \geq 0$ implies that thereafter $\dot{\phi} \geq 0$ and so (ϕ, ψ) is constrained to D and C (with no pauses in CD, see Proposition

3.3 A Problem in Resource Economics

3.3.4). Suppose that there is a switch from D to C (at $s = s_1 > s_0$). We now show:

3.3.6 Proposition

(ϕ, ψ) remains in C thereafter.

Proof. We have $\psi(s_1) = 0$; $\dot{\psi}(s)$ is given by Eq. (12c) for $s > s_1$. (Recall that ϕ is given by Eq. (16).) Solving this linear differential equation gives, for $s \geq s_1$,

$$\psi(s) = e^{-\gamma s} \int_{s_1}^{s} e^{\gamma \tau} g(\tau) \, d\tau,$$

where $g(\tau) = \gamma \phi(\tau) - \delta \beta e^{\delta(\tau - T)}$. We need only show $g(s) \geq 0$ for $s \geq s_1$, which we proceed to do.

Note that $\psi'_+(s_1) = g(s_1) \geq 0$ (since $\psi < 0$ before s_1 and $\psi > 0$ after s_1), so it suffices to prove that $g'(s) \geq 0$ for $s \geq s_1$. We calculate

$$g'(s) = \delta\{\gamma k(s) - \delta\beta e^{\delta(s-T)}\} = \delta e^{\delta(s-T)}\{\gamma(\pi r_s - \alpha) - \delta\beta\}.$$

Since the last quantity in braces is increasing in s, it suffices to show it nonnegative at $s = s_1$; that is, $\gamma k(s_1) \geq \delta \beta e^{\delta(s_1 - T)}$. Now, $g(s_1) \geq 0$ is equivalent to $\gamma \phi(s_1) \geq \delta \beta e^{\delta(s_1 - T)}$, so in fact it suffices to prove $k(s_1) \geq \phi(s_1)$.

To see this, note that for s in $(0, s_1)$ we have $\dot{\phi}(s) \leq \delta(\pi \hat{r} - \alpha) e^{\delta(s-T)}$, where $\hat{r} = r(y(t))$ at $t = T - s_1$. Integrating gives

$$\phi(s_1) \leq \phi(0) + (\pi \hat{r} - \alpha) e^{-\delta T}(e^{\delta s_1} - 1)$$

$$= k(0) + k(s_1) - (\pi \hat{r} - \alpha) e^{-\delta T}$$

$$= k(s_1) + e^{-\delta T} \pi(r_0 - \hat{r})$$

$$\leq k(s_1).$$

This completes the proof. □

We now have all the necessary information to characterize the optimal production scheme, which is known to exist by Proposition 3.2.3.

In original (unreversed) time, we have arrived at the following conclusion: the optimal production strategy consists of (in general) three segments defined by two switching times t_0 and t_1 in $[0, T]$ ($t_0 = T - s_1$, $t_1 = T - s_0$). On $[0, t_0)$ we employ $v = z(t)$, $w = L(t) - z(t)$. (This follows from Eq. (8c) since (ϕ, ψ) lies in C during this time.) Similarly we find that on (t_0, t_1) we employ $v = z(t)$, $w = 0$, while on $(t_1, T]$ we have $v = w = 0$. The "shut-down" time t_1 is characterized quite simply as the time (if any) in $[0, T]$ at which $\pi r(y(t)) - \alpha$

becomes zero. Note, however, that $y(\cdot)$ depends on $z(\cdot)$ and hence on t_0. The times t_0 and t_1 (or equivalently, s_0 and s_1) are completely determined numerically by the differential equations and boundary conditions for ϕ and ψ that we have been studying. This relationship is fairly complex—it is the "two-point boundary-value problem" in which y, z, ϕ, ψ satisfy differential equations with boundary information available for y, z at 0 and ϕ, ψ at T.

The first (or third) segment of the three-phase solution described above may in fact be absent in some cases. Here is a simple criterion which assures that no investment in additional capital is made (i.e., $t_0 = 0$):

3.3.7 Proposition

If $\gamma(\pi r(y_0) - \alpha) \leqslant \delta\beta$, then $w = 0$ on $[0, T]$.

Proof. This is equivalent to showing that the reversed function ψ remains negative (i.e., that there is no s_1 at which ψ passes from D to C).

We have

$$\pi r(y(t)) - \alpha \leqslant \pi r(y_0) - \alpha \leqslant \frac{\delta\beta}{\gamma},$$

which implies $k(s) \leqslant \delta\beta e^{\delta(s-T)}/\gamma$. From Eq. (16) we conclude

$$\phi(s) \leqslant k(0) + \frac{\delta\beta e^{-\delta T}(e^{\delta s} - 1)}{\gamma}.$$

Substituting this into Eq. (12d), gives

$$\dot{\psi}(s) \leqslant \gamma k(0) - \delta\beta e^{-\delta T}$$

$$= e^{-\delta T}\{\gamma(\pi r(y_0) - \alpha) - \delta\beta\} \leqslant 0.$$

Thus ψ does not increase and never enters C. □

3.3.8 Remark

The two inequalities $\pi r - \alpha > 0$ and $\beta > \gamma(\pi r - \alpha)/\delta$ that have been prominent in the analysis have natural interpretations in terms of marginal values. The marginal net revenue resulting from an increment dv of labour to production (assuming adequate capital) is $-\pi \, dy - \alpha \, dv = \pi r \, dv - \alpha \, dv = (\pi r - \alpha) \, dv$, and hence $\pi r - \alpha$ should be positive for any production to take place. On the other hand, assume now that all machines (capital) are being used ($v = z$). An increment in the labour applied to building capital will produce $dz = \gamma \, dw$ machines, at a cost of $\beta \, dw$, and will permit a proportional increment $dv = \gamma\, dw$ in labour used for production thereafter. The gross effect of this upon instantaneous revenue is $dR(t) = (\pi r - \alpha)\, dv = \gamma(\pi r - \alpha)\, dw$, and the *net* effect on the discounted present value when y is large is approximately given by

$$dPV = \int_0^\infty e^{-\delta t}\, dR(t)\, dt - \beta\, dw = \left\{ \frac{\gamma(\pi r - \alpha)}{\delta} - \beta \right\} dw.$$

Capital investment can only be optimal when this is positive.

Summary

The resource problem has been solved in the case in which the horizon T is finite and the extraction coefficient $r(y)$ is nondecreasing and satisfies $r(y) \leq \hat{r}y$ (so that extraction becomes more difficult as y is depleted, and actual exhaustion of the resource is not feasible). Our conclusions in this case may be summarized in the following main points.

1. We must have $\pi r(y_0) - \alpha > 0$ for any production to take place, and shutdown (i.e., $v = w = 0$ thereafter) occurs exactly if and when $\pi r(y) - \alpha$ becomes zero.
2. The optimal policy never fully capitalizes the production process: $z(t) < L(t)$ at all times.
3. Until shutdown, all capital is used at all times (machines are never idle): $v(t) = z(t)$. Thus production has priority, and is never sacrificed. One may explain this as being due to the fact that under fixed proportions a machine that is unused is completely wasted, a policy that cannot be optimal. If there is substitutability between v and z, every machine continues to contribute to production even when production labour is decreased, so that deferring current production somewhat in order to build machines for later use is more palatable. For example, if production follows a Cobb–Douglas relationship $\dot{y} = -(vz)^{1/2}$, it can be shown that for sufficiently large horizons T, early production will be sacrificed to build capital for later production.
4. If $\pi r(y_0) - \alpha > 0$, there is always a period during which some labour is idle ($v = z < L$, $w = 0$), possibly followed by shutdown.
5. The period mentioned in point 4 may be preceded by a period in which $v = z$, $w = L(t) - z$. A sufficient (but not necessary) condition for this *not* to happen is $\beta \geq \gamma(\pi r(y_0) - \alpha)/\delta$. As argued above, this last inequality is readily interpretable as saying that the marginal cost of assigning more labour to capital building exceeds the marginal revenue generated by that extra capital in the long term.
6. The optimal policy consists in general of three distinct phases, in which (v, w) is, respectively, $(z, L - z), (z, 0), (0, 0)$. The first and/or third of these may be absent; this and the times of the switches are completely determined by certain differential equations and boundary conditions.

3.4 ENDPOINT CONSTRAINTS AND PERTURBATION

We now turn to problems in which both endpoints $x(a)$, $x(b)$ are subject to explicit constraints. In fact, we consider a family of such problems, parametrized by u (in R^n). Let C_1 be a given subset of R^n.

3.4.1 Problem P_D^u

To minimize $f(x(b))$ over the arcs x which are trajectories for F and which satisfy

$$g(t, x(t)) \leq 0, \quad a \leq t \leq b$$

$$x(a) \in C_0, \quad x(b) \in C_1 + u.$$

We assume that C_1 is closed; Hypotheses 3.2.2 are retained. The problem P_D^0 is also denoted P_D (see Sections 1.3 and 1.4).

The *value function* $V: R^n \to R \cup \{\pm\infty\}$ is defined as follows: $V(u) =$ infimum in P_D^u. (Note that the infimum over the empty set is taken to be $+\infty$.) Our main interest in this section is a characterization of the generalized gradient ∂V. But first some notation.

Weak Normality

The set of solutions to P_D is denoted Y. The problem P_D is said to be *weakly normal* provided that for every x in Y, for all $r > 0$, the only multipliers $[p, \gamma, \mu, \zeta, \lambda]$ (see Definition 3.2.5) satisfying $\lambda = 0$ and hence the condition $p(b) + \int_{[a,b]} \gamma(s)\mu(ds) = 0$ are such that μ is identically zero. It is easy to see that P_D is weakly normal provided that any multiplier of the type whose existence is asserted by Theorem 3.2.6 for the free-endpoint problem involves f nonvacuously; that is, that it not have $\lambda = 0$. This is always so when the state constraint is inactive along solutions:

3.4.2 Proposition

If for every x in Y one has $g(t, x(t)) < 0$ for t in $[a, b]$, then P_D is weakly normal.

Proof. In view of Definition 3.2.5(ii), any multiplier in this situation has $\mu = 0$, so the assertion is trivially true. $\square$

We now extend the definition of multipliers given in Definition 3.2.5 to problems having C_1 different from R^n.

Multiplier Sets

Let x be an admissible arc for the problem P_D. For a given $r > 0$, the *index λ multiplier set* corresponding to x, denoted $M_r^\lambda(x)$, consists of all five-tuples

3.4 Endpoint Constraints and Perturbation

$[p, \gamma, \mu, \zeta, \lambda]$ where $p, \gamma, \mu, \zeta,$ and λ are as in Definition 3.2.5, and where (i)–(iv) hold, as well as the two following conditions, where E signifies the quantity $\lambda\zeta + p(b) + \int_{[a,b]} \gamma(s)\mu(ds)$:

(1) $\qquad p(a) \in r\{|(\lambda, E)| + \|\mu\|\} \partial d_{C_0}(x(a))$

(2) $\qquad -E \in r\{|(\lambda, E)| + \|\mu\|\} \partial d_{C_1}(x(b))$.

The reader may verify that in the case $C_1 = R^n$, P_D reduces to Problem 3.2.1, and that $[p, \gamma, \mu, \zeta, \lambda]$ belongs to $M_r^\lambda(x)$ iff $[p, \gamma, \mu, \zeta, \lambda]$ is a multiplier as defined in Definition 3.2.5 (with r of 3.2.5 being $r(\lambda + \|\mu\|)$. Because 0 belongs to $\partial d_C(u)$ whenever the point u lies in the set C (by Theorem 2.5.6), it follows that the sets $M_r^\lambda(x)$ increase as r increases.

We are ready for the main result of this section. Recall that $\partial^\infty V$ refers to the asymptotic generalized gradient defined in Section 2.9. The notation $M_r^\lambda(Y)$ refers to the set $\cup_{x \in Y} M_r^\lambda(x)$, and if $[p, \gamma, \mu, \zeta, \lambda]$ is a multiplier, $E[p, \gamma, \mu, \zeta, \lambda]$ continues to denote the quantity $\lambda\zeta + p(b) + \int_{[a,b]} \gamma(s)\mu(ds)$. Note that $E[M_r^0(Y)]$ is a cone in R^n containing 0, since for any $\alpha \geq 0$, $[\alpha p, \gamma, \alpha\mu, \zeta, 0]$ is a multiplier whenever $[p, \gamma, \mu, \zeta, 0]$ is one. A cone is said to be *pointed* if it is not possible to express 0 as a finite sum of nonzero elements of the cone. We define

$$\hat{r} = \max\{K_g, 2(K_f + 1)\}[K\ln(K) + 1],$$

where K, K_f, K_g were defined in 3.1.6 and 3.2.2.

3.4.3 Theorem

Let V be finite at 0, and let the hypotheses of Proposition 3.2.3 hold. Then the set Y of solutions to P_D is nonempty and V is lower semicontinuous. If P_D is weakly normal, then for any $r > \hat{r}$ one has

$$\partial V(0) = \overline{\text{co}}\{E[M_r^1(Y)] \cap \partial V(0) + E[M_r^0(Y)] \cap \partial^\infty V(0)\}$$

$$\partial^\infty V(0) \supset \overline{\text{co}}\{E[M_r^0(Y)] \cap \partial^\infty V(0)\}.$$

If $E[M_r^0(Y)]$ is pointed, then equality holds in the last relation, and the closure operation is superfluous in both.

Before turning to the proof of the theorem, let us note an important consequence, the extra hypothesis of which is automatically satisfied when $C_1 = R^n$ and the state constraint is inactive (as in Proposition 3.4.2).

Corollary 1

If for every r sufficiently large every multiplier $[p, \gamma, \mu, \zeta, 0]$ in $M_r^0(Y)$ has $p = 0$ and $\mu = 0$, and if $V(0)$ is finite, then V is (finite and) Lipschitz near 0.

Proof. The hypothesis assures weak normality, and also that $E[M_r^\circ(Y)] = \{0\}$. It follows from the theorem that $\partial^\infty V(0)$ reduces to $\{0\}$, which is equivalent to V being Lipschitz near 0 by Proposition 2.9.7. $\square$

The following consequence of the theorem extends the necessary conditions of Theorem 3.2.6 to the more general problem P_D. It will be derived later as part of Theorem 3.5.2.

Corollary 2

If x is in Y, and if $r > \hat{r}$, then there is a multiplier $[p, \gamma, \mu, \zeta, \lambda]$ in $M_r^\lambda(x)$ such that $\lambda + \|\mu\| + \|p\| > 0$.

Proof of Theorem 3.4.3

Step 1. If $V(0)$ is finite, then admissible arcs exist for P_D; it follows as in Proposition 3.2.3 that Y is nonempty. To see that V is lower semicontinuous, let $u_i \to u$; we wish to show that $V(u)$ is bounded above by $\liminf_{i \to \infty} V(u_i)$ which may be assumed $< +\infty$. Let x_i solve $P_D^{u_i}$ (x_i exists as a consequence of the same existence argument as above). There is no loss of generality in supposing that $\lim_{i \to \infty} V(u_i)$ exists. We invoke Theorem 3.1.7 to deduce that a subsequence of $\{x_i\}$ (we dispense with relabeling) converges uniformly to a trajectory x admissible for P_D^u. Then

$$V(u) \leq f(x(b)) = \lim_{i \to \infty} f(x_i(b))$$

$$= \liminf_{i \to \infty} V(u_i),$$

as desired.

Step 2. Since ∂V and $\partial^\infty V$ were defined through the normal cone to the epigraph of V, the relations in the theorem are ultimately statements about such a cone. It follows from the theory of convex cones (see Rockafellar (1982b) for the general result) that the relations in question result from the following:

Lemma 1

$$N_{\text{epi } V}(0, V(0)) = \overline{\text{co}} \, \{N_1 \cup N_2\},$$

where

$$N_1 := \{\alpha(E, -1) : \alpha > 0, E \in E[M_r^1(Y)] \cap \partial V(0)\}$$

$$N_2 := \{(E, 0) : E \in E[M_r^0(Y)] \cap \partial^\infty V(0)\}.$$

3.4 Endpoint Constraints and Perturbation

It is an immediate consequence of the definitions of ∂V and $\partial^\infty V$ that $N_{\text{epi }V}(0, V(0))$ contains $\overline{\text{co}}\{N_1 \cup N_2\}$. To prove the opposite inclusion, it suffices in view of Proposition 2.5.7 to prove that any limit $(v_0, -\beta_0)$ of the form $\lim_{i \to \infty}(v_i, -\beta_i)/|(v_i, -\beta_i)|$, where $(v_i, -\beta_i)$ is a perpendicular to epi V at a point (u_i, α_i), and where $(v_i, -\beta_i) \to 0$ and $(u_i, \alpha_i) \to (0, V(0))$, belongs to $N_1 \cup N_2$.

Step 3. We pause to prove the following result, which will allow us to complete the proof.

Lemma 2. Let $(v, -\beta)$ be perpendicular to epi V at (u, α). Then for any $r > \hat{r}$ there is a solution x to P_D^u and a five-tuple $[p, \gamma, \mu, \zeta, \lambda]$ with $\lambda + \|\mu\| = 1$, satisfying 3.2.5(i), (ii), and (iv), and also

$$\gamma(t) \in \partial_x^> g(t, x(t))|(v, -\beta)|$$

$$p(a) \in r|(v, -\beta)|\partial d_{C_0}(x(a))$$

$$\lambda\beta\zeta + p(b) + \int_{[a, b]} \gamma(s)\mu(ds) \in -r|(v, -\beta)|\partial d_{C_1}(x(b) - u).$$

We begin the proof of this lemma by appealing to Proposition 2.5.5, by which we conclude that for any point (u', α') in epi V one has

(3) $\quad \langle (v, -\beta), (u' - u, \alpha' - \alpha) \rangle \leq \frac{1}{2}|u' - u|^2 + \frac{1}{2}|\alpha' - \alpha|^2.$

By Proposition 3.2.3 there exists an arc x solving $P_D^u: f(x(b)) = V(u)$. Let $c = x(b) - u$, and note that if c' is any point in C_1, and if x' is any arc, then $x'(b) \in C_1 + (x'(b) - c')$. This implies that if in addition x' is a trajectory emanating from C_0 and satisfying the state constraint, then the point $(x'(b) - c', f(x'(b)) + \alpha - f(x(b)))$ lies in epi V. If we substitute this point for (u', α') in (3), $x(b) - c$ for u, and rearrange, we get

(4) $\quad \beta[f(x'(b)) - f(x(b))] - \langle v, x'(b) - x(b) - c' + c \rangle$

$\quad + \frac{1}{2}|f(x'(b)) - f(x(b))|^2 + \frac{1}{2}|x'(b) - c' - x(b) + c|^2 \geq 0.$

Notice that equality holds when $x' = x$, $c' = c$. The following definitions make explicit how to apply Theorem 3.2.6 to the solution (x, c) to a certain minimization problem. We set

$$\tilde{F}(t, x', c') = F(t, x') \times \{0\}$$

$$\tilde{C}_0 = C_0 \times C_1$$

$$\tilde{g}(t, x', c') = g(t, x')|(v, -\beta)|,$$

and we take for $\tilde{f}(x'(b), c')$ the left-hand side of (4). Then the arc (x, c) solves Problem 3.2.1 for these data. To keep x' near x and c' near c, we can employ instead of $\tilde{g}$, the function

$$\bar{g}(t, x', c') = \max\{\tilde{g}(t, x', c'), \varepsilon|c' - c| - \varepsilon^2, \varepsilon|x' - x(t)| - \varepsilon^2\}$$

for any small $\varepsilon > 0$. This has no effect on the multipliers at (x, c), but guarantees for ε small enough that $\tilde{f}$ satisfies a Lipschitz condition of rank $K_{\tilde{f}} := 2(K_f + 1)|(v, -\beta)|$ on the relevant set, and that $\bar{g}$ satisfies one of rank $K_{\bar{g}} := K_g|(v, -\beta)|$.

Applying Theorem 3.2.6, Corollary 1, to the problem described above with any $r' > \hat{r}|(v, -\beta)|$ gives a multiplier $[(p, q), \gamma, \mu, \zeta, \lambda]$ satisfying $\lambda + \|\mu\| = 1$, Conditions 3.2.5(i)–(iv), and the further conditions

$$p(a) \in r' \partial d_{C_0}(x(a))$$

$$\dot{q} = 0, \quad q(a) \in r' \partial d_{C_1}(c)$$

$$\lambda\{[-v, v] + [\beta\zeta, 0]\} + [p(b), q(b)] + \left[\int_{[a, b]} \gamma(s) \mu(ds), 0\right] = [0, 0].$$

It follows that $q(a) = q(b)$, and we deduce

$$\lambda\beta\zeta + p(b) + \int_{[a, b]} \gamma(s) \mu(ds) = \lambda v \in -r' \partial d_{C_1}(c).$$

We need only set $r' = r|(v, -\beta)|$ to get the result.

Step 4. We now suppose that $(v_i, -\beta_i)$ is a sequence of perpendiculars as given at the end of Step 2, and we let x_i be the solution to $P_D^{\mu_i}$ whose existence is asserted in Lemma 2, and $[p_i, \gamma_i, \mu_i, \zeta_i, \lambda_i]$ the multiplier. Theorem 3.1.7 implies that a subsequence of x_i converges (we dispense with relabeling) to an arc x belonging to Y. Upon dividing across by $|(v_i, -\beta_i)|$ in the conditions of Lemma 2, and upon replacing γ_i by $\tilde{\gamma}_i = \gamma_i/|(v_i, -\beta_i)|$ and p_i by $\tilde{p}_i = p_i/|(v_i, -\beta_i)|$, a further application of Theorem 3.1.7, along with Proposition 3.1.8, leads to the existence of a five-tuple $[\tilde{p}, \gamma, \tilde{\mu}, \zeta, \lambda]$ with $\lambda + \|\tilde{\mu}\| = 1$ such that conditions 3.2.5(i)–(vi) hold, as well as

$$\lambda\beta_0\zeta + \tilde{p}(b) + \int_{[a, b]} \gamma(s) \tilde{\mu}(ds) = \lambda v_0 \in -r \partial d_{C_1}(x(b)).$$

Let us note that λ is nonzero, for otherwise the weak normality of P_D would be contradicted.

Suppose first that β_0 is nonzero. Define $p = \tilde{p}/(\lambda\beta_0)$, $\mu = \tilde{\mu}/(\lambda\beta_0)$, and note that for the five-tuple $[p, \gamma, \mu, \zeta, 1]$ one has $E = v_0/\beta_0$, whence

$$|(1, E)| + \|\mu\| = \left|\left(1, \frac{v_0}{\beta_0}\right)\right| + \|\mu\| = \frac{1}{\beta_0} + \frac{\|\tilde{\mu}\|}{\lambda\beta_0}$$

(since $|(v_0, \beta_0)| = 1$)

$$= \frac{1}{\beta_0}\left\{1 + \frac{1-\lambda}{\lambda}\right\} = \frac{1}{\lambda\beta_0}.$$

It follows that $[p, \gamma, \mu, \zeta, 1]$ belongs to $M_r^1(x)$ and that v_0/β_0 belongs to $E[M_r^1(Y)]$. Of course $(v_0, -\beta_0)$ belongs to $N_{\text{epi } V}(0, V(0))$ by construction, so that $(v_0/\beta_0, -1)$ belongs to $N_{\text{epi } V}(0, V(0))$, and thus v_0/β_0 to $\partial V(0)$ by definition. Since $(v_0, -\beta_0) = \beta_0(v_0/\beta_0, -1)$, the conclusion is that $(v_0, -\beta_0)$ belongs to N_1, as required.

The case in which β_0 is zero is handled very much as above; the conclusion in that case is that $(v_0, 0)$ belongs to N_2. This completes the proof of Lemma 1 and of the theorem. □

3.5 NORMALITY AND CONTROLLABILITY

The following property proves to be of fundamental importance. The terminology below continues to be that of Section 3.4, and we continue under the basic hypotheses 3.2.2 and the assumption that C_1 is closed.

3.5.1 Definition

An arc x is said to be *normal* provided that for all $r > 0$, every multiplier $[p, \gamma, \mu, \zeta, 0]$ in $M_r^o(x)$ has p and μ identically 0. The problem P_D is termed normal if for every $r > 0$, every multiplier in $M_r^o(Y)$ has p and μ identically 0.

It follows that P_D is weakly normal if it is normal. Note that normality of an arc x is divorced from any concept of optimality; indeed, the objective function f of P_D has no bearing upon it. As we shall see, normality has important consequences for controllability and for sufficient conditions. First, however, let us derive necessary conditions for Problem P_D (see Problem 3.4.1).

3.5.2 Theorem

Let x solve P_D. Then for r sufficiently large there is a multiplier $[p, \gamma, \mu, \zeta, \lambda]$ in $M_r^\lambda(x)$ such that $\|p\| + \|\mu\| + \lambda > 0$. If P_D is normal, then this holds with $\lambda = 1$, and (under the added hypotheses of Proposition 3.2.3) with the added condition

$$\zeta + p(b) + \int_{[a,b]} \gamma(s)\mu(ds) \in \partial V(0).$$

Proof. We begin by proving the first assertion under the extra hypothesis that x is the unique solution to P_D; that is, $Y = \{x\}$. We may redefine C_0 and g without affecting the multipliers at x so that the hypotheses of Proposition 3.2.3 hold. If P_D is not weakly normal, then the assertion follows trivially. If P_D is weakly normal, then it follows from Theorem 3.4.3 and Proposition 2.9.6 that (for r large) one has

$$E[M_r^1(x)] \cup [E[M_r^\circ(x)] \setminus \{0\}] \neq \emptyset.$$

Again the result follows. Of course, if x is normal then λ is nonzero; we can arrange $\lambda = 1$ by scaling.

To treat the general case in which x is not the unique solution to P_D, we define a modified problem for which x is the unique solution. Let (y, y_0) represent points in $R^n \times R$, and set

$$\tilde{F}(t, y, y_0) = F(t, y) \times \{|y - x(t)|^2\}$$

$$\tilde{C}_0 = C_0 \times \{0\}, \qquad \tilde{C}_1 = C_1 \times R$$

$$\tilde{f}(y, y_0) = f(y) + y_0, \qquad \tilde{g}(t, y, y_0) = g(t, y).$$

The resulting problem $\tilde{P}_D$ has unique solution $(x, 0)$; that is, $\tilde{Y} = \{(x, 0)\}$. Applying the theorem to $\tilde{P}_D$ and $(x, 0)$ yields the desired result.

Let us now turn to the case in which P_D is normal. It follows from Corollary 1 to Theorem 3.4.3 that V is Lipschitz near 0. Consider the problem $\hat{P}$ of minimizing

$$\hat{f}(y(b), c) := f(y(b)) - V(y(b) - c)$$

over the admissible trajectories y for F and the points c satisfying $y(a) \in C_0$, $c \in C_1$. The objective function $\hat{f}$ is nonnegative for all such y and c by the definition of V, and $\hat{f} = 0$ when $y = x$, $c = x(b)$. Consequently, $(x, x(b))$ solves $\hat{P}$, which can be framed as a case of Problem 3.2.1 by defining $\hat{C}_0 = C_0 \times C_1$, $\hat{F}(t, y, c) = F(t, y) \times \{0\}$. (Note that $\hat{f}$ is Lipschitz near $(x(b), x(b))$ as required.) Theorem 3.2.6 applies, and the result follows after a straightforward translation of terms. □

We continue to deal with an arc x admissible for P_D.

3.5.3 Theorem

Let x be normal. Then there are constants M and $\varepsilon > 0$ such that, for all u and v in εB, there is a trajectory y for F satisfying the state constraint $g(t, y(t)) \leq 0$, the conditions

$$y(a) \in C_0 + u, \qquad y(b) \in C_1 + v,$$

3.5 Normality and Controllability

and the inequality

$$\int_a^b |x(t) - y(t)| dt \leq M|(u, v)|.$$

Proof (by bookkeeping). We begin by defining a problem based on $R \times R^n \times R^n$ (instead of R^n), in which points are denoted (y_0, y, z). The problem is also based upon $[a - 1, b]$ rather than $[a, b]$. There is no loss of generality in assuming that g is bounded below by -1 on Ω. We set

$$\tilde{C}_0 = \{(0, 0, 0)\}, \quad \tilde{C}_1 = R \times \{(c_1 - c_0, c_0) : c_0 \in C_0, c_1 \in C_1\}$$

$$\tilde{f}(y_0, y, z) = y_0$$

$$\tilde{g}(t, y_0, y, z) = \begin{cases} -1 & \text{if } t < a \\ g(t, y + z) & \text{if } t \geq a \end{cases}$$

$$\tilde{F}(t, y_0, y, z) = \begin{cases} \{[|w - x(a)|, 0, w] : w \in N\bar{B}\} & \text{if } t < a \\ \{|x(t) - y - z|\} \times F(t, y + z) \times \{0\} & \text{if } t \geq a, \end{cases}$$

where N is large enough so that $C_0 \subset NB$. There is no essential difficulty in verifying that if (y_0, y, z) is admissible for the resulting version $\tilde{P}_D$ of problem P_D, then we obtain an arc x' admissible for P_D by setting $x'(t) = y(t) + z(a)$ for $t \geq a$. Then $x'(a) = z(a)$, while

$$\tilde{f}(y_0(b), y(b), z(b)) = \int_{a-1}^a |z - x(a)| \, dt + \int_a^b |x - x'| \, dt.$$

It follows that the arc $\tilde{x}$ defined by

$$\tilde{x}(t) = \begin{cases} [0, 0, (t - a + 1)x(a)] & \text{if } a - 1 \leq t < a \\ [0, x(t) - x(a), x(a)] & \text{if } a \leq t \leq b \end{cases}$$

is the unique solution to $\tilde{P}_D$.

The next step is to verify that $\tilde{P}_D$ is normal; that is, that $\tilde{x}$ is normal. Let $[(p_0, p, q), (\gamma_0, \gamma_1, \gamma_2), \mu, \zeta, 0]$ be a multiplier of index 0 for $\tilde{x}$ (for any $r > 0$); we wish to show that (p_0, p, q) and μ are zero. Note first that $\gamma_0 = 0$, and that $\gamma_1 = \gamma_2$. The Hamiltonian is

$$\tilde{H}(t, y_0, y, z, p_0, p, q) = \begin{cases} \max\{p_0|w - x(a)| + \langle q, w \rangle : |w| \leq N\} & \text{if } t < a \\ p_0|x(t) - y - z| + H(t, y + z, p) & \text{if } t \geq a. \end{cases}$$

The Hamiltonian inclusion (see Definition 3.2.5) implies that p_0 is constant, while the transversality condition (2) of Section 3.4 yields $p_0(b) = 0$; it follows that p_0 is identically zero. The condition

$$x(a) \in \partial_q \tilde{H}(t, \tilde{x}, 0, p, q), \qquad a - 1 < t < a$$

implies that $q = 0$ on $(a - 1, a)$, since $x(a) \in \text{int}(N\bar{B})$. The Hamiltonian inclusion also gives $\dot{p} = \dot{q}$, for $t > a$, whence $p(t) = q(t) + p(a)$ for $t > a$.

With the help of Theorem 2.5.6, one can show that any element (s_0, s_1, s_2) of $\partial d_{\tilde{C}_1}(0, x(b) - x(a), x(a))$ is such that $s_0 = 0$, s_1 belongs to $\partial d_{C_1}(x(b))$, s_2 belongs to $\partial d_{C_0}(x(a)) + \{s_1\}$. The transversality conditions therefore lead to

$$p(b) + \int_{[a,b]} \gamma(s) \mu(ds) \in -r \partial d_{C_1}(x(b)), \qquad p(a) \in r \, \partial d_{C_0}(x(a)).$$

It follows that $[p, \gamma, \mu, \zeta, 0]$ belongs to $M_r^\circ(x)$. Since x is normal, we deduce $p = 0 = \mu$, whence $q = 0$ too. We conclude that $\tilde{P}_D$ is normal. By suitably redefining $\tilde{g}$ if necessary, and without affecting the multipliers at $\tilde{x}$, one can arrange to satisfy the hypotheses of Proposition 3.2.3.

Since Theorem 3.4.3, Corollary 1, is applicable, we deduce that for some constant $\tilde{M}$, for all (w, u) near 0, there is an admissible arc $[\tilde{y}_0, \tilde{y}, \tilde{z}]$ such that

$$\tilde{f}(\tilde{y}_0(b), \tilde{y}(b), \tilde{z}(b)) \leq \tilde{M}|(w, u)|$$

$$[\tilde{y}_0(b), \tilde{y}(b), \tilde{z}(b)] \in \tilde{C}_1 + (0, w, u).$$

It follows that for some $(c_0, c_1) \in C_0 \times C_1$, one has $\tilde{z}(b) = c_0 + u$, $\tilde{y}(b) = c_1 - c_0 + w$. The arc $y(t) = \tilde{y}(t) + \tilde{z}(a)$ (for $t > a$) is an admissible trajectory for F and satisfies $y(a) = c_0 + u$, $y(b) = c_1 + (w + u)$. Choosing $w = v - u$ gives the required boundary conditions, and we find

$$\int_a^b |y - x| \, dt \leq \tilde{M}|(w, u)| = \tilde{M}|(v - u, u)| \leq M|(u, v)|,$$

where $M = 2\tilde{M}$. □

The Attainable Set

The *attainable set* (from C_0), denoted $\mathcal{C}(C_0)$, is defined to be the set of all points $x(b)$ where x is a trajectory for F emanating from C_0 and satisfying the state constraint $g(t, x(t)) \leq 0$. It is a consequence of Theorem 3.1.7 that $\mathcal{C}(C_0)$ is a closed set.

3.5.4 Theorem
Let x be an admissible trajectory as above. Then, if $x(b)$ lies on the boundary of $\mathcal{C}(C_0)$, there exists a triple $[p, \gamma, \mu]$ satisfying conditions (ii)–(iv) of Definition

3.2.5 as well as $\|p\| + \|\mu\| > 0$ and the further conditions

$$p(a) \in N_{C_0}(x(a))$$

$$p(b) + \int_{[a,b]} \gamma(s)\mu(ds) \in N_{@(C_0)}(x(b)).$$

Proof. We omit the proof, whose central ideas are those used to prove theorem 5.1.2; see remark 5.1.3. □

3.6 PROBLEMS WITH FREE TIME

To this point, the time interval $[a, b]$ has been immutably given. We now consider the possibility of its being a further decision variable. Specifically, we consider arcs y defined on intervals $[a, T]$ (T depending on y) subject to the usual state and dynamic constraints:

$$\dot{y}(t) \in F(t, y(t)) \quad \text{a.e.,} \quad a \leq t \leq T$$

$$g(t, y(t)) \leq 0, \quad a \leq t \leq T.$$

We retain $y(a) \in C_0$ as the initial condition, and we adopt the terminal constraint $(T, y(T)) \in S$, where S is a given subset of $R \times R^n$. (The previous case corresponds to $S = \{b\} \times C_1$.) Our interest revolves around the minimization of $f(T, y(T))$ over this given class (notice that now f has explicit dependence on T).

We continue under Hypotheses 3.2.2, except that we now suppose Ω to be a tube on a larger interval $[a, b^+]$ for some $b^+ > b$, and we posit Lipschitz behavior of F, f, g with respect to (t, x) *jointly*. Thus the constant k now replaces the function $k(\cdot)$ of Definition 3.1.4, and acts as Lipschitz constant for the multifunction $(t, x) \to F(t, x)$ on Ω. Similarly, K_f and K_g are Lipschitz constants for f and g on Ω (with respect to (t, x)).

3.6.1 Theorem
Let x solve the above problem, where the interval of definition of x is $[a, b]$. Then, for r sufficiently large, there exist a scalar λ equal to 0 or 1, an arc p, a nonnegative Radon measure μ, a measurable function (γ_0, γ), and an element ζ of $\partial f(b, x(b))$ with $\|p\| + \|\mu\| + \lambda > 0$ such that:

(i)

$$(\gamma_0, \gamma)(t) \in \partial^> g(t, x(t)) \quad \mu\text{-a.e.,}$$

and μ is supported on the set $\{t \in [a, b]: \partial^> g(t, x) \neq \emptyset\}$.

(ii)
$$(\dot{h}, -\dot{p}, \dot{x}) \in \partial H\left(t, x, p + \int_{[a,t)} \gamma(s)\mu(ds)\right) \quad \text{a.e.,}$$

where ∂H refers to the generalized gradient of $H(t, x, p)$ with respect to (t, x, p), and where $h(\cdot)$ is an arc such that

$$h(t) = H\left(t, x(t), p(t) + \int_{[a,t)} \gamma(s)\mu(ds)\right) \quad \text{a.e.}$$

(iii) $p(a) \in r\partial c_0(x(a))$.
(iv)
$$[-h(b), p(b)] + \int_{[a,b]} (\gamma_0, \gamma)(s)\mu(ds) + \lambda \zeta \in -r\,\partial d_S(b, x(b)).$$

We shall refer to the situation in which g and F have no explicit dependence on t as the *autonomous* case.

Corollary

Let x on $[a, b]$ solve the problem in the autonomous case. Then for r sufficiently large there exist a scalar λ equal to 0 or 1, an arc p, a nonnegative Radon measure μ, a measurable function γ, and an element ζ of $\partial f(b, x(b))$ with $\|p\| + \|\mu\| + \lambda > 0$ such that:

(i)
$$\gamma(t) \in \partial^> g(x(t)) \quad \mu\text{-a.e.}$$

and μ is supported on the set $\{t : \partial^> g(x(t)) \neq \emptyset\}$.
(ii) (a)
$$(-\dot{p}, \dot{x}) \in \partial H\left(x, p + \int_{[a,t)} \gamma\,d\mu\right) \quad \text{a.e.}$$

(b) There is a constant h such that
$$H\left(x(t), p(t) + \int_{[a,t)} \gamma\,d\mu\right) = h \quad \text{a.e.}$$

(iii)
$$p(a) \in r\partial c_0(x(a)).$$

(iv)
$$\left[-h, p(b) + \int_{[a,b]} \gamma(s)\mu(ds)\right] + \lambda\zeta \in -r\,\partial d_S(b, x(b)).$$

It is evident that the corollary is merely the special (autonomous) case of the theorem. However, the latter easily follows from the corollary by the device of "absorbing" t into the state by adding another state coordinate x_0 whose derivative is 1 and initial value a. (Of course we must redefine F, and so on, along the obvious lines.) We leave these details to the reader and instead prove the corollary.

Proof of the Corollary

We shall use a transformation device to turn this "free-time" problem to a fixed-time one to which our previous results may be applied. Let us consider $(n + 1)$-dimensional arcs (y_0, y) on $[a, b]$ which are trajectories of the multifunction $\tilde{F}$ defined by

$$\tilde{F}(y_0, y) = \{(u, v) : |u| \leq \alpha, v \in (1 + u)F(y)\},$$

where α is a given number in $(0, 1)$. We define

$$\tilde{C}_0 = \{b\} \times C_0, \qquad \tilde{C}_1 = S$$

$$\tilde{g}(y_0, y) = g(y), \qquad \tilde{f}(y_0, y) = f(y_0, y).$$

Note that the arc (b, x) is feasible for the corresponding problem $\tilde{P}$, which has the form and figure of the problem P_D of bygone sections.

Lemma. (b, x) solves $\tilde{P}$ locally.

To see this, suppose that a feasible arc (y_0, y) near (b, x) exists for $\tilde{P}$ such that $f(y_0(b), y(b)) < f(b, x(b))$. We shall manufacture a contradiction by finding an arc z on an interval $[a, T]$ which is better than x at its own game.

For each t in $[a, y_0(b)]$, a unique $\tau(t)$ in $[a, b]$ is defined by the equation

$$\tau + y_0(\tau) - b = t,$$

since the function $\tau \to \tau + y_0(\tau) - b$ is monotonic with value a at a and $y_0(b)$ at b. The function $\tau(t)$ is Lipschitz by the inverse function theorem 7.1.1, so that the relation

$$z(t) = y(\tau(t))$$

defines z as an arc on $[a, T]$, where $T = y_0(b)$. We easily calculate $(d/dt)z(t) = \dot{y}(\tau)/(1 + \dot{y}_0(\tau))$, from which it follows that $z(\cdot)$ is a trajectory for F, and in fact feasible for the free-time problem. Yet $f(T, z(T)) = f(y_0(b), y(b)) < f(b, x(b))$, a contradiction which proves the lemma.

The next step in the proof is to apply the necessary conditions of Theorem 3.5.2 to the solution (b, x) of $\tilde{P}$. We derive the existence of an arc (p_0, p), a measure μ, and selection γ for $\partial^> g$, a scalar $\lambda \geq 0$, an element ζ of $\partial f(b, x(b))$, such that

$$(-\dot{p}_0, -\dot{p}, 0, \dot{x}) \in \partial \tilde{H}\left(b, x, p_0, p + \int_{[a, t)} \gamma \, d\mu\right) \quad \text{a.e.},$$

$$(E_0, E_1) := \left[p_0, p(b) + \int_{[a, b]} \gamma \, d\mu\right] + \lambda \zeta \in$$

$$-r\{|(\lambda, E_0, E_1)| + \|\mu\|\} \partial d_S(b, x(b)),$$

$$p_0(a) \in r\{|(\lambda, E_0, E_1)| + \|\mu\|\} \overline{B},$$

$$p(a) \in r\{|(\lambda, E_0, E_1)| + \|\mu\|\} \partial d_{C_0}(x(a)),$$

$$\|(p_0, p)\| + \|\mu\| + \lambda = 1 \quad \text{(by normalization)}.$$

Here of course $\tilde{H}$ is the Hamiltonian for $\tilde{F}$. We calculate

$$\tilde{H}(y_0, y, p_0, p) = \max\{p_0 u + p \cdot v : |u| \leq \alpha, v \in (1 + u)F(y)\}$$

$$= \max\{p_0 u + (1 + u)H(y, p) : |u| \leq \alpha\}$$

$$= H(y, p) + \alpha|p_0 + H(y, p)|.$$

Since H is independent of y_0, we have $\dot{p}_0 = 0$, so that p_0 is constant. This implies that, for some constant σ,

(1) $$(-\dot{p}, \dot{x}) \in \partial H\left(x, p + \int_{[a, t)} \gamma \, d\mu\right) + \alpha \sigma \overline{B} \quad \text{a.e.}$$

For almost all t, one has

$$\tilde{H}\left(b, x, p_0, p + \int_{[a,t)} \gamma\, d\mu\right) = \left\langle \dot{x}, p + \int_{[a,t)} \gamma\, d\mu \right\rangle$$

$$\leqslant H\left(x, p + \int_{[a,t)} \gamma\, d\mu\right)$$

$$\leqslant H\left(x, p + \int_{[a,t)} \gamma\, d\mu\right)$$

$$+ \alpha \left| p_0 + H\left(x, p + \int_{[a,t)} \gamma\, d\mu\right) \right|$$

$$= \tilde{H}\left(b, x, p_0, p + \int_{[a,t)} \gamma\, d\mu\right).$$

It follows that for almost all t,

(2) $$H\left(x, p + \int_{[a,t)} \gamma\, d\mu\right) = -p_0 =: h.$$

If we now pause to reflect, we find that all the assertions of the corollary are satisfied, except that (1) has the undesired term $\alpha\sigma\bar{B}$. If we go through the above for a sequence of α decreasing to 0, all the arcs, measures, and constants above depend on α (except for r and σ). A straightforward use of the now familiar technique of extracting convergent subsequences (as in Step 4 of the proof of Theorem 3.2.6) leads to the desired conclusion. (We need to note that if p, μ, and λ all vanish, then by Eq. (2) so does h, and hence also p_0, a contradiction.) □

3.7 SUFFICIENT CONDITIONS: THE HAMILTON–JACOBI EQUATION

We now turn for the first time to the important issue of *sufficient conditions*: conditions which assure that a given admissible arc (which may have been identified with the help of the necessary conditions of preceding sections) is in fact a solution of the problem. We shall explore a method whose origin lies in the classical calculus of variations, one which we call the *Hamilton–Jacobi verification technique*.

We work within the framework of the problem P_D (see Section 3.4) under the added proviso that the state constraint $g(t, x(t)) \leqslant 0$ is inactive upon local solutions x (i.e., that $g(t, x(t)) < 0$). Our hypotheses are somewhat stronger than Hypotheses 3.2.2:

3.7.1 Hypotheses

(i) F is closed, convex, and bounded on the tube Ω.
(ii) The multifunction $(t, x) \to F(t, x)$ is uniformly Lipschitz of rank k on Ω.
(iii) f is Lipschitz on Ω_b of rank K_f.
(iv) C_0 is a singleton $\{x_0\}$.

An admissible arc x for the problem is one that is a trajectory for F (on $[a, b]$) (necessarily, then, x lies in Ω), satisfying $x(a) = x_0$ and $x(b) \in C_1$. We term x a *local solution* to P_D if, for some $\varepsilon > 0$, any other admissible arc y lying in $T(x; \varepsilon)$ satisfies $f(y(b)) \geq f(x(b))$.

The classical Hamilton–Jacobi equation is the following partial differential equation for a function W of (t, x):

$$W_t(t, x) - H(t, x, -W_x(t, x)) = 0,$$

and its relationship to our problem (when H is the Hamiltonian as defined in Section 3.2) is apparent from the following:

3.7.2 Proposition

Suppose that W is a smooth function defined on an ε-tube $T(x; \varepsilon)$ about an admissible arc x, and that W satisfies the Hamilton–Jacobi equation as well as the boundary conditions

(a) $W(b, u) = f(u)$ for $u \in C_1 \cap (x(b) + \varepsilon B)$
(b) $W(a, x_0) \geq f(x(b))$.

Then x is a local solution to P_D.

Proof. Let y be any admissible arc lying in $T(x; \varepsilon)$. Then one has

$$f(y(b)) = W(b, y(b))$$

$$= W(a, y(a)) + \int_a^b \frac{d}{dt}\{W(t, y(t))\}\, dt$$

$$= W(a, x_0) + \int_a^b \{W_t(t, y) + \langle W_x(t, y), \dot{y}\rangle\}\, dt$$

$$\geq f(x(b)) + \int_a^b \{W_t(t, y) - H(t, y, -W_x(t, y))\}\, dt$$

3.7 Sufficient Conditions: the Hamilton–Jacobi Equation

(since $\dot{y} \in F(t, y)$ a.e.)
$$= f(x(b)).$$
The conclusion follows. □

How applicable is this technique in practice? Can one always hope to find a function W as above that will verify the optimality of a local solution? Here is an example to show that the answer is no.

3.7.3 Example
We set $n = 2$, $[a, b] = [0, 1]$, and we denote by (x, y) points in R^2. Let
$$F(t, x, y) = \{(0, xu) : -1 \leq u \leq 1\},$$
$x_0 = (0,0)$, $C_1 = R^2$, $f(x, y) = y$. Note that the arc $(0, 0)$ is optimal for this version of the problem. The function W of Proposition 3.7.2, if it exists, satisfies

(1) $\qquad\qquad W(0,0,0) \geq 0, \qquad W(1, x, y) = y$

(2) $\qquad\qquad W_t(t, x, y) - |xW_y(t, x, y)| = 0,$

where Eq. (2) is valid in an ε-tube about the arc $(0, 0)$. Since $W_t \geq 0$, we derive from (1) that

(3) $\qquad\qquad W(t, 0, 0) = 0.$

Again from (1), one has $W_y(1, 0, 0) = 1$, so that for some δ in $(0, \varepsilon)$ one has
$$W_y(t, x, 0) \geq \tfrac{1}{2} \quad \text{for } 1 - \delta \leq t \leq 1, \qquad |x| < \delta.$$

Combining this last conclusion with Eq. (2) yields
$$W_t(t, x, 0) \geq \frac{|x|}{2} \quad \text{for } 1 - \delta \leq t \leq 1, \qquad |x| < \delta.$$

Setting $\tau = 1 - \delta$, this leads to

(4) $\quad W(t, x, 0) \geq W(\tau, x, 0) + \dfrac{(t - \tau)|x|}{2} \quad \text{for } \tau \leq t \leq 1, \qquad |x| < \delta.$

Let t lie in $(\tau, 1)$, and consider now the quantity $q = W_x(t, 0, 0)$. One has, using (4) and (3):

$$q = \lim_{x \downarrow 0} \frac{[W(t, x, 0) - W(t, 0, 0)]}{x}$$

$$\geq \lim_{x \downarrow 0} \frac{W(\tau, x, 0) + (t - \tau)|x|/2}{x} = W_x(\tau, 0, 0) + \frac{t - \tau}{2}.$$

Similarly,
$$q = \lim_{x \downarrow 0} \frac{W(t, -x, 0) - W(t, 0, 0)}{-x} \leq W_x(\tau, 0, 0) - \frac{t - \tau}{2}.$$

This contradiction shows that no such W exists.

One way to cope with the inapplicability of the technique is to extend it to a greater level of generality, and in particular to functions W which are not smooth.

The Extended Hamilton–Jacobi Equation

Let Ω' be a tube on $[a, b]$, and let W be a function defined on Ω'. We say that W is a solution (on Ω') of the (extended) Hamilton–Jacobi equation provided:

(i) W is Lipschitz on Ω'.
(ii) For every (t, x) in Ω' with $a < t < b$, one has

$$\min\{\alpha - H(t, x, -\beta) : (\alpha, \beta) \in \partial W(t, x)\} = 0.$$

(Note that $\partial W(t, x)$ is defined for such (t, x).)

It is easy to see that (ii) reduces to the classical equation if W is C^1. The following shows that the fundamental property inherent in Proposition 3.7.2 is inherited by the extended equation.

3.7.4 Proposition

Proposition 3.7.2 is valid if W is merely Lipschitz and if the Hamilton–Jacobi equation is understood in the extended sense.

Proof. The proof of Proposition 3.7.2 goes through if one is armed with:

Lemma. For any trajectory y for F, the function $\theta(t) = W(t, y(t))$ is absolutely continuous and one has $\dot\theta(t) \geq 0$ a.e.

To see why this is so, observe first that θ is the composition of two Lipschitz maps, and hence is Lipschitz (and absolutely continuous). By Theorem 2.6.6 one has

$$\dot\theta(t) \in \operatorname{co}\{\alpha + \langle \beta, \zeta \rangle : (\alpha, \beta) \in \partial W(t, y(t)), \zeta \in \partial y(t)\},$$

so it suffices to establish that any term of the form $\alpha + \langle \beta, \zeta \rangle$ is nonnegative. But if y is a trajectory, it follows from Theorem 2.5.1 that ζ belongs to $F(t, y(t))$. Consequently,

$$\alpha + \langle \beta, \zeta \rangle = \alpha - \langle -\beta, \zeta \rangle$$
$$\geq \alpha - H(t, y(t), -\beta) \geq 0,$$

which completes the proof. □

3.7 Sufficient Conditions: the Hamilton–Jacobi Equation

The technique has now been sufficiently extended to apply to Example 3.7.3: the function $W(t, x, y) = y + (t - 1)|x|$ does the trick, as the reader may verify. Is the extended technique now adequate to treat all cases? We adduce the following to show that it is not:

3.7.5 Example

We set $n = 2$, $[a, b] = [0, 1]$, $x_0 = (0, 0)$, $C_1 = \{1\} \times R$, $f(x, y) \equiv y$, $F(t, x, y) \equiv$ closed unit ball in R^2. Note that the only admissible arc (and so the optimal one) is $(x, y)(t) \equiv (t, 0)$. A verification function W would satisfy

$$W(0, 0, 0) \geq 0, \qquad W(1, 1, y) = y,$$

and

(5) $\qquad \alpha - |(\beta, \gamma)| \geq 0 \quad \text{for all } (\alpha, \beta, \gamma) \in \partial W(t, x, y),$

where the latter holds for all (t, x, y) in an ε-tube about the arc $(t, 0)$. Now let (x', y') be any trajectory originating at $(0, 0)$ and lying in the ε-tube about $(t, 0)$. Then, by the mean-value theorem for generalized gradients, there is $\bar{t}$ in $(0, 1)$ such that

$$W(1, x'(1), y'(1)) \geq W(1, x'(1), y'(1)) - W(0, 0, 0)$$

$$\in [1, x'(1), y'(1)] \cdot \partial W(\bar{t}, \bar{t}x'(1), \bar{t}y'(1)),$$

whence $W(1, x'(1), y'(1))$ is nonnegative by condition (5). If K is a Lipschitz constant for W, one also deduces

$$y'(1) = W(1, 1, y'(1)) \geq W(1, x'(1), y'(1)) - K|1 - x'(1)|$$

$$\geq -K|1 - x'(1)|$$

in view of the preceding conclusion. Applied to the arc $x'(t) = st$, $y'(t) = -(1 - s^2)^{1/2} t$ (for any $s < 1$), this last conclusion yields

$$s \leq \frac{K^2 - 1}{K^2 + 1},$$

a contradiction which establishes the nonexistence of a Lipschitz verification function in this example.

If the verification technique still does not apply to all problems (and the example shows that it does not), then our only reason to be satisfied with it is if it applies to all "reasonable" problems. The term "reasonable" will be interpreted here to mean problems for which the necessary conditions are nondegenerate (involve the objective function f); that is, problems which are normal in the sense of Definition 3.5.1.

For the problem we are considering, any arc x for which $x(b)$ lies in the interior of C_1 is normal. It follows that the problem of Example 3.7.3 is normal. It is easy to show that $p(t) \equiv (1,0)$ is a (nontrivial) multiplier of index 0 corresponding to the optimal arc $(t,0)$ in the problem of Example 3.7.5, whence that arc fails to be normal.

3.7.6 Theorem

In order that the admissible arc x be a local solution to P_D, it is sufficient, and if all local solutions to P_D are normal also necessary, that there exist on some tube of positive radius r about x a solution W of the extended Hamilton–Jacobi equation satisfying the boundary conditions

$$W(a, x_0) = f(x(b)), \qquad W(b, y) = f(y) \quad \text{for } y \in C_1 \cap (x(b) + rB).$$

Proof. The sufficiency has already been proven in Proposition 3.7.4. Let M and δ be positive numbers, and define, for any (τ, u) in $T(x; \delta)$, the problem $P_{\tau, u}$ of minimizing the functional J_τ defined as follows:

$$J_\tau(y) := f(y(b)) + Md_{C_1}(y(b)) + M\int_\tau^b \max\{|y(t) - x(t)| - \delta, 0\}\, dt$$

over the trajectories y for F which satisfy $y(\tau) = u$ and which satisfy $|y(t) - x(t)| \leq \varepsilon_0$ for $\tau \leq t \leq b$, where $\varepsilon_0 > 0$ is such that $T(x; 2\varepsilon_0)$ is contained in Ω. Let $W(\tau, u)$ denote the value of the problem $P_{\tau, u}$.

Lemma 1. Let ε in $(0, \varepsilon_0)$ be given. Then, for any M sufficiently large, there exists δ in $(0, \varepsilon)$ such that:

(i) For all (τ, u) in $T(x; \delta)$, $P_{\tau, u}$ has a solution y satisfying $|y(t) - x(t)| < \varepsilon$ for $\tau \leq t \leq b$, and all solutions of $P_{\tau, u}$ satisfy this condition.
(ii) W is Lipschitz on $T(x; \delta)$ and satisfies there the extended Hamilton–Jacobi equation.
(iii) $W(b, y) = f(y)$ for all y in $C_1 \cap (x(b) + \delta B)$.

The proof of this lemma will not use the optimality of x. Note, however, that only one boundary condition is lacking for this W to be the function whose existence we seek to prove.

We begin by choosing δ small enough so that for any (τ, u) in $T(x; \delta)$, there is a trajectory y on $[\tau, b)$ lying in $T(x; \varepsilon_0)$. That this is possible (in fact, for $\delta < \varepsilon_0(1 + K\ln(K))^{-1}$) is a consequence of applying Theorem 3.1.6 (on $[\tau, b]$) with $x'(\cdot) := x(\cdot) + (u - x(\tau))$ and $\varepsilon' = \varepsilon_0$. By Proposition 3.2.3, $P_{\tau, u}$ admits a solution y; we wish to show that $|y(t) - x(t)| < \varepsilon$ for $\tau \leq t \leq b$ if M is large and δ small (independently of τ, u in $T(x; \delta)$).

3.7 Sufficient Conditions: the Hamilton–Jacobi Equation

There is a value τ_0 in (a, b) such that, whenever $\tau \geq \tau_0$ and y is an arc on $[\tau, b]$ with $(\tau, y(\tau))$ in $T(x; \varepsilon/2)$, one automatically has $|y(t) - x(t)| < \varepsilon$ for $\tau \leq t \leq b$; choosing $\delta < \varepsilon/2$ will suffice for such τ. There is a positive γ such that if $\tau \in [a, \tau_0]$, and if y is a trajectory for F on $[\tau, b]$ with $\max_{\tau \leq t \leq b}|y(t) - x(t)| \geq \varepsilon$, then one has

$$\int_\tau^b \max\left\{|y(t) - x(t)| - \frac{\varepsilon}{2}, 0\right\} dt \geq \gamma.$$

Suppose that for such τ the solution y to $P_{\tau, u}$ satisfies this last inequality. Then necessarily (since $\delta < \varepsilon/2$) one has

(6) $$J_\tau(y) \geq f(y(b)) + M\gamma.$$

Consider the arc $x'(t) := x(t) + u - x(\tau)$, $\tau \leq t \leq b$. It follows from Theorem 3.1.6 (since $\rho_F(x') \leq c_1 \delta$ for some constant c_1) that if $|u - x(\tau)|$ is small enough (i.e., if δ is small enough), then there exists a trajectory y' for F with $y'(\tau) = u$ such that

$$\max_{\tau \leq t \leq b} |y'(t) - x'(t)| \leq c_2 \delta,$$

where c_2 is a certain constant. Then $|y'(t) - x(t)| \leq (c_2 + 1)\delta$, and so

(7) $$J_\tau(y') \leq f(y'(b)) + M(c_2 + 1)\delta + M(b - \tau)c_2\delta.$$

For δ small enough y' will satisfy $|y'(t) - x(t)| \leq \varepsilon_0$, so that by the optimality of y one deduces $J_\tau(y) \leq J_\tau(y')$. With (6) and (7) this gives

$$f(y(b)) + M\gamma \leq f(y'(b)) + M[c_2(1 + b - \tau) + 1]\delta.$$

Since f is uniformly bounded (by N, say) on Ω_b, this will not be possible if M is first chosen larger than $2N/\gamma$ and then δ is chosen sufficiently small. This completes the proof of (i) of the lemma. Note that (iii) is immediate from the very definition of W, so there remains only (ii) to prove.

It follows as in Lemma 2, Theorem 3.2.6, that for some $c > 0$, for all (τ, u) in $T(x; \delta)$, $W(\tau, u)$ coincides with the value of the following problem. Minimize

$$J_\tau(y) + c\int_\tau^b \rho(t, y, \dot{y}) dt$$

over the arcs y satisfying $y(\tau) = u$, $|x(t) - y(t)| < \varepsilon$. We shall use this characterization of W to prove it Lipschitz. It suffices to focus attention on any product set $[T_1, T_2] \times S$ contained in $T(x; \delta)$. Further, we may prove that W is Lipschitz separately in τ and u on such a set.

Let τ and τ' in $[T_1, T_2]$ and u in S be given, and suppose that $\tau < \tau'$. Let y solve $P_{\tau, u}$, and define a new arc y' on $[\tau', b]$ by truncation of y. Then

(8) $$W(\tau', u) \leq f(y'(b)) + Md_{C_1}(y'(b))$$
$$+ M \int_{\tau'}^{b} \max\{|x(t) - y'(t)| - \delta, 0\} \, dt$$
$$+ c \int_{\tau'}^{b} \rho(t, y', \dot{y}') \, dt \leq W(\tau, u).$$

Again, let y solve $P_{\tau', u}$ and define y' on $[\tau, b]$ via

$$y'(t) = \begin{cases} u & \text{if } \tau \leq t < \tau' \\ y(t) & \text{if } \tau' \leq t \leq b. \end{cases}$$

Then

(9) $$W(\tau, u) \leq W(\tau', u) + M \int_{\tau}^{\tau'} \max\{|x(t) - u| - \delta, 0\} \, dt + c \int_{\tau}^{\tau'} \rho(t, u, 0) \, dt$$
$$\leq W(\tau', u) + \bar{c}|\tau - \tau'|$$

for an appropriate constant $\bar{c}$. It follows from inequalities (8) and (9) that $W(\cdot, u)$ is Lipschitz on $[T_1, T_2]$. We turn now to the u variable.

Let u be a point in S, and let y solve $P_{\tau, u}$. Choose any v sufficiently near u so that the arc

$$y'(t) := y(t) - u + v$$

lies in $T(x; \varepsilon)$. Then

$$W(\tau, v) \leq J_\tau(y') + c \int_{\tau}^{b} \rho(t, y', \dot{y}') \, dt$$
$$\leq c_3|u - v| + W(\tau, u)$$

for a constant c_3. Consequently we find

$$\limsup_{v \to u} \frac{\{W(\tau, v) - W(\tau, u)\}}{|u - v|} \leq c_3.$$

A (uniform) pointwise condition of this type is known to imply that $W(\tau, \cdot)$ is Lipschitz on S (of rank c_3).

Left to prove is the second assertion of (ii). We begin by showing that $\alpha - H(\tau, u, -\beta)$ is nonnegative whenever $\nabla W(\tau, u)$ exists and equals (α, β).

3.7 Sufficient Conditions: the Hamilton–Jacobi Equation

This follows from the inequality

$$\alpha + \langle \beta, v \rangle \geq 0 \quad \text{for all } v \in F(\tau, u),$$

which we proceed to prove. For any $\eta > 0$, define an arc y' on $[\tau - \eta, b]$ as follows:

$$y'(t) = \begin{cases} tv - \tau v + y(\tau) & \text{if } \tau - \eta \leq t \leq \tau \\ y(t) & \text{if } \tau \leq t \leq b, \end{cases}$$

where y solves $P_{\tau, u}$. For η small one has $(t, y'(t)) \in T(x; \varepsilon)$ for $\tau - \eta \leq t \leq \tau$, and so

$$W(\tau - \eta, u - \eta v) \leq W(\tau, u) + c \int_{\tau - \eta}^{\tau} \rho(t, y', v) \, dt.$$

Rearranging, dividing across by η, and letting $\eta \downarrow 0$, gives

$$-\alpha - \langle \beta, v \rangle \leq c\rho(\tau, u, v) = 0,$$

as required.

In light of Theorem 2.5.1, and because $H(\tau, u, \cdot)$ is convex, it follows that for any (α, β) in $\partial W(\tau, u)$, one has $\alpha - H(\tau, u, -\beta) \geq 0$. To complete the proof, we need only show that for each (τ, u) in the interior of $T(x; \delta)$, there is an element (α, β) of $\partial W(\tau, u)$ such that $\alpha - H(\tau, u, -\beta) = 0$. To see this, let y solve the problem $P_{\tau, u}$. It follows that for all $\eta > 0$ sufficiently small, the arc y restricted to $[\tau, \tau + \eta]$ remains in $T(x; \delta)$ and is a (local) solution to the free-endpoint, free-terminal-time problem of minimizing $W(T, y'(T))$ over the arcs y' on intervals $[\tau, T]$ which are trajectories for F with $y'(\tau) = u$ and which satisfy $|x(t) - y'(t)| < \delta$. Since W is Lipschitz near $(\tau + \eta, y(\tau + \eta))$, we may apply the conditions of Theorem 3.6.1 (with $\lambda = 1$) to deduce (from conditions (ii) and (iv)) that for some element $(\alpha_\eta, \beta_\eta)$ of $\partial W(\tau + \eta, y(\tau + \eta))$, one has

$$\alpha_\eta - H(\tau + \eta, y(\tau + \eta), -\beta_\eta) = 0.$$

We obtain this conclusion for a sequence of η decreasing to 0. By selecting a convergent subsequence of $(\alpha_\eta, \beta_\eta)$, we arrive at an element (α, β) of $W(\tau, u)$ with the required property, completing the proof of Lemma 1.

The function W is lacking only one property, the boundary condition $W(a, x_0) = f(x(b))$. The following result will complete the proof of the theorem. (This is where the optimality and the normality of x are used.)

Lemma 2. *There exist ε, M and δ with the properties of Lemma 1 such that x solves the problem P_{a, x_0}.*

To see this, consider the version $\tilde{P}$ of problem P_D with data

$$\tilde{F}(t, y, y_0) = F(t, y) \times \{\tilde{M} \max[|y - x(t)| - \tilde{\delta}, 0]\},$$
$$\tilde{f}(y, y_0) = f(y) + y_0, \qquad \tilde{g}(t, y, y_0) = |y - x(t)| - \varepsilon_0,$$
$$\tilde{C}_0 = \{x_0\} \times \{0\}, \qquad \tilde{C}_1 = C_1 \times R.$$

Let $\tilde{V}(u)$ be the value of this problem when C_1 is replaced by $C_1 + u$. As shown in Lemma 1, for any $\tilde{\varepsilon} > 0$, there exist $\tilde{M}$ and $\tilde{\delta}$ with the property that $\tilde{P}$ has solutions, all of which lie in $T(x; \tilde{\varepsilon})$. If $\tilde{\varepsilon}$ is sufficiently small, then $\tilde{V}(0) = f(x(b))$ (since x is a local solution to P_D). We now claim that $\tilde{V}$ is Lipschitz in a neighborhood of 0. Let (p, q) be a multiplier of index 0 corresponding to a solution (y, y_0) of $\tilde{P}$. It follows that $y_0 = 0$ and that y is a local solution of P_D (with $f(y(b)) = f(x(b))$). The definition of multiplier implies in this case

$$(-\dot{p}, \dot{y}) \in \partial H(t, y, p) + (\tilde{M}q\bar{B} \times \{0\})$$
$$\dot{q} = 0, \qquad q(b) = 0.$$

In consequence we find that p is a multiplier for y of index 0; by hypothesis therefore, $p = 0$. We have shown that $\tilde{P}$ is normal, which implies that $\tilde{V}$ is Lipschitz (of rank $\tilde{k}$, say) on a neighborhood rB of 0 (Corollary 1, Theorem 3.4.3) for some $r > 0$.

Now let $\tilde{\varepsilon}$, $\tilde{M}$, $\tilde{\delta}$ be as above, and choose $\varepsilon < r$, $M > \max\{\tilde{M}, \tilde{k}\}$, $\delta < \tilde{\delta}$ with the properties of Lemma 1. We claim that x solves the resulting problem P_{a, x_0}. To see this, let y be any solution to P_{a, x_0}. Then y lies in $T(x; \varepsilon)$ by Lemma 1, whence

$$d_{C_1}(y(b)) \leq |y(b) - x(b)| < \varepsilon < r.$$

It follows that for some point u in rB, one has $y(b) \in C_1 + u$ and $d_{C_1}(y(b)) = |u|$. We calculate

$$J_a(y) = f(y(b)) + Md_{C_1}(y(b)) + M\int_a^b \max\{|x - y| - \delta, 0\} \, dt$$
$$\geq f(y(b)) + M|u| + \tilde{M}\int_a^b \max\{|x + y| - \tilde{\delta}, 0\} \, dt$$
$$\geq \tilde{V}(u) + M|u| \geq \tilde{V}(u) + \tilde{k}|u|$$
$$\geq \tilde{V}(0) = f(x(b)) = J_a(x).$$

It follows that x solves P_{a, x_0} and the proof is complete. $\square$

We remark that in the proof of the theorem, the requirement that all local solutions be normal can be replaced by the requirement that x be normal and that x be a local unique solution.

Chapter Four

The Calculus of Variations

We must welcome the future, remembering that soon it will be the past; and we must respect the past, remembering that once it was all that was humanly possible.

GEORGE SANTAYANA

We now explore the second of our paradigms for dynamic optimization, the generalized problem of Bolza. The methods used in the analysis are modern, but there is a wealth of history guiding our steps. The reader who has read about the past in Sections 1.3 and 1.4 will better be able to appreciate the present.

4.1 THE GENERALIZED PROBLEM OF BOLZA

The object of our attention is the problem (dubbed P_B) of minimizing the functional

$$J(y) := l(y(a), y(b)) + \int_a^b L(t, y(t), \dot{y}(t))\, dt$$

over all arcs y (absolutely continuous functions y from $[a, b]$ to R^n). This has the appearance of a classical problem of Bolza in the calculus of variations. The big difference is that we shall allow l and L to be extended-valued (i.e., to have values in $R \cup \{+\infty\}$).

Many apparently different problems in dynamic optimization can be reduced to P_B. Consider for example the problem P_D defined in 3.4.1. If we define l and L via

$$l(y_0, y_1) := \psi_{C_0}(y_0) + \psi_{C_1}(y_1) + f(y_1)$$

$$L(t, s, v) := \begin{cases} 0 & \text{if } v \in F(t, s) \text{ and } g(t, s) \leq 0 \\ +\infty & \text{otherwise} \end{cases}$$

(where ψ_C denotes the indicator function of the set C) the resulting problem P_B is P_D. Similar (somewhat deeper) equivalences hold for other types of constrained problems, and in particular the control problems treated in the next chapter. Such results hinge in large part upon measurability and selection results, which we review briefly in this section.

$\mathcal{L} \times \mathcal{B}$ *Measurability*

To ensure that P_B is well defined, we wish as a first step to guarantee that for each arc y the composite function $t \to L(t, y(t), \dot{y}(t))$ is (Lebesgue) measurable. In the calculus of variations this follows from the usual requirement that L be continuous (say). In the present extended-valued setting, the appropriate (weakest) measurability requirement guaranteeing the required property is that L be measurable with respect to $\mathcal{L} \times \mathcal{B}$, where $\mathcal{L} \times \mathcal{B}$ denotes the σ-algebra of subsets of $[a, b] \times R^{2n}$ generated by product sets $M \times N$, where M is a Lebesgue measurable subset of $[a, b]$ and N is a Borel subset of R^{2n}. In practice, when P_B is derived from an initial problem having different structure, it follows under mild hypotheses that L possesses this measurability property; Rockafellar (1975) proves some useful general results to this effect.

Measurability in the above sense also arises in the issue of measurable selections, especially for multifunctions not having closed values. The following result of Aumann (1967) (see also related results discussed in Wagner, 1977) is useful:

4.1.1 Theorem
Let Γ be a multifunction mapping a measurable subset S of R^m to the nonempty subsets of R^n, and suppose that the graph of Γ (i.e., the set $\{(s, \gamma) : s \in S, \gamma \in \Gamma(s)\}$) is $\mathcal{L} \times \mathcal{B}$ measurable. Then Γ admits a measurable selection.

(In the above, of course, $\mathcal{L} \times \mathcal{B}$ is generated by products of Lebesgue sets in R^m and Borel sets in R^n.) This result can be viewed as an extension of Theorem 3.1.1, since when Γ is closed, the measurability of Γ implies that the graph of Γ is $\mathcal{L} \times \mathcal{B}$ measurable.

If for an arc y the integral appearing in the Bolza functional J fails to be well defined and greater than $-\infty$, we adopt the convention that $J(y)$ is $+\infty$. In all other cases the functional J has an unambiguous value (possibly $+\infty$).

4.1.2 Basic Hypotheses

Throughout the chapter we shall assume the following:

(i) L is $\mathcal{L} \times \mathcal{B}$ measurable.
(ii) l is lower semicontinuous.
(iii) For each t in $[a, b]$, the function $L(t, \cdot, \cdot)$ is lower semicontinuous.

It follows from the above that the multifunction F defined by

$$F(t, x) = \text{epi } L(t, x, \cdot)$$

is measurable in t, a fact which will be of use to us later on.

The Hamiltonian

The Hamiltonian H is defined on $[a, b] \times R^n \times R^n$ as follows:

$$H(t, s, p) = \sup\{\langle p, v \rangle - L(t, s, v) : v \in R^n\}.$$

A central theme in this chapter is that of analyzing P_B through the use of the Hamiltonian. This is possible to a great extent only if H "captures" L adequately (i.e., if knowing H is equivalent to knowing L). The condition guaranteeing this, which we label the *convexity condition* (on L), is that for each (t, s) the function $L(t, s, \cdot)$ be convex. When this holds, conjugacy theory of convex analysis provides the formula

$$L(t, s, v) = \sup\{\langle p, v \rangle - H(t, s, p) : p \in R^n\}.$$

It has been known since the time of Tonelli that the convexity condition plays a crucial role in the semicontinuity of the Bolza functional, and hence in the theory of existence of solutions to P_B. The latter has been studied via the Hamiltonian by Rockafellar (1975); the following is taken from his work.

4.1.3 · Theorem

Let L satisfy the convexity condition and the basic hypotheses, and let l and H satisfy the following growth conditions:

$$H(t, x, p) \leq \mu(t, p) + |x|(\sigma(t) + \rho(t)|p|)$$

$$l(s_0, s_1) \geq l_0(s_0) + l_1(s_1),$$

where $\sigma(t)$, $\rho(t)$, and $\mu(t, p)$ are finite and summable as functions of t (with σ

and ρ nonnegative), and where l_0, l_1 are bounded below on bounded sets, with

$$\liminf_{|s|\to\infty} \frac{l_0(s)}{|s|} = +\infty$$

$$\liminf_{|s|\to\infty} \frac{l_1(s)}{|s|} > -\infty.$$

Then if the Bolza functional $J(y)$ is finite for at least one arc y, the problem P_B admits a solution.

4.2 NECESSARY CONDITIONS

We now turn our attention to finding necessary conditions for arcs x which solve the problem P_B. We shall require a constraint qualification in our first result. To describe this condition, let $V_0(s)$ denote the infimum in the modified version of P_B in which the function l is replaced by its perturbation $l_{s,0}$ defined by

$$l_{s,0}(u, v) := l(u + s, v).$$

Similarly, $V_1(s)$ is the infimum where l is replaced by

$$l_{0,s}(u, v) := l(u, v + s).$$

We shall call P_B *calm* provided, for j equal to either 0 or 1, one has

$$\liminf_{s\to 0} \frac{V_j(s) - V_j(0)}{|s|} > -\infty.$$

Note that P_B is certainly calm if l is Lipschitz in either of its variables, and in particular if P_B is a free-endpoint problem (one in which l depends only upon $y(a)$).

Given an arc x on $[a, b]$ we shall say that H satisfies the *strong Lipschitz condition* near x provided that there exists an $\varepsilon > 0$ and an integrable function $k(\cdot)$ on $[a, b]$ such that, for all p in R^n, for all (t, y_1) and (t, y_2) in the tube $T(x; \varepsilon)$ one has

$$|H(t, y_1, p) - H(t, y_2, p)| \le k(t)\{1 + |p|\}|y_1 - y_2|.$$

4.2.1 Remark
This condition requires tacitly that H be finite near x. This implies that for every y in $x(t) + \varepsilon B$ there is at least one value of v such that $L(t, y, v) < +\infty$. Thus explicit state constraints such as $g(x(t)) \le 0$ are ruled out by this assumption.

4.2.2 Theorem

Let L satisfy the convexity condition and the basic hypotheses. Suppose that x solves P_B, that P_B is calm, and that H satisfies the strong Lipschitz condition near x. Then there exists an arc p such that

$$\begin{bmatrix} -\dot{p}(t) \\ \dot{x}(t) \end{bmatrix} \in \partial H(t, x, p) \quad \text{a.e.}$$

$$\begin{bmatrix} p(a) \\ -p(b) \end{bmatrix} \in \partial l(x(a), x(b)).$$

If L (or H) has no explicit dependence on t, then furthermore $H(x(t), p(t)) =$ constant, and $\dot{x}$ is essentially bounded.

Proof.

Step 1. Our approach is to reformulate the problem into one to which the results of Chapter 3 will apply. We shall use overbars (e.g., $\bar{y}$) to signify points $\bar{y} = (y_1, y_2, y_3, y_4)$ belonging to $R^n \times R^n \times R \times R$. We define

$$C_0 := \{\bar{y}: y_4 \geq l(y_1, y_2), y_3 = 0\}$$

$$C_1 := \{\bar{y}: y_1 = y_2\}$$

$$F_i(t, \bar{y}) := \{\bar{v}: v_2 = 0, v_4 = 0, (v_1, v_3)$$

$$\in [\text{epi } L(t, y_1, \cdot)] \cap [\hat{e}(t) + ik(t)\bar{B}]\},$$

where $\hat{e}(t) := (\dot{x}(t), L(t, x(t), \dot{x}(t)))$,

$$f(\bar{y}) := y_3 + y_4.$$

It is simple to check that, for any positive integer i, the arc $\bar{x}$ defined by

$$\bar{x}(t) = \left(x(t), x(b), \int_a^t L(s, x, \dot{x})\, ds, l(x(a), x(b))\right)$$

solves the problem of minimizing $f(\bar{y}(b))$ over the arcs $\bar{y}$ which satisfy $\bar{y}(b) \in C_1$ and

(1) $$\bar{y}(a) \in C_0$$

(2) $$\dot{\bar{y}} \in F_i(t, \bar{y}) \quad \text{a.e.}$$

Lemma 1. There are positive numbers M and ε such that, for every i, $\bar{x}$ minimizes $f(\bar{y}(b)) + Md_{C_i}(\bar{y}(b))$ over the arcs $\bar{y}$ in $T(\bar{x}; \varepsilon)$ satisfying the constraints (1) and (2).

By the calmness hypothesis, there exist positive numbers M_1 and ε_1 such that, for j equal to either 0 or 1, one has

$$V_j(s) + M_1|s| \geq V_j(0)$$

whenever s lies in $\varepsilon_1 B$. Set $\varepsilon = \varepsilon_1/2$ and $M = \sqrt{2} M_1$. Suppose that the assertion of the lemma is false. Then there is an arc $\bar{y}$ as in the statement of the lemma such that

(3) $$f(\bar{y}(b)) + Md_{C_1}(\bar{y}(b)) < f(\bar{x}(b)) = V_j(0).$$

Let $\bar{u}$ be the point in C_1 closest to $\bar{y}(b)$. Then $u_1 = u_2 =: u$ and furthermore

$$d_{C_1}(\bar{y}(b)) = |(y_1(b) - u, y_2(b) - u)|$$

$$= \sqrt{2}|y_1(b) - u| = \sqrt{2}|y_2(b) - u|,$$

and we have $2u = y_1(b) + y_2(b)$.

Suppose first that $j = 1$. It follows that we may rewrite (3) as follows, where $y := y_1$ and $s := y_2(b) - y_1(b)$:

$$l(y(a), y(b) + s) + \int_a^b L(t, y, \dot{y})\, dt + \left(\frac{M}{\sqrt{2}}\right)|s| < V_1(0).$$

But then

$$V_1(s) + M_1|s| < V_1(0)$$

and $|s| < \varepsilon_1$, which is a contradiction. The other case, in which $j = 0$, is handled the same way. Lemma 1 is proved.

Step 2. We now wish to apply the necessary conditions of Chapter 3 to the problem described in Lemma 1. All Hypotheses 3.2.2 are present to permit this, as is clear except for the requirement that F_i be Lipschitz near $\bar{x}$. We proceed to verify this.

Lemma 2. F_i is Lipschitz near $\bar{x}$, with Lipschitz function $10k(t)$, for each i.

Let $F(t, y)$ denote epi $L(t, y, \cdot)$. We begin by proving that for each t, $F(t, \cdot)$ is Lipschitz within ε of $x(t)$, with Lipschitz constant $2k(t)$. Let s_1, s_2 in $x(t) + \varepsilon B$ and w_1 in $F(t, s_1)$ be given. It suffices to exhibit w_2 in $F(t, s_2)$ satisfying $|w_1 - w_2| \leq 2k(t)|s_2 - s_1|$.

4.2 Necessary Conditions

If there is no such w_2, then there would be a unit vector (p, q) and a positive number δ such that, for all w_2 in $F(t, s_2)$, one has

(4) $\qquad (p, q) \cdot (w_1 - w_2) \geq 2k(t)|s_1 - s_2| + \delta.$

Since $F(t, s_2)$ is an epigraph, it follows that $q \leq 0$. Suppose first that q is negative. Since one has $w_1 = [v_1, r_1]$ for some $r_1 \geq L(t, s_1, v_1)$, (4) implies

(5) $\qquad \tilde{p} \cdot v_1 - L(t, s_1, v_1) \geq \tilde{p} \cdot v_2 - L(t, s_2, v_2) + \dfrac{(2k|s_1 - s_2| + \delta)}{|q|}$

for all v_2 in R^n, where $\tilde{p} = p/|q|$. Since $|q| + |p| < 2$, it follows that $2/|q|$ is no less than $1 + |\tilde{p}|$, and hence (5) yields

$$H(t, s_1, \tilde{p}) > H(t, s_2, \tilde{p}) + k(1 + |\tilde{p}|)|s_1 - s_2|,$$

the desired contradiction. There remains the case $q = 0$. Since $L(t, s_2, v_2) \geq -H(t, s_2, 0)$ for all v_2, we may choose λ in $(0, 1)$ such that $\lambda L(t, s_2, v_2) > -\delta/3$ for all v_2 and such that $\lambda L(t, s_1, v_1) < \delta/3$. Then for all v_2 one has

$$p \cdot v_1 - \lambda L(t, s_1, v_1) \geq p \cdot v_2 - \lambda L(t, s_2, v_2) + 2k|s_1 - s_2| + \dfrac{\delta}{3},$$

which leads to

$$H\!\left(t, s_1, \dfrac{p}{\lambda}\right) - H\!\left(t, s_2, \dfrac{p}{\lambda}\right) \geq \dfrac{2k|s_1 - s_2|}{\lambda} + \dfrac{\delta}{3\lambda}$$

$$> k\!\left(1 + \dfrac{|p|}{\lambda}\right)|s_1 - s_2|,$$

again a contradiction.

This establishes the Lipschitz property of F; we proceed to look at F_i. Let $\hat{v} = \dot{x}(t), \hat{r} = L(t, x, \dot{x})$. Observe that

$$[F(t, s) - (\hat{v}, \hat{r})] \cap ik(t)\overline{B} = (F(t, s) \cap [(\hat{v}, \hat{r}) + ik(t)\overline{B}]) - (\hat{v}, \hat{r})$$

$$= \{(v_1, v_3) : (v_1, 0, v_3, 0) \in F_i(t, \bar{s})\} - (\hat{v}, \hat{r}).$$

To complete the proof of Lemma 2, it suffices to show that, for some $\delta > 0$, for almost all t, the multifunction

$$G(s) := \{F(t, x(t) + s) - (\hat{v}, \hat{r})\} \cap ik(t)\overline{B}$$

is Lipschitz on δB with constant $10k(t)$. This will follow from the following general result.

Lemma 3. Let Γ be a multifunction which, on δB, is closed, convex, nonempty, and Lipschitz of rank M. Suppose that for some number r exceeding $2M\delta$, $\Gamma(s) \cap rB$ is nonempty for s in δB. Then the multifunction $\tilde{\Gamma}(s) := \Gamma(s) \cap 3r\bar{B}$ is Lipschitz on δB of rank $5M$.

To see this, let x, y in δB and v in $\tilde{\Gamma}(x)$ be given. We wish to exhibit w in $\tilde{\Gamma}(y)$ such that $|v - w| \leq 5M|x - y|$. Suppose first that $|v|$ is bounded above by the (positive) quantity $3r - M|x - y|$. By the Lipschitz property of Γ, there exists w in $\Gamma(y)$ such that $|w - v| \leq M|x - y|$. Then

$$|w| \leq M|x - y| + |v| \leq 3r,$$

whence w belongs to $\tilde{\Gamma}(y)$ and satisfies our requirements. Suppose now that $|v|$ exceeds the quantity $3r - M|x - y|$, which we label N. Let ζ be an element of $\Gamma(x) \cap rB$, and set

$$\lambda := \frac{N - |\zeta|}{|v| - |\zeta|}, \qquad \phi := \lambda v + (1 - \lambda)\zeta.$$

Note that λ lies between 0 and 1, so that ϕ belongs to $\Gamma(x)$ by convexity. We find

$$|\phi - v| \leq (1 - \lambda)|v - \zeta| \leq \frac{|v| + |\zeta|}{N - |\zeta|}\{|v| - N\}$$

$$\leq \frac{4r}{N - r}\{|v| - 3r + M|x - y|\} \leq 4M|x - y| \quad (\text{since } N - r \geq r).$$

By the Lipschitz property of Γ, there is w in $\Gamma(y)$ such that $|w - \phi|$ is bounded above by $M|x - y|$. We calculate

$$|w| \leq |\phi| + M|x - y| \leq N + M|x - y| \leq 3r,$$

so that w belongs to $\tilde{\Gamma}(y)$. Finally we confirm

$$|w - v| \leq M|x - y| + 4M|x - y| = 5M|x - y|.$$

This completes the proof of the lemma.

It is a simple matter to complete the proof of Lemma 2 by applying the preceding to the multifunction G, with $M := 2k(t)$, $r := ik(t)/3$, and with any positive δ less than both ε and $i/12$.

4.2 Necessary Conditions

We now proceed to apply Theorem 3.2.6 to the problem described in Lemma 1. We deduce the existence of an arc p_i such that

(6) $\quad\quad\quad (-\dot{p}_i, \dot{x}, \hat{L}) \in \partial h_i(t, x, p_i, -1) \quad$ a.e.

(7) $\quad\quad\quad (p_i(a), -p_i(b)) \in \partial l(x(a), x(b))$

(8) $\quad\quad\quad |p_i(a)| \leq r,$

where $\hat{L}$ signifies $L(t, x, \dot{x})$, $h_i(t, x, \cdot, \cdot)$ is the support function of $F_i(t, \bar{x})$, and r is a constant not depending on i (see Corollary 1 to Theorem 3.2.6). (Note that conditions like (6) are written horizontally for convenience.)

Lemma 4. One has

(9) $\quad\quad\quad (-\dot{p}_i, \dot{x}) \in \partial H_i(t, x, p_i) \quad$ a.e.,

where H_i is given by

$$H_i(t, y, p) = \max\{ p \cdot v - L(t, y, v) : (v, L(t, y, v)) \in (\dot{x}, \hat{L}) + ik(t)\bar{B} \}.$$

To prove this, note that if f_1 is the function

$$f_1(y, p, q) = h_i(t, y, p, q),$$

and if f_2 is defined by

$$f_2(y, p) = h_i(t, y, p, -1),$$

then

$$f_1(y, p, q) = |q| f_2\left(y, \frac{p}{|q|} \right) \quad \text{for } q < 0.$$

It follows that whenever ∇f_1 exists, with

$$[\alpha, \beta, \gamma] = \nabla f_1(y, p, q),$$

then one has

$$\alpha = |q| \nabla_y f_2\left(y, \frac{p}{|q|} \right), \beta = \nabla_p f_2\left(y, \frac{p}{|q|} \right).$$

It now follows from Theorem 2.5.1 that any element (α, β, γ) of $\partial f_1(y, p, -1)$ is such that (α, β) belongs to $\partial f_2(y, p)$. We have shown that (9) follows from (6), which is the assertion of Lemma 4.

Lemma 5. There is a constant M_0 such that, for all i and for all t in $[a, b]$, one has

$$|p_i(t)| < M_0.$$

By Proposition 3.2.4(b) and (6), one has

$$|\dot{p}_i| \leq 10k(t)|(p_i, -1)|.$$

Combined with (8), this leads to the required conclusion.

Step 3

Lemma 6. For almost every t, there exists $N(t)$ such that, for all $i \geq N(t)$, one has $H_i(t, \cdot, \cdot) = H(t, \cdot, \cdot)$ in the neighborhood of $(x(t), p_i(t))$.

Let t be such that (9) holds. The multifunction

$$(y, p) \to \partial_p H(t, y, p)$$

is upper semicontinuous and compact-valued, and so the image of a compact set is compact. Note that $\dot{x}(t) \in \partial_p H(t, x(t), p_i(t))$ by Proposition 2.5.3. It follows from this and Lemma 5 that for some constant c (depending in general on t) one has

$$\partial_p H(t, y, p) \subset \dot{x}(t) + cB$$

for all (y, p) in a neighborhood of $(x(t), p_i(t))$.

For all v in $\partial_p H(t, y, p)$, one has (by the definition of subgradient)

$$\langle p, v \rangle - L(t, y, v) = H(t, y, p).$$

Since $H(t, \cdot, \cdot)$ is continuous, it follows that for (y, p) sufficiently near $(x(t), p_i(t))$ one has, for such v,

$$|L(t, y, v) - L(t, x, \dot{x})| \leq |\langle p, v \rangle - \langle p_i, \dot{x} \rangle| + |H(t, y, p) - H(t, x, p_i)|$$

$$\leq |\langle p - p_i, \dot{x} \rangle| + |\langle p, v - \dot{x} \rangle| + 1$$

$$\leq M_0\{2|\dot{x}(t)| + c\} + 1 =: c_1.$$

If now we choose $i > c + c_1$, it follows that $H(t, \cdot, \cdot)$ and $H_i(t, \cdot, \cdot)$ agree near (x, p_i) (since, always, $H_i \leq H$). This proves the lemma.

We may assume that the function $N(\cdot)$ of Lemma 6 is measurable. Let us now define

$$A_i = \{t : N(t) \leq i\}.$$

4.2 Necessary Conditions

We have, by construction,

$$(-\dot{p}_i, \dot{x}) \in \partial H(t, x, p_i) \quad \text{for } t \in A_i.$$

Note that meas(A_i) → ($b - a$) as $i \to \infty$, and that the arcs p_i are such that $|\dot{p}_i|$ is uniformly integrably bounded on all of $[a, b]$, and such that $|p_i(a)|$ is bounded. We are justified in applying Theorem 3.1.7 and (in view of (7)) the conclusions of the theorem follow.

Now if L (and thus H) has no dependence on t, the Hamiltonian inclusion clearly implies that $\dot{x}$ is essentially bounded, since H is uniformly Lipschitz on compact subsets of $R^n \times R^n$. The relation $\langle p, \dot{x} \rangle - L(x, \dot{x}) = H(x, p)$ then implies that $L(x(t), \dot{x}(t))$ is also essentially bounded. Because of these facts, it is possible to redo the proof using a slightly different multifunction F_i:

$$F_i(\bar{y}) := \{(v_1, 0, v_3, 0) : (v_1, v_3) \in \text{epi } L(y_i, \cdot) \cap i\bar{B}\}.$$

The proof goes through just as before, but now F_i is independent of t, so that the corollary to Theorem 3.6.1 may be applied after Lemma 2 (rather than Theorem 3.2.6). We gain the added conclusion $H_i(x(t), p_i(t)) = $ constant. Upon completing the proof, we get this conclusion for $H(x, p)$ as desired. □

4.2.3 Remark

It is possible to prove a version of the theorem in which one dispenses with the convexity condition. One may posit (along with calmness) directly that epi $L(y, \cdot)$ is Lipschitz, and that H is finite. The proof goes through as above, except that one shows as a preliminary step that the problem can be "relaxed" (i.e., that x also solves (with the same value) the problem in which L is replaced by the function $\tilde{L}$ such that, for each (t, y), $\tilde{L}(t, y, \cdot)$ is the greatest convex function majorized by $L(t, y, \cdot)$). Details are in Clarke (1977a).

On The Calmness Condition

There are several cases in which the calmness hypothesis of Theorem 4.2.1 is automatically satisfied. Among these are the case in which l is locally Lipschitz (in either variable). Notable in this regard is the free-endpoint case in which l is actually independent of $x(b)$. Another is the case in which L is Lipschitz (a situation to be explored presently). A third example is provided by the case (see Section 4.1) in which

$$L(t, s, v) = \psi_{F(t, s)}(v), \, l(s_0, s_1) = \psi_{C_0}(s_0) + \psi_{C_1}(s_1) + f(s_1).$$

The problem of Bolza P_B then reduces to the problem P_D of Chapter 3 (see Problem 3.4.1), and if P_D is normal, then, under certain conditions, P_D is calm by Theorem 3.4.3, Corollary 1.

Another result in a similar vein is the following:

4.2.4 Proposition

Let l and L have the form

$$L(t, s, v) = f(t, s, v) + \psi_V(v)$$

$$l(s_0, s_1) = \psi_T(s_0, s_1),$$

where f is a locally Lipschitz function and V, T are compact convex sets in R^n, $R^n \times R^n$, respectively. Suppose that

$$0 \in \text{int}\{(b - a)V - \Delta\},$$

where Δ is the set $\{s_1 - s_0 : (s_0, s_1) \in T\}$. Then P_B is calm and H satisfies the strong Lipschitz condition.

Proof. There is a uniform bound M on any admissible x, since $x(a)$ and $\dot{x}$ are bounded. For points y_1, y_2 in $x(t) + B$, one has

$$H(t, y_1, p) - H(t, y_2, p) \leq \sup\{f(t, y_2, v) - f(t, y_1, v) : v \in V\}$$

$$\leq K|y_1 - y_2|,$$

where K is a Lipschitz constant for f on the relevant (bounded) set. The required Lipschitz property for H follows.

We shall derive the calmness of P_B from a general stability result in mathematical programming (see Theorem 6.3.2, Corollary 1). Let X be the Banach space of arcs x on $[a, b]$, with the norm

$$\|x\| := |x(a)| + \int_a^b |\dot{x}|\, dt.$$

Define the functionals $\tilde{f} \colon X \to R$ and $A \colon X \to R^{2n}$ via

$$\tilde{f}(x) = \int_a^b f(t, x, \dot{x})\, dt$$

$$A(x) = [x(a), x(b)],$$

and the set S via

$$S := \{x \in X : \dot{x}(t) \in V \text{ a.e.}, |A(x)| \leq 2r\},$$

where r is a scalar such that rB contains T. It is easy to show that $\tilde{f}$ is Lipschitz on S, that A is a continuous linear functional on X, and that S is closed, convex, and bounded.

Lemma. $0 \in \text{int}\{AS - T\}$.

We wish to show that for every (σ_1, σ_2) near 0 in R^{2n}, there is an element x of S such that, for some (s_1, s_2) in T, one has

(10) $$\sigma_1 = x(a) - s_1, \quad \sigma_2 = x(b) - s_2.$$

We know that (for (σ_1, σ_2) small enough) there exists v in V and (s_1, s_2) in T such that

$$\sigma_2 - \sigma_1 = (b - a)v + s_1 - s_2.$$

To derive Eq. (10), it suffices to take $x(\cdot)$ to be the arc

$$x(t) = s_1 + \sigma_1 + (t - a)v.$$

It follows from Theorem 6.3.2, Corollary 1, that the function

$$W(\sigma_1, \sigma_2) := \inf\{\tilde{f}(x) : x \in S, (x(a), x(b)) \in T + (\sigma_1, \sigma_2)\}$$

is Lipschitz in a neighborhood of $(0,0)$. This implies that the functions V_0, V_1 are Lipschitz near 0, so that P_B is calm. □

4.3 SUFFICIENT CONDITIONS

As a rule, necessary conditions for optimality become sufficient as well when the problems in question are convex. In the case of P_B, for example, the conclusions of Theorem 4.2.2 guarantee that x is a solution when l is convex and $L(t, \cdot, \cdot)$ is convex for each t. The latter is equivalent to $H(t, y, p)$ being concave in y (as well as convex in p, which is always the case). (A formal statement of this fact will be given presently.)

In what appears to be quite a different vein, the classical calculus of variations exhibits sufficient conditions involving the notion of conjugate points, which are determined via a certain second-order differential equation. Zeidan (1982; 1983) has demonstrated through the context of P_B that both these approaches can be viewed as manifestations of the same technique (in fact, so can the Hamilton–Jacobi technique discussed in Chapter 3). We shall limit ourselves here to a simple yet powerful special case of her results. We posit Basic Hypotheses 4.1.2.

4.3.1 Theorem
Suppose that x and p are arcs on $[a, b]$ with

(1) $$L(t, x, \dot{x} + v) - L(t, x, \dot{x}) \geq \langle p, v \rangle \quad \text{for } v \text{ in } R^n, \text{ a.e.,}$$

and suppose that for some $\varepsilon > 0$, for almost all t in $[a, b]$, for all y in εB, one has

(2) $H(t, x(t) + y, p(t) - Q(t)y) - H(t, x(t), p(t))$
$$\leqslant -\langle \dot{p}(t) + Q(t)\dot{x}(t), y \rangle + \tfrac{1}{2}\langle y, \dot{Q}(t)y \rangle,$$

and that for all u, v in εB one has

(3) $l(x(a) + u, x(b) + v) - l(x(a), x(b))$
$$\geqslant -\langle p(b), v \rangle + \langle p(a), u \rangle - \tfrac{1}{2}\langle u, Q(a)u \rangle + \tfrac{1}{2}\langle v, Q(b)v \rangle,$$

where $Q(\cdot)$ is a Lipschitz $n \times n$ symmetric matrix-valued function on $[a, b]$. Then x solves P_B locally (i.e., relative to the tube $T(x; \varepsilon)$).

4.3.2 Remark

When $Q = 0$, the conditions (1)–(3) of the theorem imply the Hamiltonian inclusion and the transversality condition of Theorem 4.2.2. In this light they can be viewed as a (local) strengthening of the necessary conditions.

Proof. It is implicit that x lies in Ω and assigns a finite value to the Bolza functional. Let us define a function W as follows:

$$W(t, y) := -\tfrac{1}{2}\langle y - x(t), Q(t)(y - x(t)) \rangle + \langle p(t), y \rangle.$$

Then, for almost all t in $[a, b]$, for all y in $x(t) + \varepsilon B$, one has

$$W_t(t, y) + H(t, y, W_y(t, y)) \leqslant W_t(t, x(t)) + H(t, x(t), W_y(t, x(t))),$$

since this is merely inequality (2) rewritten.

As in Remark 4.2.3, we denote by $\tilde{L}(t, y, \cdot)$ the convex hull of the function $L(t, y, \cdot)$. Condition (1) implies $\tilde{L}(t, x, \dot{x}) = L(t, x, \dot{x})$ a.e.

Now let $y(\cdot)$ be any arc in the ε-tube about x for which $l(y(a), y(b))$ is finite. Then

$$L(t, y, \dot{y}) - L(t, x, \dot{x}) \geqslant \tilde{L}(t, y, \dot{y}) - \tilde{L}(t, x, \dot{x})$$
$$= \sup_q \{\langle \dot{y}, q \rangle - H(t, y, q)\} - \langle p, \dot{x} \rangle + H(t, x, p)$$
$$\geqslant \langle \dot{y}, W_y(t, y) \rangle - H(t, y, W_y(t, y)) - \langle p, \dot{x} \rangle + H(t, x, W_y(t, x))$$
$$\geqslant W_t(t, y) + \langle \dot{y}, W_y(t, y) \rangle - W_t(t, x) - \langle p, \dot{x} \rangle$$
$$= \frac{d}{dt} W(t, y(t)) - \frac{d}{dt} W(t, x(t)).$$

4.3 Sufficient Conditions

The above implies (recall that the Bolza functional is denoted by J)

$$J(y) - J(x) \geq l(y(a), y(b)) - l(x(a), x(b))$$
$$+ W(b, y(b)) - W(a, y(a)) - W(b, x(b)) + W(a, x(a)).$$

The right-hand side above is non-negative by (3). It follows that x solves P_B relative to $T(x; \varepsilon)$. □

Here is the promised result for the "convex case."

Corollary

Let L satisfy the convexity condition, and suppose that there is an arc p satisfying

$$\begin{bmatrix} -\dot{p}(t) \\ \dot{x}(t) \end{bmatrix} \in \partial H(t, x(t), p(t)) \quad \text{a.e.}$$

$$\begin{bmatrix} p(a) \\ -p(b) \end{bmatrix} \in \partial l(x(a), x(b)),$$

where l is convex, and where, for each t, the function $y \to H(t, y, p(t))$ is concave. Then x solves P_B.

Proof. Apply the theorem with $\varepsilon = \infty$, $Q(t) \equiv 0$. □

Conjugate Points

The classical theorem of Jacobi in the calculus of variations affirms that when L is C^3 (and under other technical assumptions which we shall not list in this comment, and ignoring l), an extremal x (i.e., an arc x which, together with some arc p, solves the Hamiltonian equations) is a local solution if there is no point c in $(a, b]$ *conjugate* to a; that is, such that the following boundary-value problem admits a nontrivial solution:

$$\frac{d}{dt}\{L_{vv}\dot{h} + L_{vx}h\} - L_{xv}\dot{h} - L_{xx}h = 0$$

$$h(a) = h(c) = 0$$

(where all derivatives of L are evaluated at $(t, x(t), \dot{x}(t))$). The nonexistence of such c is called "the (strong) Jacobi condition." Using the theory of such boundary-value problems, it is possible to express Jacobi's condition in the following Hamiltonian terms:

There exists a symmetric matrix-valued function Q such that, on $[a, b]$, one has

(4) $$\dot{Q} - QH_{pp}Q + H_{xp}Q + QH_{px} - H_{xx} > 0$$

(derivatives of H evaluated at $(t, x(t), p(t))$; ≥ 0 indicates positive semidefinite). Now condition (2) of Theorem 4.3.1 stipulates that, for each t, the function

$$\theta(y) := \tfrac{1}{2}\langle y, Qy \rangle - \langle \dot{p} + Q\dot{x}, y \rangle - H(t, x(t) + y, p(t) - Qy)$$

attains a local minimum at $y = 0$. One way to guarantee this is the condition $\theta_{yy}(0) > 0$, which is precisely (4). (The condition $\theta_y(0) = 0$ is a consequence of Hamilton's equations.) This is how the theorem is linked to the classical theory; we refer the reader to the cited works of Zeidan for these and other matters.

4.4 FINITE LAGRANGIANS

We now present one of several versions of the Hamiltonian necessary conditions that can be derived when more stringent requirements are imposed upon the Lagrangian $L(t, y, v)$.

For simplicity, we shall deal in this section with the autonomous problem (i.e., L has no explicit dependence on t). Our assumptions will imply that L is locally Lipschitz (and so, finite) as a function of (y, v). More specifically, we require that the growth of L in (y, v) be at most exponential, in the following sense: there are constants k_0, k_1 and c_0 such that for all y and v in R^n, one has

$$|\partial L(y, v)| \leq k_0 |L(y, v)| + k_1 |(y, v)| + c_0.$$

In addition, we suppose that for every y in R^n, one has

$$\lim_{|v| \to +\infty} \frac{L(y, v)}{|v|} = +\infty.$$

(This is easily seen to imply that the Hamiltonian too is finite.) Because assumptions are being made directly on L (in contrast with Theorem 4.2.2), it will not be necessary to assume that L is convex as a function of v.

We shall suppose that l is of the form

$$l(s_0, s_1) = f(s_1) + \psi_{C_0}(s_0) + \psi_{C_1}(s_1),$$

where f is locally Lipschitz and C_0, C_1 are closed.

4.4.1 Theorem

Let x solve P_B under the above hypotheses on L. Then $\dot{x}$ is essentially bounded, H is Lipschitz on bounded sets, and there exists an arc p such that

$$\begin{bmatrix} -\dot{p}(t) \\ \dot{x}(t) \end{bmatrix} \in \partial H(x(t), p(t)) \quad \text{a.e.}$$

$$p(a) \in N_{C_0}(x(a)), \quad -p(b) \in \partial f(x(b)) + N_{C_1}(x(b))$$

$$H(x(t), p(t)) = \text{constant}.$$

4.4 Finite Lagrangians

Proof. Preliminary Lemmas. We begin with a few technical results.

Lemma 1. If $f: R^m \to R$ satisfies

$$|\partial f(x)| \leq k_0 |f(x)| + k_1 |x| + c_0 \quad \text{for all } x,$$

then one has, for every x and for every y in $x + iB$,

$$|f(y) - f(x)| \leq \alpha_i [k_0 |f(x)| + c_0 + k_1 \max\{|x|, |y|\}]|y - x|$$

where $\alpha_i := (e^{k_0 i} - 1)/(k_0 i)$. (Here i can be any positive real number.)

To see this, let us set $g(t) := |f(x + tv) - f(x)|$ for $v = y - x$. Then the hypothesis implies the inequality

$$|g'(t)| \leq |v|\{k_0 g(t) + k_0 |f(x)| + c_0 + k_1 \max\{|x|, |y|\}\} \quad \text{a.e.}$$

Integration leads to

$$g(1) \leq \left[|f(x)| + \frac{c_0}{k_0} + k_1 \frac{\max\{|x|, |y|\}}{k_0}\right][e^{k_0 |v|} - 1].$$

We have $e^{k_0 |v|} - 1 \leq \alpha_i k_0 |v|$ for $|v| \leq i$, so that the preceding inequality immediately yields the desired result.

Lemma 2. For each positive integer i there exists an integrable function $k_i(\cdot)$ with the property that, for all $(y_1, v_1), (y_2, v_2)$ in $(x(t), \dot{x}(t)) + iB$, one has

$$|L(y_1, v_1) - L(y_2, v_2)| \leq k_i(t)|(y_1 - y_2, v_1 - v_2)|.$$

If $L(t)$ signifies $L(x(t), \dot{x}(t))$ and $\beta(t)$ signifies $|(x(t), \dot{x}(t))|$, then we may apply Lemma 1 once to deduce

$$|L(y_1, v_1) - L(t)| \leq \alpha_i [k_0 |L(t)| + c_0 + k_1(i + \beta(t))]i,$$

and so

$$|L(y_1, v_1)| \leq (i\alpha_i k_0 + 1)|L(t)| + i\alpha_i [c_0 + k_1(i + \beta(t))] =: \tilde{k}_i(t).$$

Note that $\tilde{k}_i(\cdot)$ is integrable, since $L(\cdot)$ is integrable by assumption. Applying Lemma 1 again (for $2i$) with this bound gives

$$|L(y_2, v_2) - L(y_1, v_1)|$$
$$\leq \alpha_{2i} [k_0 \tilde{k}_i(t) + c_0 + k_1(i + \beta(t))]|(y_2 - y_1, v_2 - v_1)|,$$

so that the required result follows, with

$$k_i(t) := \alpha_{2i}[k_0 \tilde{k}_i(t) + c_0 + k_1(i + \beta(t))].$$

Lemma 3. *H is Lipschitz on bounded subsets of $R^n \times R^n$.*

It suffices to prove that H is Lipschitz near any point (x_0, p_0). Choose any $M > 3|p_0|$, and let $\varepsilon > 0$ be such that $8\varepsilon k_0 < 1, 8\varepsilon k_1 < M$, and such that $[\exp(k_0 \varepsilon) - 1]/(k_0 \varepsilon) < 2$. By hypothesis there exists a constant c such that, for all v in R^n, one has

(1) $$L(x_0, v) \geq c + M|v|.$$

Now let y be any point in $x_0 + \varepsilon B$. Then, by Lemma 1,

(2) $$L(y, v) \geq L(x_0, v) - 2[k_0|L(x_0, v)| + c_0 + k_1(|v| + |x_0| + \varepsilon)]\varepsilon.$$

It follows from (1) that $|L(x_0, v)|$ is bounded above by $L(x_0, v) + 2|c|$. Putting this into (2) gives

(3)
$$L(y, v) \geq (1 - 2\varepsilon k_0)L(x_0, v) - 2\varepsilon(2|c|k_0 + c_0 + k_1|x_0| + k_1 \varepsilon) - 2\varepsilon k_1|v|$$
$$\geq \frac{M}{2}|v| + c_1$$

for an appropriate constant c_1 not depending on y and v.

If p satisfies $3|p| \leq M$, then

$$\langle p, 0 \rangle - L(y, 0) \leq H(y, p) = \sup_v \{\langle p, v \rangle - L(y, v)\}.$$

If v is such that

$$\langle p, v \rangle - L(y, v) \geq H(y, p) - 1,$$

it follows that, for an appropriate constant c_2 not depending on y, v, or p, one has

$$L(y, v) \leq c_2 + \frac{M}{3}|v|,$$

which, combined with (3), gives $|v| \leq 6(c_2 - c_1)/M =: c_3$. It follows that for any y in $x_0 + \varepsilon B$ and p in $(M/3)B$, one has

$$H(y, p) = \max\{\langle p, v \rangle - L(y, v) : v \in c_3 \bar{B}\}.$$

4.4 Finite Lagrangians

Because L is Lipschitz on bounded sets, it follows easily from this characterization that H is Lipschitz on the set in question.

Let us note that in the proof we also showed:

Lemma 4. For any bounded subset Y of R^n, for any $M > 0$, there exists a constant c such that, for all y in Y and v in R^n, one has

$$c + M|v| \leq L(y, v) \leq |L(y, v)| \leq L(y, v) + 2|c|.$$

Proof of Theorem 4.4.1

Step 1. There is no essential change in the proof in the special case $f \equiv 0$, which we shall treat for simplicity. We claim that for some ε and $r > 0$, x minimizes

$$r d_{C_1}(y(b)) + \int_a^b L(y, \dot{y})\, dt$$

over all arcs y in $T(x; \varepsilon)$ satisfying $y(a) \in C_0$. If this is false, then there is a sequence x_i of arcs converging uniformly to x such that $x_i(a) \in C_0$ and

$$(4) \qquad i d_{C_1}(x_i(b)) + \int_a^b L(x_i, \dot{x}_i)\, dt < \int_a^b L(x, \dot{x})\, dt.$$

Because $\int_a^b L(x_i, \dot{x}_i)\, dt$ is bounded below (as a consequence of Lemma 4), it follows that $d_{C_1}(x_i(b))$ goes to 0. Let c_i in C_1 satisfy $|x_i(b) - c_i| = d_{C_1}(x_i(b))$, and set $y_i(t) := x_i(t) - (t - a)(x_i(b) - c_i)/(b - a)$. Note that $|y_i - x_i|$ and $|\dot{y}_i - \dot{x}_i|$ are bounded by $c_1 d_{C_1}(x_i(b))$, where $c_1 = \max\{1, 1/(b - a)\}$.

Invoking Lemmas 1 and 4, we can infer that (for i large enough) $L(y_i, \dot{y}_i)$ is bounded above by (letting L_i denote $L(x_i, \dot{x}_i)$)

$$L_i + \alpha_1 [k_0 |L_i| + c_0 + k_1(2 + |x_i| + |\dot{x}_i|)] |(y_i - x_i, \dot{y}_i - \dot{x}_i)|$$

$$\leq L_i + \alpha_1 [k_0 L_i + c_0 + 2k_0 |c| + k_1(3 + |x| + |\dot{x}_i|)] 2 c_1 d_{C_1}(x_i(b)).$$

If this is combined with the inequalities $L(x_i, \dot{x}_i) \geq c + M|\dot{x}_i|$ and (from inequality (4)) $\int_a^b L(x_i, \dot{x}_i)\, dt < \int_a^b L(x, \dot{x})\, dt$, one concludes that $\int_a^b L(y_i, \dot{y}_i)\, dt$ is bounded above by

$$\int_a^b L(x_i, \dot{x}_i) + 2\alpha_1 c_1 \left[\left(k_0 + \frac{k_1}{M} \right) \int_a^b L(x, \dot{x})\, dt + c_2 \right] d_{C_1}(x_i(b))$$

(where $c_2 := \{2k_0 |c| + c_0 + k_1(3 + \|x\|) - c/M\}(b - a)$), which (for i large) is strictly less than $\int_a^b L(x, \dot{x})\, dt$. This contradicts the optimality of x, since y_i satisfies $y_i(b) \in C_1$, and proves the claim.

Step 2. With an eye to applying Theorem 4.2.2, we proceed to prove that, in the problem described in Step 1, we can convexify L in v and preserve the optimality of x. Let us define a multifunction F_i by setting $F_i(t, \bar{y})$ equal to

$$\text{epi } L(y, \cdot) \cap \{(v, \alpha) : |v - \dot{x}(t)| \leq i, |\alpha| \leq c_i(t)\},$$

where $\bar{y} = [y, y_0]$, and where $c_i(t)$ is an integrable function satisfying

$$|L(y, v)| + 1 \leq c_i(t)$$

for all y in $x(t) + \varepsilon\bar{B}$ and v in $\dot{x}(t) + i\bar{B}$ (the existence of c_i follows easily from Lemma 2). Note that the arc $\bar{x}(t) := [x(t), \int_a^t L(x, \dot{x}) \, d\tau]$ solves the problem of minimizing

$$r d_{C_1}(y(b)) + y_0(b)$$

over all arcs $\bar{y} = [y, y_0]$ which lie near $\bar{x}$, which are trajectories for F_i, and which satisfy $\bar{y}(a) \in C_0 \times \{0\}$. Since F_i is measurably Lipschitz and integrably bounded, it follows from the Corollary to Theorem 3.1.6 that $\bar{x}$ continues to solve this problem (with the same value for the minimum) if F_i is replaced by co F_i. Returning to Lagrangian form, this says that for any large integer i, x solves (over $T(x; \varepsilon)$) the problem of minimizing

(5) $$r d_{C_1}(y(b)) + \int_a^b L_i(t, y, \dot{y}) \, dt$$

over the arcs y satisfying $y(a) \in C_0$, where $L_i(t, y, \cdot)$ is the "convexification" of $L(y, \cdot) + \psi_{\dot{x}+i\bar{B}}(\cdot)$ (i.e., the greatest convex function which is majorized by $L(y, \cdot)$ on $\dot{x} + i\bar{B}$). Further, $L_i(t, x, \dot{x}) = L(x, \dot{x})$ a.e. If $\tilde{L}$ denotes the convex hull of L in v (i.e., for each y, $\tilde{L}(y, \cdot)$ is the greatest convex function majorized by $L(y, \cdot)$), the preceding implies $\tilde{L}(x, \dot{x}) = L(x, \dot{x})$ a.e.

Step 3. Let us note that $L_i(t, y, v) = +\infty$ for $|v - \dot{x}(t)| > i$, and that the Hamiltonian H_i corresponding to L_i may be written

$$H_i(t, y, p) = \max\{\langle p, v \rangle - L(y, v) : |v - \dot{x}(t)| \leq i\}.$$

H_i satisfies the strong Lipschitz condition by Lemma 2. We apply Theorem 4.2.2 to the problem (5) to deduce the existence of an arc p_i satisfying

(6) $$(-\dot{p}_i, \dot{x}) \in \partial H_i(t, x, p_i) \quad \text{a.e.}$$

(7) $$p_i(a) \in N_{C_0}(x(b)), -p_i(b) \in r\partial d_{C_1}(x(b)) \subset N_{C_1}(x(b)).$$

Step 4. From Proposition 2.5.3 we know that (6) implies that $\dot{x}$ belongs to $\partial_p H_i(t, x, p_i)$, which in turn implies that the concave function

$$v \to \langle p_i, v \rangle - L_i(t, x, v)$$

achieves a maximum at $v = \dot{x}$. Since $L_i(t, \cdot, \cdot) \geq \tilde{L}(\cdot, \cdot)$ (with equality at $(x, \dot{x})$), it follows that the concave function

$$v \to \langle p_i, v \rangle - \tilde{L}(x, v)$$

attains a maximum over v in $\dot{x}(t) + iB$ at $v = \dot{x}$. For concave functions local and global maxima coincide, and we deduce

$$\langle p_i, \dot{x} \rangle - L(x, \dot{x}) = \langle p_i, \dot{x} \rangle - \tilde{L}(x, \dot{x})$$
$$\geq -\tilde{L}(x, 0) \geq -L(x, 0).$$

Let $M > 2\|p_i\|$ and apply Lemma 4 to deduce the existence of a constant c such that, for all v, one has

$$L(x, v) \geq c + M|v|.$$

Then

$$\frac{M|\dot{x}|}{2} \geq |p_i||\dot{x}| \geq \langle p_i, \dot{x} \rangle$$
$$\geq L(x, \dot{x}) - L(x, 0) \geq c + M|\dot{x}| - L(x, 0).$$

This easily implies that $\dot{x}$ is essentially bounded as claimed in the theorem.

Armed with this fact, we can redo Steps 2 and 3 of the proof with a slightly different multifunction F_i:

$$F_i(y) := \text{epi } L(y, \cdot) \cap \{(v, \alpha) : |v| \leq i, |\alpha| \leq c_i\}.$$

All goes through unchanged, and we obtain the counterpart of (6) and (7):

(8) $\qquad (-\dot{p}_i, \dot{x}) \in \partial H_i(x, p_i)$ a.e.

(9) $\qquad p_i(a) \in N_{C_0}(x(b)), -p_i(b) \in r\, \partial d_{C_1}(x(b)) \subset N_{C_1}(x(b)).$

Because now F_i is independent of t, we also derive (by Theorem 3.6.1, Corollary)

(10) $\qquad H_i(x(t), p_i(t)) = \text{constant}.$

Step 5. Condition (8) yields $p_i \in \partial_v \tilde{L}(x, \dot{x})$ a.e. We note

$$L(x, \dot{x} + v) - L(x, \dot{x}) \geq \tilde{L}(x, \dot{x} + v) - \tilde{L}(x, \dot{x})$$
$$\geq \langle p_i, v \rangle \quad \text{for all } v \text{ in } R^n.$$

It follows that $L_v^\circ(x, \dot{x}; v)$ majorizes $\langle p_i, v \rangle$ for all v, whence p_i belongs to $\partial_v L(x, \dot{x})$. Consequently

$$|p_i| \leq k_0(|L(x, \dot{x})| + k_1|(x, \dot{x})| + c_0),$$

and since L is bounded on bounded sets, we derive an a priori bound (M, say) on $\|p_i\|$.

Let Δ denote the quantity

$$\sup\{\operatorname{diam} \partial_p H(y, p) : |y| \leq \|x\|, |p| \leq M\},$$

where diam denotes diameter (Δ is finite by Lemma 3). We claim that for $i > \|\dot{x}\| + \Delta + 1$, for each t, there is a neighborhood of $(x(t), p_i(t))$ upon which H_i and H coincide. Clearly, the theorem follows from this, in view of the relations (8)–(10).

It is easy to see that the multifunction $(y, p) \to \partial_p H(y, p)$ is upper semicontinuous, so that for each t there is a neighborhood S of $(x(t), p_i(t))$ such that

$$\partial_p H(y, p) \subset \partial_p H(x, p_i) + B \quad \text{for } (y, p) \text{ in } S.$$

Since $\dot{x}$ belongs to $\partial_p H(x, p_i)$, we have

$$\partial_p H(x, p_i) \subset \dot{x} + \Delta \bar{B},$$

and so

$$\partial_p H(y, p) \subset \dot{x} + (\Delta + 1)B \quad \text{for } (y, p) \text{ in } S.$$

It follows that for (y, p) in S the v for which $\langle p, v \rangle - L(y, v)$ equals $H(y, p)$ belong to iB (for $i > \|\dot{x}\| + \Delta + 1$), which implies that $H(y, p) = H_i(y, p)$. □

4.4.2 Remark

The proof makes clear that the hypotheses on $L(y, v)$ need only hold for y in a neighborhood of the set $\{x(t) : a \leq t \leq b\}$ (but for all v).

Necessary Conditions in Lagrangian Form

When L does not possess the growth properties necessary for H to be finite, it is convenient to be able to express the necessary conditions in terms of L itself. As an illustration, and to facilitate comparison with the classical necessary conditions, consider the following:

4.4.3 Theorem

Let x solve P_B in the case in which $L(s, v)$ is a locally Lipschitz function independent of t, and suppose that $\dot{x}(t)$ is essentially bounded. Then there is an arc p such that (writing $L(t)$ for $L(x(t), \dot{x}(t))$)

(11) $$\dot{p}(t) \in \partial_s L(t), \, p(t) \in \partial_v L(t) \quad \text{a.e.}$$

(12) $$L(t) - \langle p(t), \dot{x}(t) \rangle = \text{constant a.e.}$$

(13) $\quad L(x(t), \dot{x}(t) + v) \geq L(x(t), \dot{x}(t)) + \langle p(t), v \rangle \quad \text{for all } v, \text{ a.e.}$

(14) $$\begin{bmatrix} p(a) \\ -p(b) \end{bmatrix} \in \partial l(x(a), x(b)).$$

Those who are familiar with the first-order necessary conditions of the classical calculus of variations will recognize in (11) the counterpart of the Euler–Lagrange equation $(d/dt)\nabla_v L = \nabla_s L$. The first Erdmann condition corresponds to the continuity of p, and the second to Eq. (12). The Weierstrass necessary condition is in (13), while (14) reflects the transversality (or normal boundary) conditions. The theorem is derivable as a special case of the maximum principle, Theorem 5.2.1. (See Clarke, 1980a.)

More general results in this vein dealing with cases in which L is extended-valued appear in Clarke (1976b).

4.5 THE MULTIPLIER RULE FOR INEQUALITY CONSTRAINTS

The very nature of the problem P_B is such as to make all constraints implicit (by suitably defining l and L). In the early part of this century, which was a time of great activity for the calculus of variations, the thrust was towards multiplier rules in which constraints appear explicitly. We shall digress somewhat in this section to present a result in this vein. As in the preceding section, the results of Chapter 3 lie at the heart of things.

The Problem

We consider the following problem (of Mayer). To minimize $f(x(1))$ over the arcs x on $[0, 1]$ which satisfy the boundary conditions

$$x(0) \in C_0, \quad x(1) \in C_1$$

and the inequality constraint

$$\phi(x(t), \dot{x}(t)) \leq 0 \quad \text{a.e.,}$$

where ϕ maps $R^n \times R^n$ to R. We assume that C_0, C_1 are closed and that f and ϕ are locally Lipschitz. An arc x is a *weak local minimum* if, for some $\varepsilon > 0$, it solves this problem relative to the arcs y satisfying $|x(t) - y(t)| < \varepsilon$, $|\dot{x}(t) - \dot{y}(t)| < \varepsilon$ a.e.

The arc x is termed *piecewise-smooth* if there is a partition $0 = t_0 < t_1 < \cdots < t_k = 1$ of $[0, 1]$ such that $\dot{x}$ exists and is continuous on (t_{i-1}, t_i) ($i = 1, 2, \ldots, k$) and admits finite limits at both t_{i-1} (from the right) and t_i (from the left). These limits are denoted $\dot{x}(t_{i-1}^+)$ and $\dot{x}(t_i^-)$, respectively. A point t at which x fails to be differentiable is termed a *corner point*.

We shall say that x satisfies the *rank condition* provided that for any t such that $\phi(x(t), \dot{x}(t))$ is zero, for any element (ζ_1, ζ_2) of $\partial\phi(x(t), \dot{x}(t))$, one has $\zeta_2 \neq 0$. If t is a corner point, this condition is understood to hold with $\dot{x}(t)$ replaced by both $\dot{x}(t^+)$ and $\dot{x}(t^-)$.

4.5.1 Theorem

Suppose that the piecewise-smooth arc x provides a weak local minimum for the preceding problem, and that x satisfies the rank condition. Then there exist an arc p, a measurable function $\lambda: [0, 1] \to R$, and a scalar λ_0 equal to 0 or 1, such that:

(i)
$$\begin{bmatrix} \dot{p}(t) \\ p(t) \end{bmatrix} \in \lambda(t) \partial\phi(x(t), \dot{x}(t)) \ a.e.$$

(ii) $\lambda(t) \geq 0$ a.e., $\lambda(t) = 0$ when $\phi(x(t), \dot{x}(t)) < 0$.

(iii) $p(0) \in N_{C_0}(x(0))$.

(iv) $-p(1) \in \lambda_0 \partial f(x(1)) + N_{C_1}(x(1))$.

(v) $|p(t)| + \lambda_0$ is never 0.

4.5.2 Remark

The case of several inequality constraints $\phi_i(x, \dot{x}) \leq 0$ is implicit in the above, since we may define $\phi = \sup \phi_i$. The classical theory uses Valentine's method of "slack variables" to reduce the problem to one with equality constraints. The classical rank condition in that smooth setting is the requirement that the vectors $\{D_{\dot{x}}\phi_i(x, \dot{x})\}$ be linearly independent (so that perforce there are at most n inequality constraints). When we treat a problem having r constraints $\phi_i \leq 0$ ($i = 1, 2, \ldots, r$) by setting $\phi = \max_{1 \leq i \leq r} \phi_i$, the rank condition above (if the ϕ_i are smooth, for example) amounts to requiring that the vectors $\{D_{\dot{x}}\phi(x, \dot{x})\}$ be *positively* linearly independent (this is a consequence of Proposition 2.3.12). Theorem 4.5.1 thus extends the classical multiplier rule for inequality constraints even in the case of smooth data.

Let us note formally:

Corollary 1

Let the piecewise-smooth arc x minimize $f(y(1))$ over the arcs y satisfying $y(0) \in C_0$, $y(1) \in C_1$ and the r constraints $\phi_i(y, \dot{y}) \leq 0$ a.e. ($i = 1, 2, \ldots, r$),

4.5 The Multiplier Rule for Inequality Constraints

where f, ϕ_i are locally Lipschitz. Suppose that for all t, for all (ζ_1, ζ_2) in the set

$$\mathrm{co}\{\partial \phi_i(x, \dot{x}) : i \text{ such that } \phi_i(x, \dot{x}) = 0\},$$

one has $\zeta_2 \neq 0$. Then there exist an arc p, measurable functions λ_i, and a scalar λ_0 equal to 0 or 1 such that (iii), (iv), and (v) hold, as well as

(i)′
$$\begin{bmatrix} \dot{p}(t) \\ p(t) \end{bmatrix} \in \sum_{i=1}^{r} \lambda_i(t) \partial \phi_i(x, \dot{x}) \quad a.e.$$

(ii)′ $\lambda_i \geq 0$, $\lambda_i(t) = 0$ if $\phi_i(x, \dot{x}) < 0$.

Proof of the theorem. For ease of notation, we denote $\phi(x(t), \dot{x}(t))$ and $\partial \phi(x(t), \dot{x}(t))$ by $\phi(t)$ and $\partial \phi(t)$, respectively. When t is a corner point, there will be occasions when $\phi(t)$ is to be interpreted as $\phi(x(t), x(t^+))$ or $\phi(x(t), x(t^-))$, but the context will make this evident.

Lemma 1. There is a constant M with the following property: given any t in $[0, 1]$ and (s, v) in $(x(t), \dot{x}(t)) + B$, then for all ζ in $\partial \phi(s, v)$ we have $|\zeta| \leq M$.

This follows from the hypothesis that ϕ is Lipschitz on bounded sets.

Lemma 2. There exist positive numbers δ_1 and δ_2 such that, for any t in $[0, 1]$, for any (s, v) in $(x(t), \dot{x}(t)) + \delta_1 B$, for any (α, β) in $\partial \phi(s, v)$, we have $|\beta| \geq \delta_2$.

Suppose the lemma false. Then for $i = 1, 2, \ldots$, there exist t_i in $[0, 1]$, (s_i, v_i) in $(x(t_i), \dot{x}(t_i)) + (1/i)B$, and (α_i, β_i) in $\partial \phi(s_i, v_i)$ such that $|\beta_i| < 1/i$. By taking subsequences we may assume that, for some t in $[0, 1]$, for some (s, v) and $(\alpha, 0)$ in R^{2n}, we have $t_i \to t$, $(s_i, v_i) \to (s, v)$, and $(\alpha_i, \beta_i) \to (\alpha, 0)$. It follows that $(s, v) = (x(t), \dot{x}(t))$. Furthermore, by the upper semicontinuity of the generalized gradient, we know that $(\alpha, 0)$ belongs to $\partial \phi(t)$. This contradicts the rank condition. (An evident modification of this argument handles the case in which t is a corner point.)

Now let any positive integer K be given, and choose ε_K so that, for any t in $[0, 1]$, the inequality

$$|(s, v) - (x(t), \dot{x}(t))| < \varepsilon_K$$

implies

$$\phi(s, v) \leq \phi(x(t), \dot{x}(t)) + \frac{1}{K}.$$

Such a choice is possible because ϕ is uniformly continuous on compact sets. We may suppose that ε_K is less than $1/K$, and also less than the ε occurring in the definition of weak local minimum.

Let us set

$$A_K(t) = \left\{ \zeta : \zeta \in \partial \phi(s, v), |(s, v) - (x(t), \dot{x}(t))| < \frac{1}{K} \right\},$$

and define, for t such that $\phi(t) > -1/K$,

$$G_K(t) := A_K(t)^* := \{ \gamma : \gamma \cdot \zeta \leq 0 \text{ for all } \zeta \text{ in } A_K(t) \}.$$

For t such that $\phi(t) \leq -1/K$, set $G_K(t) = R^{2n}$.

Now let K be larger than $1/\delta_1$. The following result then follows from Lemmas 1 and 2:

Lemma 3. There is a constant $N > 1$ such that the convex cone $G_K(t)$ has the following property for each t: given any s in R^n, there exists v in R^n such that $|v| \leq N|s|$ and $(s, v) \in G_K(t)$.

We now define a multifunction F_K from $[0, 1] \times R^n$ to R^n as follows:

$$F_K(t, s) = \left\{ v : |v| \leq \frac{\varepsilon_K}{2}, (s, v) \in G_K(t) \right\}.$$

It follows that for $|s| < \varepsilon_K/(2N)$, the multifunction $F_K(t, s)$ is nonempty, convex, compact, integrably bounded, and measurably Lipschitz with Lipschitz constant N.

Lemma 4. The arc $z(t) \equiv 0$ minimizes

$$f(x(1) + z(1))$$

over all arcs z satisfying $|z(t)| < \varepsilon_K/(2N)$ and the constraints

$$z(0) \in C_0 - x(0),$$

(1) $$z(1) \in C_1 - x(1)$$

$$\dot{z}(t) \in F_K(t, z(t)) \quad \text{a.e.}$$

Let any such z be given. Notice that it suffices to prove the inequality

(2) $$\phi(x + z, \dot{x} + \dot{z}) \leq 0 \quad \text{a.e.},$$

since then the fact that x is optimal for our original problem over a class of

4.5 The Multiplier Rule for Inequality Constraints

arcs including $x + z$ yields

$$f(x(1)) \leq f(x(1) + z(1)).$$

In proving (2) consider first any t such that $\phi(t) \leq -1/K$. Then (2) follows from the choice of ε_K, since we have

$$|(z(t), \dot{z}(t))| < \varepsilon_K.$$

Now let us consider any t such that $\phi(t) > -1/K$. We have

(3) $$\phi(x(t) + z(t), \dot{x}(t) + \dot{z}(t)) = \phi(t) + \int_0^1 Dg(\lambda) \, d\lambda,$$

where the Lipschitz function g is defined by

$$g(\lambda) = \phi(x(t) + \lambda z(t), \dot{x}(t) + \lambda \dot{z}(t)),$$

and $Dg(\lambda)$ exists a.e. It now suffices to prove that $Dg(\lambda)$ is nonpositive for λ in $[0, 1]$, since then Eq. (3) implies

$$\phi(x(t) + z(t), \dot{x}(t) + \dot{z}(t)) \leq \phi(t) \leq 0.$$

Let S be the set defined as follows:

$$S = \partial \phi(x(t) + \lambda z(t), \dot{x}(t) + \lambda \dot{z}(t)) \cdot (z(t), \dot{z}(t)).$$

According to Proposition 2.1.2(b), $\max\{\sigma : \sigma \in S\}$ is the lim sup of the quantity

$$\frac{\phi(x + \lambda z + h + \delta z, \dot{x} + \lambda \dot{z} + h' + \delta \dot{z}) - \phi(x + \lambda z + h, \dot{x} + \lambda \dot{z} + h')}{\delta},$$

as h and h' converge to 0 in R^n and δ decreases to 0. By definition, $Dg(\lambda)$ is equal to

$$\lim_{\delta \downarrow 0} \frac{\phi(x + \lambda z + \delta z, \dot{x} + \lambda \dot{z} + \delta \dot{z}) - \phi(x + \lambda z, \dot{x} + \lambda \dot{z})}{\delta},$$

whence

$$Dg(\lambda) \leq \max\{\sigma : \sigma \in S\} \leq 0.$$

This completes the proof of the lemma.

We now apply Theorem 3.5.2 to the problem in the statement of Lemma 4. If the function $H: [0, 1] \times R^n \times R^n \to R$ is defined as follows:

$$H(t, s, p) = \max\{p \cdot v : v \in F_K(t, s)\},$$

we deduce that an arc p_K and a scalar λ_K equal to 0 or 1 exist such that:
(a) $(-\dot{p}_K, 0) \in \partial H(t, 0, p_K)$ a.e.
(b) $p_K(0)$ is normal to $C_0 - x(0)$ at 0.
(c) For some vector ζ_K in $\partial f(x(1))$, $-p_K(1) - \lambda_K \zeta_K$ is normal to $C_1 - x(1)$ at 0.
(d) $|p_K(t)| + \lambda_K$ is never zero.

Lemma 5. For almost all t,

(4) $$\dot{p}_K \cdot s + p_K \cdot v \leq 0 \quad \text{for all } (s, v) \in G_K(t).$$

It suffices to show this for $|v|$ small, since $G_K(t)$ is a cone. Let t be such that (a) holds. Then we may suppose that v belongs to $F_K(t, s)$, and consequently

(5) $$H(t, s, p_K) \geq p_K \cdot v.$$

It is elementary to verify that the function $H(t, x, p)$ is concave in x. Along with (a), this implies that $-\dot{p}_K$ belongs to the super-differential at 0 of the concave function $x \to H(t, x, p_K)$. From this we deduce:

(6) $$H(t, s, p_K) - H(t, 0, p_K) \leq -\dot{p}_K \cdot s.$$

It follows from (a) that 0 belongs to $\partial_p H(t, 0, p_k)$, whence

(7) $$H(t, 0, p_K) = 0.$$

Now we combine (5)–(7) to obtain condition (4) and prove the lemma.

For future reference, let us note that from Lemma 5 and the definition of $G_K(t)$ we deduce:

(8) $$\dot{p}_K(t) \text{ and } p_K(t) \text{ are zero when } \phi(t) < -\frac{1}{K}.$$

We shall now be considering all the above as the integer K increases to infinity. By taking subsequences, we may assume that the λ_K are either all 0 or all equal to 1, and that the ζ_K converge to a vector ζ. From the easily proven fact that the function $x \to H(t, x, p)$ is Lipschitz with constant $N|p|$, along with (a), we deduce:

(9) $$|\dot{p}_K| \leq N|p_K| \quad \text{a.e.,}$$

where the constant N is independent of K. (Since G_K increases with K, N can only decrease as K increases.)

4.5 The Multiplier Rule for Inequality Constraints

Lemma 6. There exists an arc p and a scalar λ_0 equal to 0 or 1 satisfying (iii)(iv)(v) as well as:

(10) $\qquad \dot{p} \cdot s + p \cdot v \leqslant 0$ for all $(s, v) \in \partial\phi(t)^*$ a.e.

(11) $\qquad \dot{p}$ and p equal 0 when $\phi(t) < 0$.

Case 1. The λ_K are all 0. By scaling, we may assume that all the p_K are nonvanishing and $\|p_K\| = 1$, where the rescaled functions continue to satisfy conditions (8) and (9) and (b) and (c) (with $\lambda_K = 0$). In view of (9), the Dunford–Pettis criterion implies that $\{\dot{p}_K\}$ admits a subsequence converging weakly in L^1 to $\dot{p}$ (say). It follows for suitable subsequences that $\dot{p}$ is the derivative of an arc p to which p_K converges uniformly. Since p satisfies (9) and $\|p\| = 1$, (v) holds (with $\lambda_0 = 0$), as well as (iii) and (iv). Relation (11) is an immediate consequence of (8). In order to prove (10), note first that $G_K(t)$ increases to $\partial\phi(t)^*$ for any t such that $\phi(t) = 0$ (this uses the upper semicontinuity of $\partial\phi$). Furthermore, weak convergence preserves linear inequalities such as (4); the result follows.

Case 2. The λ_K are all equal to 1, and $\|p_K\|$ is bounded. In this case the argument is unchanged, except that the need to rescale initially is eliminated. The conclusions (iv)(v) hold with $\lambda_0 = 1$.

Case 3. The λ_K are all equal to 1, and $\|p_K\|$ is unbounded. We may assume that $\|p_K\|$ increases to infinity. We rescale the arcs p_K by dividing by $\|p_K\|$ (which is certainly nonzero for K large). The argument then continues as in Case 1, and we get conditions (iv) and (v) with $\lambda_0 = 0$, since $\lambda_K/\|p_K\|$ converges to 0. This proves the lemma.

In order to complete the proof of the theorem, it now suffices to infer (i) and (ii) from (10) and (11). The condition (10) says that $(\dot{p}, p)$ belongs to $(\partial\phi(t)^*)^*$, which is the closed convex cone generated by $\partial\phi(t)$. This has the following characterization, for any t such that $\phi(t) = 0$,

$$(\partial\phi(t)^*)^* = \{\lambda\zeta : \lambda \geqslant 0, \zeta \in \partial\phi(t)\},$$

because $\partial\phi(t)$ is a compact convex set not containing zero. Invoking the measurable selection theorem, we obtain (i) when $\phi(t) = 0$, and (ii) follows by simply setting $\lambda(t) = 0$ when $\phi(t) < 0$ and using (11). □

4.5.3 Remark
The case in which ϕ has an explicit dependence on t may be treated exactly as above with the additional hypotheses:
(i) $\phi(t, x, v)$ is a measurable function of t for each (x, v).
(ii) $\partial\phi(t, x, v)$ is an upper semicontinuous multifunction. (Here $\partial\phi$ refers to the generalized gradient with respect to (x, v).)

Both these hypotheses are automatically satisfied when ϕ is independent of t. In the case of t-dependence, (i) is required to ensure that the multifunction F_K constructed in the proof is measurable in t, while (ii) is necessary for the conclusions of Lemmas 2 and 6.

4.5.4 Example: Queen Dido and the Badlands

Queen Dido is given a length of rope with which to enclose a region along the shore, the latter being represented by the line $x = 0$ in the $t - x$ plane (see Figure 4.1). In doing this, she seeks to join the point $(0, 0)$ to the point $(1, 0)$ by a curve of length L lying in the half-plane $x \geqslant 0$ so as to maximize the area between the curve and the t axis. The problem as described to this point is classical, but let us now suppose that for a given positive α, the terrain $x > \alpha$ is inferior, and worth only half as much as the terrain $x < \alpha$. The return corresponding to a choice of border function $x(t)$ is then

(12) $$\int_0^1 g(x(t))\, dt,$$

where

$$g(x) = \begin{cases} x & \text{if } x \leqslant \alpha \\ (x + \alpha)/2 & \text{if } x \geqslant \alpha. \end{cases}$$

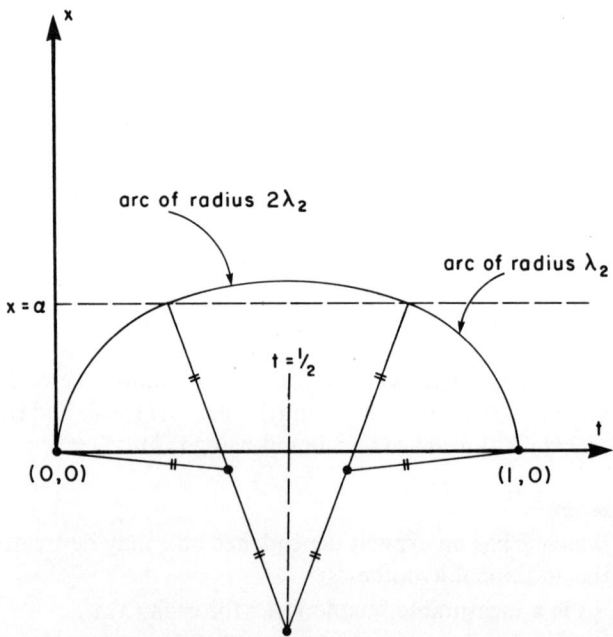

Figure 4.1 The solution to Queen Dido's problem.

4.5 The Multiplier Rule for Inequality Constraints

Her majesty is seeking to maximize the integral (12) (or minimize its negative) subject to the constraints

(13) $$x(0) = 0, \quad x(1) = 0$$

(14) $$\int_0^1 (1 + \dot{x}^2)^{1/2} \, dt = L.$$

Note that g is Lipschitz and nondifferentiable.

We proceed to place this problem within the framework of the problem treated in this section. We consider the two additional variables y and z and the constraints

(15) $$\phi_1(x, y, z, \dot{x}, \dot{y}, \dot{z}) = -\dot{y} - g(x) \leq 0$$

(16) $$\phi_2(x, y, z, \dot{x}, \dot{y}, \dot{z}) = -\dot{z} + (1 + \dot{x}^2)^{1/2} \leq 0$$

(17) $$x(0) = 0, \quad y(0) = 0, \quad z(0) = 0, \quad x(1) = 0, \quad z(1) = L,$$

and we define

(18) $$f(x(1), y(1), z(1)) = y(1).$$

It is not difficult to see that the problem of minimizing the quantity (18) subject to the constraints (15)–(17) is equivalent to Queen Dido's. The equality (14) has been replaced by

$$\int_0^1 (1 + \dot{x}^2)^{1/2} \, dt \leq L$$

in this transition, which makes no difference in as much as all the available cord will be used. In fact, it is clear from the nature of the problem that both constraints (15) and (16) will be active at all times.

In applying Corollary 1, note that the vector x is here replaced by (x, y, z), that $n = 3$ and $r = 2$. The sets C_0 and C_1 are $\{(0, 0, 0)\}$ and $\{0\} \times R \times \{L\}$, respectively. The functions involved are Lipschitz as required, and the sets $\partial \phi_1$ and $\partial \phi_2$ are seen to be:

$$\partial \phi_1(x, y, z, \dot{x}, \dot{y}, \dot{z}) = \{(\zeta, 0, 0, 0, -1, 0) : -\zeta \in \partial g(x)\}$$

$$\partial \phi_2(x, y, z, \dot{x}, \dot{y}, \dot{z}) = \left\{ \left(0, 0, 0, \frac{\dot{x}}{(1 + \dot{x}^2)^{1/2}}, 0, -1\right) \right\},$$

from which we infer that the conclusions of Corollary 1 are available to us for any piecewise-smooth solution, which we shall denote (x, y, z). We deduce the existence of nonnegative functions λ_1 and λ_2 such that the function $p(t)$ defined by $p(t) = [\lambda_2 \dot{x}/(1 + \dot{x}^2)^{1/2}, -\lambda_1, -\lambda_2]$ is absolutely continuous, and

(19) $$\dot{p}(t) \in \{-\lambda_1(t) \partial g(x)\} \times \{0\} \times \{0\}.$$

It follows that λ_1 and λ_2 are constant. From (iv) of the theorem we obtain:

$$\lambda_1 - \lambda_0 = 0.$$

If λ_0 is zero, then λ_1 is zero also, and it follows from (v) of the theorem that λ_2 must be strictly positive. But then (19) implies that the sign of $\dot{x}$ is constant, which is not possible except in the degenerate case $L = 1$.

We may thus suppose $\lambda_0 = 1 = \lambda_1$. Now if λ_2 were zero, (19) would yield $0 \in \partial g(x)$, which is never the case. Thus λ_2 is positive.

We have arrived at the following conclusions: $\dot{x}$ is continuous and satisfies the equation

(20) $$\frac{d}{dt}\left\{\frac{\dot{x}}{(1 + \dot{x}^2)^{1/2}}\right\} = \begin{cases} -\dfrac{1}{\lambda_2} & \text{if } x < \alpha \\ -\dfrac{1}{2\lambda_2} & \text{if } x > \alpha. \end{cases}$$

Note that $x(t)$ cannot equal α on any interval, since zero does not belong to $\partial g(\alpha)$.

The solutions to the two separate cases in Eq. (20) are well known, since each case is the type of equation that arises in the classical version of Queen Dido's problem. We find with no difficulty that x describes an arc of a circle of radius λ_2 for $x < \alpha$, and an arc of a circle of radius $2\lambda_2$ for $x > \alpha$. The requirement that these arcs meet with a common tangent (at $x = \alpha$) assures that to each λ_2 there corresponds at most one such configuration (see Figure 4.1).

Consequently, the optimal arc x is uniquely specified once λ_2 is known; λ_2 is determined by the condition that x is of given length L. Once the nature of x is known to be as described above, it is an easy exercise to obtain (implicit) equations for λ_2 (and the other parameters of the solution). These relations could then be used to calculate explicitly the solution x.

It is interesting to determine the nature of the information contributed by the new multiplier rule. Based on the known classical solution, one might expect the solution to the present problem to consist of an amalgam of circular arcs on either side of the line $x = \alpha$ (as indeed it does). The multiplier rule has served to rule out the possibility that x lies along the line $x = \alpha$ for any length of time, and has yielded the crucial facts that the radii of the upper and lower arcs are in the ratio of 2:1, and that these three pieces are smoothly joined. Thus the information obtained from its use is essentially global.

4.6 MULTIPLE INTEGRALS

Let Ω be a domain in R^m, and let $L: R \times R^m \to R$ be given. An important problem in the calculus of variations involves the minimization of the functional

$$\text{(1)} \qquad \int_\Omega L(x(\omega), \nabla x(\omega))\, d\omega$$

over a certain class X of functions $x(\cdot)$ defined on Ω. Specific examples of this problem (several involving nondifferentiability) are given in Ekeland and Temam (1976); see also Morrey (1966). We shall derive the first-order necessary condition for this problem under the following hypotheses:

(i) L satisfies the growth condition of Section 4.4, namely,

$$|\partial L(y, v)| \leq k_0 |L(y, v)| + k_1 |(y, v)| + c_0.$$

(ii) X is a subset of the Sobolev space $H_1^1(\Omega)$ which is closed under addition by elements of $C_0^\infty(\Omega)$, the set of C^∞ functions having compact support in Ω; that is, $x + \phi$ belongs to X for every x in X and ϕ in $C_0^\infty(\Omega)$.

The import of hypothesis (ii) is essentially that the functional (1) is being minimized over a subclass of functions x in $H_1^1(\Omega)$ having prescribed values on the boundary of Ω. Below, div signifies divergence. A weak solution $\hat{x}$ is one that solves the problem with the added constraint $|x(\omega) - \hat{x}(\omega)| + |\nabla x(\omega) - \nabla \hat{x}(\omega)| < \varepsilon$, for some $\varepsilon > 0$.

4.6.1 Theorem
Suppose that $\hat{x}$ is a weak solution to the above problem. Then there exists a function $p: \Omega \to R^m$ such that each component of p belongs to $L^1(\Omega)$, div p belongs to $L^1(\Omega)$, and

$$(\text{div } p(\omega), p(\omega)) \in \partial L(\hat{x}(\omega), \nabla \hat{x}(\omega)) \quad a.e. \text{ in } \Omega.$$

Proof. Let us define A to be the set of all measurable functions (ϕ, ζ) defined on Ω such that

$$(\phi(\omega), \zeta(\omega)) \in \partial L(\hat{x}(\omega), \nabla \hat{x}(\omega)) \quad \text{a.e. in } \Omega.$$

It follows from Lemma 2 of the proof of Theorem 4.4.1 that the set $\partial L(\hat{x}(\omega), \nabla \hat{x}(\omega))$ is bounded by an integrable function $k(\omega)$, so that A is a subset of $L^1(\Omega)^{m+1}$. Since A is convex and rather evidently strongly closed in $L^1(\Omega)^{m+1}$, A is also weakly closed. The Dunford–Pettis criterion implies that A is weakly compact, and the measurable selection theorem implies that A is nonempty. We summarize:

Lemma 1. *A is a nonempty convex weakly compact subset of $L^1(\Omega)^{m+1}$.*

Now let C be the (convex) set $C_0^\infty(\Omega)$ defined above, and define $f: A \times C \to R$ as follows (for $a = (\phi, \zeta)$):

$$f(a, c) := \int_\Omega \{\phi(\omega)c(\omega) + \langle \zeta(\omega), \nabla c(\omega)\rangle\}\, d\omega.$$

Note that f is well defined since ϕ and ζ are integrable. It is evident that for each c, $f(\cdot, c)$ is continuous on A in the weak topology.

Lemma 2.

$$\inf_{c \in C} \sup_{a \in A} f(a, c) = 0.$$

Since $f(\cdot, 0)$ is identically zero, it suffices to prove that for any given c in C, the following quantity is nonnegative:

(2) $$\sup_{a \in A} f(a, c).$$

We proceed to prove this. From the measurable selection theorem it follows that expression (2) is equal to

$$\int_\Omega \sup\{\phi c(\omega) + \langle \zeta, \nabla c(\omega)\rangle : (\phi, \zeta) \in \partial L(\hat{x}, \nabla \hat{x})\}\, d\omega$$

$$\geq \int_\Omega \limsup_{\lambda \downarrow 0} \frac{L(\hat{x} + \lambda c, \nabla \hat{x} + \lambda \nabla c) - L(\hat{x}, \nabla \hat{x})}{\lambda}\, d\omega,$$

which in turn (by Fatou's Lemma) majorizes

$$\limsup_{\lambda \downarrow 0} \frac{\left\{\int_\Omega L(\hat{x} + \lambda c, \nabla \hat{x} + \lambda \nabla c)\, d\omega - \int_\Omega L(\hat{x}, \nabla \hat{x})\, d\omega\right\}}{\lambda}$$

But this last expression is certainly nonnegative since (for λ small) $\hat{x}$ assigns a value to the functional (1) no higher than that assigned by $x := \hat{x} + \lambda c$. This completes the proof of the lemma.

We have set the stage for applying Aubin's lop-sided minimax theorem, which yields the existence of an element $a = (\phi, \zeta)$ of A such that

(3) $$\int_\Omega \{\phi c + \langle \zeta, \nabla c\rangle\}\, d\omega \geq 0 \quad \text{for all } c \text{ in } C.$$

Since $C = -C$, equality holds in (3) for all c. By definition, then, $\phi = \text{div }\zeta$. □

Chapter Five

Optimal Control

Tradition is the forgetting of the origins.

EDMUND HUSSERL, *The Origin of Geometry*

What I tell you three times is true.

LEWIS CARROLL, *The Hunting of the Snark*

Most applications of dynamic optimization today take place in the context of our third main paradigm, the optimal control problem in standard or Pontryagin form. It is a formulation that has proven to be a natural one in the modeling of a variety of physical, economic, and engineering problems. Most of the existing theory of optimal control has been developed in this context; for the most part it deals with a (pseudo-)Hamiltonian distinct from the one that Chapters 3 and 4 (for an appropriate reformulation of the given problem) would consider. An advantage of the approach of this chapter is that when the data of the problem are smooth, it is possible to express the results entirely in terms of derivatives and ordinary differential equations. This is the case in particular for the best known result in the theory, the set of necessary conditions that is known as the (Pontryagin) maximum principle. The context of this chapter will be clearer to the reader who has read Sections 1.3 and 1.4.

We shall develop in this chapter general results on existence, on necessary and sufficient conditions, on controllability and sensitivity, within the standard framework. It will be clear that there are strong links with the preceding chapters, whose developments are in fact the base from which the results of this one are derived. We shall illustrate both the abstract and specific advantages that can result from being able to study a problem from the points of view of its various formulations.

5.1 CONTROLLABILITY

The Dynamics

We shall be concerned with a given interval $[a, b]$ in R. We are given a multifunction U mapping $[a, b]$ to the nonempty subsets of R^m. A *control* is a measurable selection $u(\cdot)$ for U; that is, a measurable function u satisfying

$$u(t) \in U(t) \quad \text{a.e.}$$

We are given a function $\phi: [a, b] \times R^n \times R^m \to R^n$. An arc x on $[a, b]$ which satisfies

$$\dot{x}(t) = \phi(t, x(t), u(t)) \quad \text{a.e.}$$

is called a *trajectory* (corresponding to the control u). We subject trajectories to the following *state constraint*:

$$g(t, x(t)) \leq 0 \quad \text{a.e.}$$

In addition, we limit trajectories to those lying in a given tube Ω (see Section 3.1). Trajectories x which lie in Ω and satisfy the state constraint are termed *admissible*.

5.1.1 Basic Hypotheses

We impose the following conditions on the data (cf. Hypotheses 3.2.2):
  (i) For each x in R^n, the function $\phi(\cdot, x, \cdot)$ is $\mathcal{L} \times \mathcal{B}$ measurable. There exists an $\mathcal{L} \times \mathcal{B}$ measurable real-valued function k defined on $\text{Gr}(U)$ such that, for each (t, u) in $\text{Gr}(U)$, the function $\phi(t, \cdot, u)$ is Lipschitz on Ω_t, of rank $k(t, u)$.
  (ii) $\text{Gr}(U)$ is $\mathcal{L} \times \mathcal{B}$ measurable (see Section 4.1).
  (iii) g is upper semicontinuous on Ω, and for each t in $[a, b]$, the function $g(t, \cdot)$ is Lipschitz on Ω_t of rank K_g (not depending on t).

The Attainable Set

Let C be a subset of R^n. The *attainable set* from C, denoted $\mathcal{C}[C]$, is the set of all points $x(b)$, where x is an admissible trajectory corresponding to some control u, and where x satisfies the initial condition $x(a) \in C$.

The following gives necessary conditions for x to be a "boundary trajectory." The difficulty of the result stems largely from the fact that the attainable set

5.1 Controllability

need not be closed (or convex). (Recall the set $\bar{\partial}_x g$ which was defined for Corollary 2 of Theorem 3.2.6.) Let the set C be closed.

5.1.2 Theorem

Let the admissible trajectory x satisfy $x(a) \in C$, where x corresponds to the control u. Assume that the function $t \to k(t, u(t))$ is integrable. Let $\theta: R^n \to R^d$ be a function which is Lipschitz near $x(b)$. If $\theta(x(b))$ lies in the boundary of $\theta(\mathcal{R}[C])$, then there exist an arc p, a nonnegative Radon measure μ, a measurable function γ, and a vector v in R^d such that:

(i)
$$-\dot{p}(t) \in \partial_x \phi(t, x(t), u(t))^* \left[p(t) + \int_{[a,t)} \gamma(s) \mu(ds) \right] \quad \text{a.e.}$$

(where * denotes transpose).

(ii)
$$\max\left\{ \left\langle p(t) + \int_{[a,t)} \gamma(s) \mu(ds), \phi(t, x(t), w) \right\rangle : w \in U(t) \right\}$$
$$= \left\langle p(t) + \int_{[a,t)} \gamma(s) \mu(ds), \phi(t, x(t), u(t)) \right\rangle \quad \text{a.e.}$$

(iii) $\gamma(t) \in \bar{\partial}_x g(t, x(t))$ μ-a.e., and μ is supported on the set $\{t \in [a, b] : g(t, x(t)) = 0\}$.
(iv) $p(a) \in N_C(x(a))$.
(v) $p(b) + \int_{[a,b]} \gamma(s) \mu(ds) \in \partial\theta(x(b))^* v$.
(vi) $\|\mu\| + |v| = 1$.

Proof of Theorem 5.1.2. We shall proceed under two interim hypotheses whose removal will be the final step in the proof:

Interim Hypotheses

(i) For each t, $U(t)$ consists of finitely many points.
(ii) There is an integrable function $c(\cdot)$ such that for almost all t, for all w in $U(t)$,
$$|\phi(t, x(t), w)| \leq c(t), \quad k(t, w) \leq c(t),$$
where $k(\cdot, \cdot)$ is the function whose existence was postulated in Hypotheses 5.1.1.

Step 1. Let i be any positive integer, and let ζ be a point in $\theta(x(b)) + (1/i^2)B$ such that $\zeta \notin \theta(\mathcal{R}[C])$. (Such a point ζ exists because $\theta(x(b))$ is a boundary point.) Let $\varepsilon > 0$ be such that the tube $T(x; 2\varepsilon)$ is contained in Ω, and let us define W as the set of all pairs (w, z), where w is an admissible control, z a point in C, and where the initial-value problem

$$\dot{y} = \phi(t, y, w(t)), \quad y(a) = z$$

admits a solution y on $[a, b]$ lying in the closed ε-tube about x and satisfying

$$g(t, y(t)) \leq 0 \quad \text{a.e.}$$

Note that W is nonempty. In fact, $(u, x(a))$ belongs to W. Note also that since the function

$$y \to \phi(t, y, w(t))$$

is Lipschitz with constant $c(t)$, the arc y (when it exists) is unique. Let us define $F: W \to R$ as follows:

$$F(w, z) = |\zeta - \theta(y(b))|,$$

where y is the state corresponding to w and z.

We define a metric Δ on W via

$$\Delta((w_1, z_1), (w_2, z_2)) = \text{meas}\{t : w_1(t) \neq w_2(t)\} + |z_1 - z_2|.$$

Lemma 1. Δ is a metric, and relative to Δ, W is a complete metric space and F a continuous functional. If y_j and y_0 are the arcs corresponding to (w_j, z_j) and (w_0, z_0), and if (w_j, z_j) converges in W to (w_0, z_0), then $\|y_j - y_0\|$ converges to 0.

That Δ is a metric is left as an exercise. Let us show that a Cauchy sequence $\{(w_j, z_j)\}$ in W necessarily converges to a limit (w_0, z_0) in W. It suffices to prove that this is so for a subsequence, since the sequence is Cauchy, so there is no loss of generality in supposing (by relabeling) that

$$\Delta((w_j, z_j), (w_{j+1}, z_{j+1})) \leq 2^{-j}.$$

Set

$$E_j = \bigcup_{k \geq j} \{t : w_k(t) \neq w_{k+1}(t)\}.$$

We have $E_{j+1} \subset E_j$ and

$$\text{meas}(E_j) \leq \sum_{k \geq j} 2^{-k} = 2^{1-j}.$$

5.1 Controllability

Define w_0 by

$$w_0(t) = w_k(t) \quad \text{for } t \notin E_k.$$

It follows easily that w_0 is well defined a.e., and is a control. We also have $|z_j - z_{j+1}| \leq 2^{-j}$, whence z_j converges to an element z_0 belonging to C (since R^n is complete and C is closed). To show that (w_0, z_0) belongs to W, we need to verify that there is a trajectory y_0 on $[a, b]$ corresponding to (w_0, z_0) which lies in the closed ε-tube about x and which satisfies the state constraint. To this end, define the multifunction $\Gamma(t, y) := \{\phi(t, y, w_0(t))\}$, and let y_j be the trajectory corresponding to (w_j, z_j). If A_j is the set $\{t : w_j(t) = w_0(t)\}$, then note that

$$\dot{y}_j(t) \in \Gamma(t, y_j(t)) \quad \text{for } t \text{ in } A_j$$

and that $\text{meas}(A_j) \to (b - a)$. Theorem 3.1.7 applies to yield the desired trajectory y_0 corresponding to (w_0, z_0).

We turn now to the other assertions of Lemma 1. It clearly suffices to prove the last one. Let Γ be as above, and ρ its associated function (see 3.1). Note that $\rho(t, y_j, \dot{y}_j) = 0$ when t belongs to A_j, while $\rho(t, y_j, \dot{y}_j)$ is bounded by $2c(t)[1 + \varepsilon]$ (in view of the interim hypotheses) at all times. Since $\text{meas}(A_j) \to (b - a)$, it follows that for any positive δ, for all j sufficiently large, one has $\rho_\Gamma(y_j) < \delta$. We now apply Theorem 3.1.6 to deduce the existence of a trajectory $\tilde{y}_j$ for Γ such that $\tilde{y}_j(a) = y_j(a) = z_j$ and such that $|y_j(t) - \tilde{y}_j(t)| \leq K\delta$. Observe the inequality

$$|\dot{y}_0(t) - \dot{\tilde{y}}_j(t)| = |\phi(t, y_0, w_0) - \phi(t, \tilde{y}_j, w_0)|$$

$$\leq c(t)|y_0(t) - \tilde{y}_j(t)|.$$

Together with $|y_0(a) - \tilde{y}_j(a)| = |y_0(a) - z_j| = |z_0 - z_j|$, this yields

$$|y_0(t) - \tilde{y}_j(t)| \leq |z_0 - z_j| \exp\left(\int_a^b c(\tau)\, d\tau\right) = :c_1|z_0 - z_j|.$$

Thus, for large j, one has

$$|y_0(t) - y_j(t)| \leq |y_0(t) - \tilde{y}_j(t)| + |\tilde{y}_j(t) - y_j(t)|$$

$$\leq c_1|z_0 - z_j| + K\delta,$$

which demonstrates the convergence of y_j to y_0, and proves Lemma 1.

Observe that $F(u, x(a)) < i^{-2}$; since F is nonnegative, it follows that

$$F(u, x(a)) < \inf_W F + i^{-2}.$$

Lemma 1 justifies the application of Theorem 7.5.1 to conclude that there exists an element $(\hat{w}, \hat{z})$ of W such that

(1) $$\Delta((\hat{w}, \hat{z}), (u, x(a))) \leq i^{-1}$$

and such that $(\hat{w}, \hat{z})$ minimizes

$$F(w, z) + i^{-1}\Delta((w, z), (\hat{w}, \hat{z}))$$

over W. Let $\hat{y}$ be the state corresponding to $(\hat{w}, \hat{z})$. Then the last assertion may be reworded as follows:

Lemma 2. Let (w, y) be any admissible control/state pair on $[a, b]$ for which

$$y(a) \in C, \qquad g(t, y(t)) \leq 0$$

$$|y(t) - x(t)| \leq \varepsilon.$$

Then

$$|\zeta - \theta(y(b))| + i^{-1}\operatorname{meas}\{t : w \neq \hat{w}\} + i^{-1}|y(a) - \hat{z}| \geq |\zeta - \theta(\hat{y}(b))|.$$

Step 2. We now proceed to interpret our present situation as one to which the necessary conditions of Theorem 3.2.6 may be applied. To this end, some complicated bookkeeping will be required. We shall consider states Y having components $[y^1, y^2, y^3]$ in $R^m \times R \times R^n$, the last of which will correspond to the one we already have. We define a multifunction

$$\hat{F}(t, Y) := \left\{ \left[\frac{w}{1 + |w|}, \chi_t(w), \phi(t, y^3, w) \right] : w \in U(t) \right\},$$

where $\chi_t(w) = 1$ if $w \neq \hat{w}(t)$, and 0 otherwise. The set C_0 is defined as follows:

$$C_0 := \{[y^1, y^2, y^3] : y^3 \in C\},$$

Note that any trajectory $Y = [y^1, y^2, y^3]$ for $\hat{F}$ emanating from C_0 is of the form

(2) $$Y(t) = \left[\beta_1 + \int_a^t \frac{w}{1 + |w|} ds, \beta_2 + \int_a^t \chi_s(w) \, ds, y(t) \right],$$

where (w, y) is an admissible control/state pair and $y(a) \in C$. Let f_0, f_1 be defined as follows:

$$f_0(Y) = i^{-1}|y^3 - \hat{z}| - i^{-1}y^2$$

$$f_1(Y) = |\zeta - \theta(y^3)| + i^{-1}y^2.$$

5.1 Controllability

It follows then from Lemma 2 that the trajectory $\hat{Y} = [\hat{y}^1, \hat{y}^2, \hat{y}^3]$ given by Eq. (2) with $w = \hat{w}$, $y = \hat{y}$, $\beta_1 = 0$, $\beta_2 = 0$ minimizes

$$f_1(Y(b)) + f_0(Y(a))$$

over the trajectories for $\hat{F}$ which originate in C_0 and which satisfy

(3) $$g(t, y^3(t)) \leq 0$$

and

(4) $$|y^3(t) - x(t)| \leq \varepsilon.$$

Let $\hat{f} := f_1(\hat{Y}(b)) + f_0(\hat{Y}(a)) = |\zeta - \theta(\hat{y}(b))|$, and for any arc Y, set

$$G(Y) := \max_t g(t, y^3(t)).$$

Then it follows that $\hat{Y}(\cdot)$ minimizes

(5) $$\max\{f_1(Y(b)) + f_0(Y(a)) - \hat{f}, G(Y)\}$$

over the trajectories Y for $\hat{F}$ such that

(6) $$Y(a) \in C_0$$

and which satisfy the constraints (3) and (4).

When i is sufficiently large, (1) implies that any y near $\hat{y}$ will automatically satisfy (4), so that we may conclude that $\hat{Y}$ provides a strong local minimum (i.e., a local minimum with respect to the supremum norm) for the functional (5) subject to the constraint (6) (and subject of course to being a trajectory for $\hat{F}$). Because any trajectory for co $\hat{F}$ can be uniformly approximated by one for $\hat{F}$ having the same initial point (see Corollary, Theorem 3.1.6), this remains true if $\hat{F}$ is replaced by co $\hat{F}$.

Step 3. We have just seen that $\hat{Y}$ provides a strong local minimum for (5) over the trajectories for co $\hat{F}$ satisfying (6). This is precisely the situation for which Corollary 2, Theorem 3.2.6, was designed. Its hypotheses are clearly present. (Note that co $\hat{F}$ is closed because $\hat{F}(t, Y)$ consists of finitely many points.

We shall denote by P points of the form $[p^1, p^2, p^3]$, in agreement with our preceding notation. The Hamiltonian $\tilde{H}$ for the problem in question is

$$\tilde{H}(t, Y, P) := \max_{w \in U(t)} \left\{ \frac{\langle p^1, w \rangle}{1 + |w|} + \chi_t(w) p^2 + \langle p^3, \phi(t, y^3, w) \rangle \right\}.$$

A straightforward analysis of the Hamiltonian inclusion of Theorem 3.2.6 with the aid of Theorem 2.8.6 yields the conclusion that the quantity $[-\dot{p}^3, 0]$ (we have looked at only the y^3 and p^2 components of $\partial \tilde{H}$) is expressible a.e. as a convex combination of points of the form $[A^*\{p^3 + \int_{[a,t)} \gamma \, d\mu\}, r]$, where

$$r = \lim_{j \to \infty} \chi_t(w_j)$$

$$A = \lim_{j \to \infty} D_y \phi(t, y_j, w_j),$$

and where w_j converges to a point in $U(t)$ providing the maximum defining $\tilde{H}$, and where y_j converges to $y(t)$. Since the various values of r must combine in the convex combination in question to give 0, the only relevant sequences $\{w_j\}$ are those which are equal to $\hat{w}(t)$ for large j. We deduce (setting $p^3 = p$)

(7) $$-\dot{p} \in \partial_y \phi(t, \hat{y}, \hat{w})^* \left\{ p + \int_{[a,t)} \gamma \, d\mu \right\}.$$

The transversality conditions of Theorem 3.2.6 and the other components of the differential inclusion imply that p^1 is identically 0, and that p^2 is a constant whose absolute value is at most $1/i$. We deduce from the maximization cited above that

(8)
$$\left\langle p + \int_{[a,t)} \gamma \, d\mu, \hat{y} \right\rangle \geq \max \left\{ \left\langle p + \int_{[a,t)} \gamma \, d\mu, \phi(t, \hat{y}, w) \right\rangle : w \in U(t) \right\} \quad \text{a.e.}$$

In addition, there is a nonnegative scalar λ such that

$$p^3(b) + \int_{[a,b]} \gamma \, d\mu \in -\lambda \, \partial l(\hat{y}(b)),$$

where

$$l(y) := |\zeta - \theta(y)|.$$

Note that $\theta(\hat{y}(b)) \neq \zeta$, since ζ does not belong to $\theta(\mathcal{C}[C])$ by choice. This allows us to conclude from the above (and from Theorem 2.6.6) that

(9) $$p(b) + \int_{[a,b]} \gamma \, d\mu \in \partial \theta(\hat{y}(b))^* v,$$

where $-v = \lambda(\theta(\hat{y}(b)) - \zeta)/|\theta(\hat{y}(b)) - \zeta|$. Note that since $\|\mu\| + \lambda = 1$, we have

(10) $$\|\mu\| + |v| = 1.$$

5.1 Controllability

The final condition involves C_0:

$$P(a) \in r\, \partial d_{C_0}(\hat{Y}(a)) + \lambda \partial f_0(\hat{Y}(a)),$$

which implies

(11) $$p(a) \in r\partial d_C(\hat{y}(a)) + i^{-1}B.$$

It is now time to recall that all the above has been obtained for any (sufficiently large) integer i, so that in the key relations (7)–(11), the quantities $p, \gamma, \mu, \hat{y}, v, \hat{w}, \lambda$ (but not r) all actually depend on i. Note that as $i \to \infty$, the measure of the set

$$\{t: \hat{w}(t) \neq u(t)\}$$

goes to 0 by (1), as does $|\hat{y}(a) - x(a)|$. It follows from Lemma 1 that $\hat{y}$ converges uniformly to x. Of course, we may assume by taking subsequences that v_i converges with i to a vector v in R^d. Invoking Theorem 3.1.7 and Proposition 3.1.8 as in Step 4 of the proof of Theorem 3.2.6, we may choose further subsequences so that (7)–(10) hold in the limit (with $\hat{y} = x, \hat{w} = u$), and also

(12) $$p(a) \in r\partial d_C(x(a)).$$

Note that these give the conclusions of the theorem.

Step 4. It now remains to remove the Interim Hypotheses. Let us begin by showing how the validity of the theorem as proven above implies the same result in the absence of Interim Hypothesis (i) (i.e., we continue to assume (ii) for now). Note that in the previous step it follows that there is an a priori bound (M, say) on $|p(t) + \int_{[a,t)} \gamma\, d\mu|$ independent of U.

Let S_j be an increasing family of finite subsets of MB such that $S_j + j^{-1}B \supset MB$. For each element s of S_j, choose an element w_s of $U(t)$ such that

$$h(t, x(t), s) \leq \langle s, \phi(t, x(t), w_s) \rangle + j^{-1},$$

where

$$h(t, y, s) := \sup\{\langle s, \phi(t, y, w) \rangle : w \in U(t)\}.$$

Aumann's Selection Theorem 4.1.1 implies that we can choose w_s as a measurable function of t. Let $U_j(t) = \{w_s(t): s \in S_j\} \cup \{u(t)\}$. Note that all the hypotheses of the theorem are satisfied if U_j replaces U, including Interim Hypotheses (i) and (ii). In consequence, we apply the result already proven to obtain the existence of p, γ, μ, v (all depending on j) such that (for $U = U_j$)

the conclusions of the theorem hold, and in particular

$$\left\langle p + \int_{[a,t)} \gamma\, d\mu,\, \dot{x} \right\rangle = h_j\left(t, x, p + \int_{[a,t)} \gamma\, d\mu\right) \quad \text{a.e.}$$

where h_j is defined as

$$h_j(t, y, s) = \max\{\langle s, \phi(t, y, w)\rangle : w \in U_j(t)\}.$$

Let s be an element of S_j such that

$$p + \int_{[a,t)} \gamma\, d\mu \in s + j^{-1} B,$$

and note that $c(t)$ is a Lipschitz constant for $h(t, x(t), \cdot)$ and $h_j(t, x(t), \cdot)$. Thus we have

$$h_j\left(t, x, p + \int_{[a,t)} \gamma\, d\mu\right) \geq h_j(t, x, s) - j^{-1} c(t)$$

$$\geq \langle s, \phi(t, x, w_s)\rangle - j^{-1} c(t)$$

$$\geq h(t, x, s) - j^{-1} - j^{-1} c(t)$$

$$\geq h\left(t, x, p + \int_{[a,t)} \gamma\, d\mu\right) - 2 j^{-1} c(t) - j^{-1}.$$

It follows that

$$(13) \quad \left| \left\langle p + \int_{[a,t)} \gamma\, d\mu,\, \dot{x} \right\rangle - h\left(t, x, p + \int_{[a,t)} \gamma\, d\mu\right) \right| \leq j^{-1} [2c(t) + 1].$$

Since (v) of the theorem implies an a priori bound on $p(b)$, another straightforward application of Theorem 3.1.7 leads to p, γ, μ, v satisfying the required conditions in the limit as $j \to \infty$ (it is (13) that yields (ii) of the theorem).

The final touch is to show that Interim Hypothesis (ii) can be deleted. For any integer j, define

$$U_j(t) = \{w \in U(t) : k(t, w) \leq k(t, u(t)) + j, |\phi(t, x(t), w)| \leq |\dot{x}(t)| + j\}.$$

Note that $U_j(\cdot)$ is an increasing sequence of multifunctions, and that any element of $U(t)$ belongs to $U_j(t)$ for j large enough.

If we replace U by U_j, all the hypotheses of the theorem are satisfied, and also Interim Hypothesis (ii). We may therefore apply the theorem as proven

5.1 Controllability

above to deduce the existence of p, γ, μ, and v (all depending on j), satisfying the conclusions of the theorem for $U = U_j$, and in particular

(14)
$$\left\langle p + \int_{[a,t)} \gamma\, d\mu, \dot{x} \right\rangle \geq \left\langle p + \int_{[a,t)} \gamma\, d\mu, \phi(t, x, w) \right\rangle \quad \text{for all } w \text{ in } U_j(t), \text{ a.e.}$$

As in the preceding, Theorem 3.1.7 applies and yields p, γ, μ, and v which continue to satisfy (i) (iii) (iv) (v) (vi) of the theorem. Any w in $U(t)$ eventually belongs to $U_j(t)$ for all j large enough. This fact, together with (14), implies (ii) for the limiting data. □

5.1.3 Remark
When $\theta(\mathcal{R}[C])$ is closed, our argument involving perpendiculars can be used to derive the further conclusion that the vector V is normal to $\theta(\mathcal{R}[C])$ at $\theta(x(b))$.

Null Controllability

Let us consider the dynamics in the *autonomous* case (i.e., the case in which ϕ, g, and U have no dependence on t). To simplify matters, let us further suppose that 0 belongs to U, that $\phi(0, 0) = 0$, and that $g(0)$ is negative. It follows that the arc $x = 0$ is an admissible trajectory for the control $u = 0$, and that the state constraint is inactive along x.

The set $\mathcal{N}$ of null controllability is the set of all points α such that, for some $T > 0$, for some control w on $[0, T]$ and its corresponding admissible state y, one has $y(0) = \alpha$, $y(T) = 0$. Thus $\mathcal{N}$ is the set of points which can be steered to the origin in finite time; note that 0 belongs to $\mathcal{N}$. In applications, it is frequently of interest to know whether $\mathcal{N}$ contains a neighborhood of zero. Theorem 5.1.2 can be used to construct conditions assuring this. To illustrate, let us suppose that ϕ is Borel measurable and strictly differentiable as a function of the state variable, and let us set $A = D_x\phi(0, 0)$. Let U be a Borel set in R^m.

5.1.4 Proposition
Suppose that for some $T > 0$, one has

$$0 \in \text{int co}\{e^{At}\phi(0, w) : w \in U, 0 \leq t \leq T\}.$$

Then $0 \in \text{int } \mathcal{N}$.

Proof. If the conclusion is false, then (for $[a, b] = [0, T]$) the set $\mathcal{R}[\{0\}]$ for the "time-reversed" system $\dot{y} = -\phi(y, w)$ admits 0 as a boundary point (for note that this set is a subset of $\mathcal{N}$). The hypotheses of Theorem 5.1.2 (with

θ = identity, $x = u = 0$) are present, so we conclude that a nonvanishing arc p exists such that

$$-\dot{p} = -A^*p \quad \text{a.e.}$$

$$\min\{\langle p(t), \phi(0, w)\rangle : w \in U\} = 0, \quad 0 \leq t \leq T \text{ a.e.}$$

We deduce that $p(t) = e^{A^*t}\alpha$ for some nonzero vector α, and that, for all t in $[0, T]$ and w in U, one has

$$\langle \alpha, e^{At}\phi(0, w)\rangle = \langle e^{A^*t}\alpha, \phi(0, w)\rangle \geq 0.$$

This is the required contradiction. □

The following is a version of a classic result in control theory.

Corollary

In addition to the hypotheses of the proposition, suppose that $\phi(0, \cdot)$ is differentiable at 0 and that 0 belongs to int U. Let $B = D_u\phi(0, 0)$. Then, if the rank of the $n \times nm$ matrix

$$\begin{bmatrix} B & AB & A^2B & \cdots & A^{n-1}B \end{bmatrix}$$

is maximal, 0 belongs to int $\mathfrak{N}$.

Proof. We shall show that the interiority hypothesis of Proposition 5.1.4 must hold. If it did not, there would be a nonzero vector v in R^n such that

$$\langle v, e^{At}\phi(0, w)\rangle \leq 0 \quad \text{for } 0 \leq t \leq T, \quad \text{for } w \in U.$$

Consequently, the function $w \to \langle v, e^{At}\phi(0, w)\rangle$ attains a local maximum at $w = 0$, whence $v^*e^{At}B = 0$. By alternately setting $t = 0$ and differentiating in this identity, one derives $v^*A^iB = 0$ for any nonnegative integer i. It follows that v annihilates every column of the matrix defined in the corollary, contradicting the fact that it has maximum rank. □

5.2 THE MAXIMUM PRINCIPLE

We now pose the standard optimal control problem as it is best known to the world, and we derive a general version of the well-known necessary conditions which go under the label "the (Pontryagin) maximum principle."

The Optimal Control Problem P_C

Given the dynamics, control, and state constraints described in the preceding section, and functions $f: R^n \to R$ and $F: [a, b] \times R^n \times R^m \to R$, the problem

5.2 The Maximum Principle

P_C is that of minimizing

$$f(x(b)) + \int_a^b F(t, x(t), u(t))\, dt$$

over the controls u and corresponding admissible trajectories x satisfying

$$x(a) \in C_0, \quad x(b) \in C_1.$$

Hypotheses 5.1.1 remain in force. As for the new data, we assume that C_0, C_1 are closed, that f is locally Lipschitz, and that F satisfies the same condition as ϕ (i.e., 5.1.1(i)).

The pseudo (or Pontryagin) Hamiltonian $H_P: [a, b] \times R^n \times R^n \times R^m \times R \to R$ is the function

$$H_P(t, x, p, u, \lambda) := \langle p, \phi(t, x, u) \rangle - \lambda F(t, x, u).$$

(In the following theorem, when the state constraint is inactive along the solution x (i.e., when $g(t, x(t)) < 0$), the statement of the theorem simplifies due to the fact that all mention of μ (and the corresponding integrals) may be excised. This may be a useful reduction for first reading.)

5.2.1 Theorem (Maximum Principle)

Let (x, u) solve P_C, and assume that the function $t \to k(t, u(t))$ (see 5.1.1(i)) is integrable. Then there exist a scalar λ equal to 0 or 1, an arc p, a measurable function γ, and a nonnegative measure μ on $[a, b]$ such that:

(i) p satisfies the "adjoint equation"

$$-\dot{p}(t) \in \partial_x H_P\left(t, x(t), p(t) + \int_{[a,t)} \gamma(s)\mu(ds), u(t), \lambda\right) \quad \text{a.e.}$$

(ii) H_P is maximized at $u(t)$:

$$\max\left\{ H_P\left(t, x(t), p(t) + \int_{[a,t)} \gamma(s)\mu(ds), w, \lambda\right) : w \in U(t) \right\}$$

$$= H_P\left(t, x(t), p(t) + \int_{[a,t)} \gamma(s)\mu(ds), u(t), \lambda\right) \quad \text{a.e.}$$

(iii) $\gamma(t) \in \bar{\partial}_x g(t, x(t))$ μ-a.e., and μ is supported on the set

$$\{t: g(t, x(t)) = 0\}.$$

(iv) *For some ζ in $\partial f(x(b))$, the following transversality conditions hold*:

$$p(a) \in N_{C_0}(x(a)), \qquad -\lambda\zeta - p(b) - \int_{[a,b]} \gamma(s)\mu(ds) \in N_{C_1}(x(b)).$$

(v) $\|p\| + \|\mu\| + \lambda > 0$.

5.2.2 Remark

The measurability of k and the integrability of $k(t, u(t))$ can be guaranteed by other (perhaps more familiar) types of hypotheses. For example, these properties follow if ϕ and F possess derivatives ϕ_x and F_x which are continuous in (t, x, u), and if $u(\cdot)$ is essentially bounded (these are standard hypotheses for smooth versions of the maximum principle). A second alternate hypothesis would be the growth conditions (see Section 4.4):

$$|\partial_x \phi| \leq k|\phi| + c, \qquad |\partial_x F| \leq k|F| + c.$$

The maximum principle does not apply to cases in which the control set U depends explicitly upon the state: $U = U(t, x)$. Such problems can be formulated as differential inclusions (see Filippov's Lemma, Section 5.5).

Proof. We shall derive the result as an immediate consequence of Theorem 5.1.2. Let $\tilde{s} = [s_0, s_1, s_2]$ denote points in $R \times R^n \times R^n$,

$$C = \{\tilde{s} : s_0 = 0, s_1 \in C_0, s_2 \in C_1\}$$

$$\tilde{\phi}(t, \tilde{s}, w) = [F(t, s_1, w), \phi(t, s_1, w), 0]$$

$$\tilde{g}(t, \tilde{s}) = g(t, s_1)$$

$$\theta(\tilde{s}) = [s_0 + f(s_1), s_2 - s_1].$$

In the notation of Theorem 5.1.2, we claim that the admissible trajectory $\tilde{x}(t) = [\int_a^t F(\tau, x, u) \, d\tau, x(t), x(b)]$ is such that $\theta(\tilde{x}(b))$ lies in the boundary of $\theta(\mathcal{C}[C])$. If this were false, there would exist an admissible trajectory $\tilde{y}$ (and corresponding control v) such that for some $\varepsilon > 0$, one has $y_0(b) + f(y_1(b)) = x_0(b) + f(x_1(b)) - \varepsilon$ and $y_1(b) = y_2(b)$. It follows easily that $y = y_1$ would then be an admissible trajectory for P_C which is better for P_C than x, a contradiction.

Theorem 5.1.2 therefore applies to the situation above. A straightforward translation of its conclusions yields the maximum principle. □

Variants of the Maximum Principle

Additional conclusions can be added to those of Theorem 5.2.1 if more is assumed about the nature of the problem and the data. Notable examples are the cases in which (1) the data depend upon t in a more than merely measurable way, and (2) the interval $[a, b]$ is itself a choice variable. The transformation techniques introduced in Section 3.6 for the differential inclusion problem adapt in a straightforward manner to the control problem, so that appropriate versions of the maximum principle for problems such as the above can be derived directly from Theorem 5.2.1.

As an illustration, we state the result for the autonomous problem in which a is fixed but b is free, subject to the constraint $(b, x(b)) \in S$, where S is a given closed subset of $R \times R^n$. (The earlier case is that in which $S = \{b\} \times C_1$.) Thus ϕ, F, g, and U have no explicit dependence on t, and we seek to minimize

$$f(b^1, x(b^1)) + \int_a^{b^1} F(x(t), u(t))\, dt$$

over the state/control pairs (x, u) for (ϕ, U) defined on an interval $[a, b^1]$, subject to the constraints $g(x(t)) \leq 0$ ($a \leq t \leq b^1$), $x(a) \in C_0$, $(b^1, x(b^1)) \in S$. Let (x, u) solve this problem, where x and u are defined on $[a, b]$, and suppose that the hypotheses of Theorem 5.2.1 hold, where we also assume Lipschitz behavior of f in the b^1 variable.

5.2.3 Theorem

There exist p, γ, μ, and λ (equal to 0 or 1) satisfying conditions 5.2.1(i), (ii), (iii), and (v), as well as the further conditions:

(i) *There is a constant h such that*

$$H_P\left(x(t), p(t) + \int_{[a,t)} \gamma\, d\mu, u(t), \lambda\right) = h \quad \text{a.e.}$$

(ii)

$$\left[h, -p(b) - \int_{[a,b]} \gamma(s)\, \mu(ds)\right] \in \lambda\, \partial f(b, x(b)) + N_S(b, x(b)).$$

(iii) $p(a) \in N_{C_0}(x(a))$.

5.2.4 Remark

When $S = [a, \infty) \times C_1$ and f is time-free, it follows from the above that the constant value of H_P is 0.

5.3 EXAMPLE: LINEAR REGULATOR WITH DIODE

A type of optimal control problem that has found wide application in engineering design is that known as the *linear regulator*. The model in question has linear dynamics $\dot{x} = Ax + Bu$ which may arise from any of a variety of electrical, mechanical, or thermal systems. It is desired to find a control u which brings the system from an initial state value x_0 at $t = 0$ to a target value x_T at $t = T$, and having the property that the cost of the transformation is a minimum. This cost is measured by the positive definite quadratic functional $\int_0^T \langle Qu, u \rangle \, dt$. This is clearly a special case of the problem P_C, one which has smooth data. The maximum principle, Theorem 5.2.1, leads rather easily to a characterization of the unique solution.

Many simple models lead to dynamics which are not only nonlinear but even nonsmooth. One illustration of this is the resource economics problem of Section 3.3. Another example, one which is in the same spirit as the classical linear regulator problem, is provided by the case in which the underlying system is a simple electrical circuit consisting of a diode, a capacitor, and an impressed voltage (see Figure 1.3). As shown in McClamroch (1980), the governing differential equation of the circuit with impressed voltage u is given by

$$(1) \qquad \dot{x} = \begin{cases} \alpha(u - x) & \text{if } x \leqslant u \\ -\beta(x - u) & \text{if } x \geqslant u, \end{cases}$$

where α, β are positive constants (with $\alpha > \beta$) and x is the voltage across the capacitor. We wish to choose $u(\cdot)$ on $[0,T]$ to minimize $\frac{1}{2}\int_0^T u(t)^2 \, dt$, and so that the state x satisfies prescribed boundary conditions $x(0) = x_0$, $x(T) = x_T$.

We will show that this nonsmooth problem may be formulated as a special case of the control problem P_C of Section 5.2, or as a special case of the Bolza problem P_B of Section 4.1. It turns out that the latter formulation is superior in this case, which illustrates the value of having alternative methods at hand.

Formulation as P_B

By solving in Eq. (1) for u in terms of $\dot{x}$, one sees that the problem may be recast as a problem of Bolza P_B by defining

$$L(x, v) = \begin{cases} \dfrac{(x + v/\alpha)^2}{2} & \text{if } v \geqslant 0 \\ \dfrac{(x + v/\beta)^2}{2} & \text{if } v \leqslant 0 \end{cases}$$

$$l(s_0, s_1) = \begin{cases} 0 & \text{if } s_0 = x_0, s_1 = x_T \\ +\infty & \text{otherwise.} \end{cases}$$

It is easy to see that Basic Hypotheses 4.1.2 are satisfied. Note that $L(x, \cdot)$ fails to be convex when x is positive.

Let us proceed to calculate the Hamiltonian. We have

$$(2) \qquad H(x, p) = \max_{v} \{ pv - L(x, v) \}.$$

5.3 Example: Linear Regulator with Diode

Those v for which the maximum is attained satisfy the condition $p \in \partial_v L(x, v)$. We find (by using Theorem 2.5.1, for example):

$$\partial_v L(x, v) = \begin{cases} \left\{\frac{1}{\alpha}\left(x + \frac{v}{\alpha}\right)\right\} & \text{if } v > 0 \\ \left\{\frac{1}{\beta}\left(x + \frac{v}{\beta}\right)\right\} & \text{if } v < 0 \\ \left\{tx : \frac{1}{\alpha} \leq t \leq \frac{1}{\beta}\right\} & \text{if } v = 0. \end{cases}$$

This narrows the range of possibilities for the maximizing v in Eq. (2) to the following: $v = \alpha(\alpha p - x)$ (when $\alpha p > x$); $v = \beta(\beta p - x)$ (when $\beta p < x$); $v = 0$ (when x lies between αp and βp). By comparison we find the maximizing v and calculate H. We find

$$H(x, p) = \alpha p\left(\frac{\alpha p}{2} - x\right) \quad \text{when } x \leq \min(\alpha p, 0),$$

$$\qquad \text{and when } 0 \leq x \leq \frac{\alpha + \beta}{2}p$$

$$= \frac{-x^2}{2} \quad \text{when } \alpha p \leq x \leq \beta p \leq 0$$

$$= \beta p\left(\frac{\beta p}{2} - x\right) \quad \text{in all other cases.}$$

The three regions identified above are labelled A, B, and C, respectively (see Figure 5.1(a)).

Formulation as P_C

The dynamics (1) may be written $\dot{x} = \phi(x, u)$ where $\phi(x, u) := \max\{\alpha(u - x), \beta(u - x)\}$. The control set U is the set of reals, and we set $F(t, x, u) = u^2/2$, $f \equiv 0$. The pseudo-Hamiltonian H_P is given by

$$H_P(t, x, p, u, \lambda) = p \max\{\alpha(u - x), \beta(u - x)\} - \lambda \frac{u^2}{2}.$$

There is no state constraint, so we could simply set $g \equiv -1$. Alternatively, to consider local solutions x, we could define

$$g(t, s) = |s - x(t)| - \varepsilon.$$

In either case, g (and μ) would not figure in the necessary conditions. Of course, we take $C_0 = \{x_0\}$, $C_1 = \{x_T\}$.

216   Optimal Control

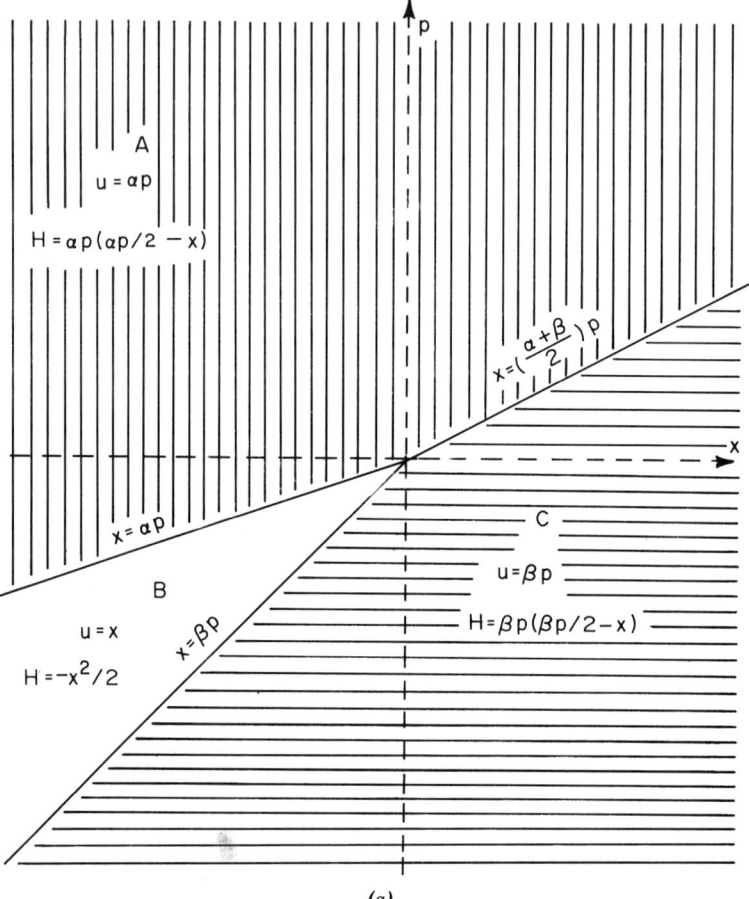

Figure 5.1(a) The three regions of definition for the Hamiltonian.

Necessary Conditions

Suppose x solves the problem described above (at least locally). Let us apply the necessary conditions in order to identify x. Beginning with the P_B formulation we invoke Theorem 4.4.1 to deduce the existence of an arc p satisfying $(-\dot{p}, \dot{x}) \in \partial H(x, p)$ a.e., $H(x, p) =$ constant. It is clear from the nature of H that in the interior of the regions A, B, and C this gives rise to ordinary differential equations whose solutions trace out level curves of H. (Figure 5.1 (b) depicts a few such directed level curves.) Let us show that the (x, p) trajectories of the Hamiltonian inclusion cannot "pause" in crossing from one region to another. Simple calculation shows that $(0, 0) \notin \partial H(x, p)$ for any (x, p) on the boundary of the regions, with one exception: we have $(0, 0) \in$

5.3 Example: Linear Regulator with Diode

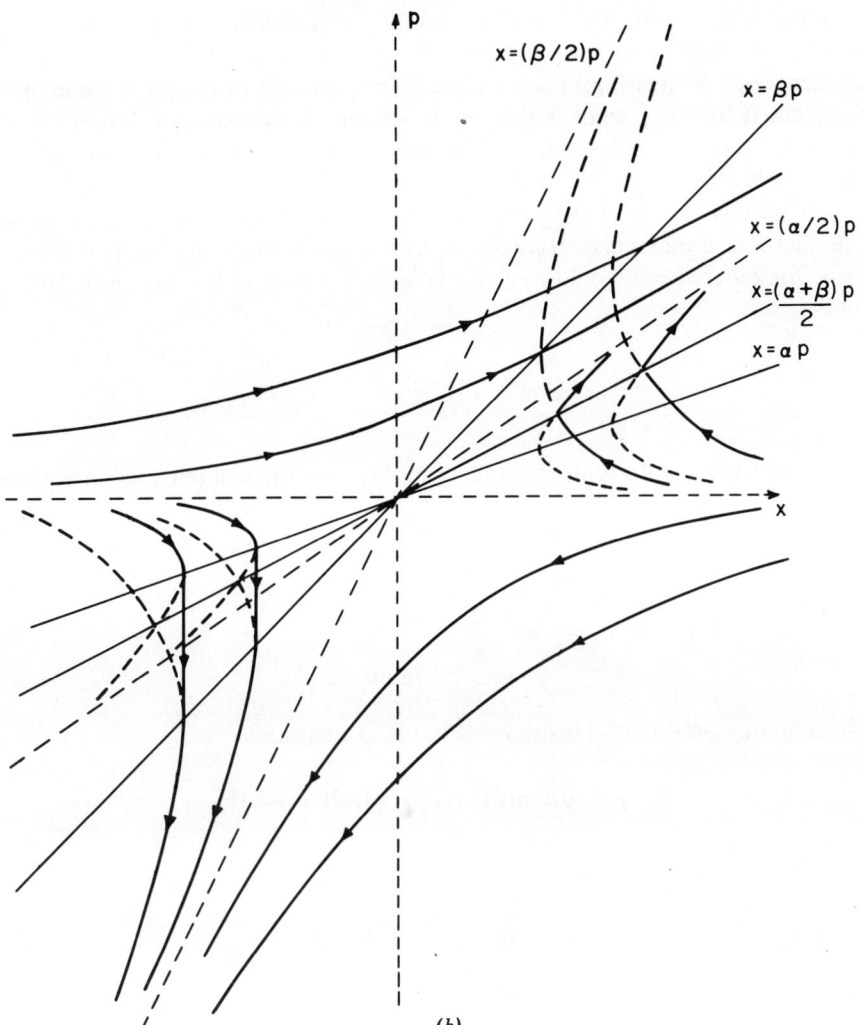

Figure 5.1(b) Optimal flow lines in the phase plane.

$\partial H(0,0)$. The level set $H = 0$ contains three parts: (a) $p = 0$, (b) $\beta p = 2x < 0$, (c) $\alpha p = 2x > 0$. Along (a), one has $\dot{x} = -\alpha x$ or $-\beta x$, whence $|\dot{x}| \leq k|x|$ for some constant k. A similar argument derives the same conclusion along (b) and (c), and it follows that if (x, p) is ever at the origin, then x is identically zero. By ruling out the trivial case $x_0 = x_T = 0$ we can assume that $(x(t), p(t))$ lies in the boundary of the regions A, B, or C for only finitely many values of t.

We now have an accurate qualitative picture of x in all cases. For example, in the case $x_T < x_0 < 0$ there are three segments to the solution: a period of time during which $\dot{x} = \alpha(\alpha p - x) > 0$, $\dot{p} = \alpha p < 0$ (in region A), followed by a period with $\dot{x} = 0$, $\dot{p} = x < 0$ (region B), and finally a spell in C during which $\dot{x} = \beta(\beta p - x) < 0$, $\dot{p} = \beta p < 0$. (Of course, one or more of these segments will be missing in some cases.) The behavior of (x, p) is completely determined by $p(0)$, since $x(0) = x_0$ is known. A numerical solution is obtained by choosing $p(0)$ so that $x(T) = x_T$ results (i.e., a two-point boundary-value problem).

Let us now examine the necessary conditions of the maximum principle. The fact that u maximizes $H_P(x, p, w, \lambda)$ over all w easily implies that λ must be 1, for otherwise p would be 0 along with λ, which is not the case. Thus u maximizes

$$p \max\{\alpha(w - x), \beta(w - x)\} - \frac{w^2}{2}.$$

The reader may verify that, as a function of (x, p), the maximizing u are given by

$$u = \alpha p \quad \text{in int } A$$
$$= \beta p \quad \text{in int } C$$
$$= x \quad \text{in int } B.$$

The adjoint equation of Theorem 5.2.1(i) is in this case

$$-\dot{p} \in p \partial_x \max\{\alpha(u - x), \beta(u - x)\},$$

which gives in B the following:

$$\dot{p} = \alpha p \quad \text{if } u > x$$
$$= \beta p \quad \text{if } u < x$$
$$\in [\alpha p, \beta p] \quad \text{if } u = x.$$

The maximum principle gives $\dot{p} \in [\alpha p, \beta p]$ in int B (in contrast to $\dot{p} = x$ from the previous analysis), so that no unique p (and hence x) is identified by its application.

We have managed to find the solution to the problem if one exists. In the next section we shall illustrate the sufficient conditions by using them to prove that the candidate identified above is in fact optimal.

The moral, as always, is that a diversity of techniques is a valuable asset. The Hamiltonian approach of Chapters 3 and 4 maximizes first (to find H) and then "differentiates" (to get the Hamiltonian inclusion), while the maximum principle can be thought of as performing these operations in reverse order. Although related, the two approaches are different, and neither is

5.4 SUFFICIENT CONDITIONS AND EXISTENCE

A useful approach to finding sufficient conditions for the control problem P_C is to formulate it as a Bolza problem and interpret the result of Zeidan proven in Section 4.3. There are several ways to write P_C in the form P_B. Here is a general one. Set

$$L(t, x, v) = \begin{cases} \inf\{F(t, x, u) : \phi(t, x, u) = v, u \in U(t)\} & \text{if } g(t, x) \leq 0 \\ +\infty & \text{if } g(t, x) > 0 \end{cases}$$

$$l(s_0, s_1) = \begin{cases} f(s_1) & \text{if } s_0 \in C_0, s_1 \in C_1 \\ +\infty & \text{otherwise.} \end{cases}$$

(Note that the infimum over the empty set is taken to be $+\infty$.) A moment's thought reveals that it is reasonable to expect that an arc x solves the resulting P_B iff, for some corresponding control u, (x, u) solves P_C. Under mild assumptions, it also follows that L satisfies the basic hypotheses of Chapter 4, as shown by the following equivalence theorem drawn from Rockafellar (1975). (We omit the proof, which is based upon measurable selection theory.)

5.4.1 Theorem (Equivalence)

Let the data of the problem P_C satisfy the following conditions:

(i) $F(t, x, u)$ is $\mathcal{L} \times \mathcal{B}$ measurable (see Section 4.1) and lower semicontinuous in (x, u).
(ii) $\phi(t, x, u)$ is measurable in t and continuous in (x, u).
(iii) $U(\cdot)$ is closed-valued and its graph is $\mathcal{L} \times \mathcal{B}$ measurable.
(iv) $g(t, x)$ is $\mathcal{L} \times \mathcal{B}$ measurable and lower semicontinuous in x.
(v) C_0 and C_1 are closed; f is lower semicontinuous.
(vi) For each t, for every bounded subset S of $R^n \times R^n$, the following set is bounded:

$$\{u \in U(t) : \text{for some } (x, v) \text{ in } S, v = \phi(t, x, u)\}.$$

Then the data l, L of the associated problem P_B of Bolza satisfy Basic Hypotheses 4.1.2, and an arc x solves P_B iff there is a control u corresponding to x such that (x, u) solves P_C.

5.4.2 Theorem (Sufficient Conditions)

Let the data of the problem P_C satisfy the hypotheses of Theorem 5.4.1. Let (x, u) be an admissible state/control pair for P_C, and suppose that there exist an arc p, a Lipschitz $n \times n$ symmetric matrix-valued function Q, and a positive ε such that the following hold:

(i) For almost all t, for all y in $\{y : g(t, x(t) + y) \leq 0\} \cap \varepsilon B$, and for all w in $U(t)$, one has

$$H_P(t, x(t) + y, p(t) - Q(t)y, w, 1) \leq$$
$$H_P(t, x(t), p(t), u(t), 1) - \langle \dot{p}(t) + Q(t)\dot{x}(t), y \rangle$$
$$+ \tfrac{1}{2}\langle y, \dot{Q}(t)y \rangle.$$

(ii) For all α, β in εB such that $x(a) + \alpha \in C_0$ and $x(b) + \beta \in C_1$, one has

$$\langle p(a), \alpha \rangle - \langle p(b), \beta \rangle - \tfrac{1}{2}\langle \alpha, Q(a)\alpha \rangle + \tfrac{1}{2}\langle \beta, Q(b)\beta \rangle$$
$$\leq f(x(b) + \beta) - f(x(b)).$$

Then (x, u) solves P_C relative to the tube $T(x; \varepsilon)$.

Proof. We verify the hypotheses of Theorem 4.3.1 for the associated Bolza problem. The condition (i) above gives

(1) $\quad F(t, x, w) \geq \langle p, \phi(t, x, w) - \dot{x} \rangle + F(t, x, u) \quad$ for $w \in U(t)$.

This immediately gives condition (1) of Theorem 4.3.1 in view of how L is defined. Condition (2) of Theorem 4.3.1 follows upon recalling that $H(t, y, p)$ coincides with $\max\{H_P(t, y, p, w, 1) : w \in U(t)\}$, and (3) is evident. It follows that x solves P_B restricted to $T(x; \varepsilon)$. Theorem 5.4.1 implies that there exists $\tilde{u}$ for x such that $(x, \tilde{u})$ solves P_C. We can take $\tilde{u} = u$ in this conclusion, since $L(t, x, \dot{x}) = F(t, x, u)$, as follows from (1). □

The following extends a well-known result of Mangasarian (1966).

Corollary

Let the data of the problem P_C satisfy the hypotheses of Theorem 5.4.1, and assume in addition that C_0 and C_1 are convex, that $U(\cdot)$ is convex-valued, that f is convex, and that ϕ and F are differentiable in the state variable. Let (x, u) be admissible for P_C, and suppose that there is an arc p satisfying:

(i) For almost all t, the function $w \to H_P(t, x(t), p(t), w, 1)$ attains a maximum over $U(t)$ at $w = u(t)$.
(ii) $-\dot{p}(t) = \nabla_x H_P(t, x(t), p(t), u(t), 1)$ a.e.
(iii) $p(a) \in N_{C_0}(x(a))$, $\quad -p(b) \in N_{C_1}(x(b)) + \partial f(x(b))$.
(iv) For each t, the function $(y, w) \to H_P(t, y, p(t), w, 1) - \psi_{U(t)}(w)$ is concave.

Then (x, u) solves P_C.

5.4 Sufficient Conditions and Existence

Proof. The superdifferential at $(x(t), u(t))$ of the concave function of (iv) contains the point $(-p(t), 0)$. Expressing the defining inequality for this supergradient gives exactly condition (i) of Theorem 5.4.2 (with $Q \equiv 0$); (ii) of Theorem 5.4.2 is trivially satisfied, and the theorem applies (with $\varepsilon = \infty$) to give the result. □

The corollary is appealing because its hypotheses resemble so closely the necessary conditions of the maximum principle (strengthened by concavity). It is not well adapted, however, to problems which are nondifferentiable in x. For example, it does not suffice to simply replace condition (ii) by $-\dot{p} \in \partial_x H_p$.

We now illustrate the use of the theorem with the problem of Section 5.3 (to which the corollary does not apply).

5.4.3 Example

We return to the problem of Section 5.3. Recall that a unique candidate (x, u) was identified, and that for the corresponding adjoint arc p (of the true Hamiltonian inclusion) we had (see Figure 5.1 (a)) $u = \alpha p$ in A, $u = \beta p$ in C, and $u = x$ in B. Condition (ii) of Theorem 5.4.2 certainly holds when $Q \equiv 0$. Let us verify condition (i) of Theorem 5.4.2 for $Q \equiv 0$ and $\varepsilon = +\infty$. This amounts to proving the inequality

$$(2) \quad p \max\{\alpha(w - x - y), \beta(w - x - y)\} - \frac{w^2}{2} \leq H(x, p) - \dot{p}y$$

for all y and w. In region A this reduces to

$$(3) \quad p \max\{\alpha(w - x - y), \beta(w - x - y)\} - \frac{w^2}{2} \leq \alpha p\left(\frac{\alpha p}{2} - x - y\right)$$

since in A we have $\dot{p} = \alpha p$, $H(x, p) = \alpha p(\alpha p/2 - x)$. Consider the case $w \geq x + y$, and set $w = x + y + \delta$ for $\delta \geq 0$. Then (setting $x + y = z$) (3) is equivalent to

$$z^2 + 2\delta z + \delta^2 + \alpha p(\alpha p - 2z - 2\delta) \geq 0.$$

The minimum over z occurs at $z = \alpha p - \delta$, and the left-hand side vanishes there. This proves (3) for $w \geq x + y$; the other case is similar.

The proof of inequality (2) within regions B and C is no more difficult than that for A. The hypotheses of Theorem 5.4.2 are all present, and it follows that (x, u) provides in all cases a (global) solution to the original problem.

It can be shown that under the hypotheses of the corollary to Theorem 5.4.2, the function $y \to H(y, p(t))$ is concave. The Hamiltonian H of the problem above fails to be concave in y, however, as an examination of points (y, p) on the common boundary of the regions A and C will reveal.

Existence

The original motivation for formulating control problems in Bolza form was to obtain general existence theorems. The following is an immediate consequence of existence theorem 4.1.3 and equivalence theorem 5.4.1. (Note that the set in (i) below coincides with epi $L(t, x, \cdot)$.)

5.4.4 Theorem (Existence)

Let the data of the problem P_C satisfy the hypotheses of Theorem 5.4.1 and suppose that in addition:

(i) For each t and x with $g(t, x) \leq 0$, the following set is convex:

$$\{[\phi(t, x, u), F(t, x, u) + \delta] : u \in U(t), \delta \geq 0\}.$$

(ii) There exist functions $\sigma(t)$, $\rho(t)$, and $\mu(t, p)$ finite and summable in t (with σ and ρ nonnegative) such that for all t, x satisfying $g(t, x) \leq 0$, for all w in $U(t)$, and for all p one has

$$H_P(t, x, p, w, 1) \leq \mu(t, p) + |x|(\sigma(t) + \rho(t)|p|).$$

(iii) The set C_0 is compact; $f(s)$ is bounded below on C_1 by some function of the form $c - M|s|$.

Then, if there is at least one admissible (x, u) for P_C giving a finite value to the cost functional, there is a solution to P_C.

5.5 THE RELAXED CONTROL PROBLEM

The Mayer Problem

We deal in this section with the autonomous control problem of Mayer, which is the special case of the problem P_C of Section 5.2 in which $F \equiv 0$ and ϕ, g, and U have no explicit dependence on t. Thus we seek to minimize $f(x(b))$ over the state/control pairs (x, u) satisfying

$$\dot{x}(t) = \phi(x(t), u(t)), \quad u(t) \in U \quad \text{a.e.}$$

$$g(x(t)) \leq 0 \quad \text{a.e.}, \quad x(a) \in C_0, \quad x(b) \in C_1.$$

We label this problem P_M.

Our goal is to interpret some of the controllability and sensitivity results of Chapter 3 in terms of the structure of P_M (or, more accurately, its "relaxation," which will be defined below). The simplified setting used here will make the connection particularly clear. We assume throughout that U and C_0 are compact and C_1 is closed, that f and g are locally Lipschitz, and that the set

$\{x : g(x) \leq 0\}$ is compact. As for ϕ, we assume it is continuous and admits a derivative $\nabla_x \phi(\cdot, \cdot)$ which is continuous.

The Relaxed Problem

Let F be the multifunction defined by $F(x) = \phi(x, U)$. As a consequence of the measurable selection theorem (or as a special case of Theorem 5.4.1) we deduce that x solves the resulting version of P_D (see Problem 3.4.1) iff (x, u) solves P_M for some control u. (This is often referred to as *Filippov's Lemma*.)

It is easy to verify that F, f, g, and C_0 satisfy on any sufficiently large bounded tube Ω Basic Hypotheses 3.2.2 of Chapter 3, with one exception. The one requirement that does not follow is that F be convex-valued. This condition is familiar in existence theory; note that it coincides in this case with condition (i) of Theorem 5.4.4. We saw in the corollary to Theorem 3.1.6 that trajectories of co F can be uniformly approximated to an arbitrary tolerance by trajectories for co F, and it follows from Theorem 3.1.7 that the trajectories for co F are "sequentially compact." It is therefore reasonable to consider the problem with F replaced by co F as the *closure* of the original problem.

It is possible to define an associated problem with data $\tilde{\phi}, \tilde{U}$ such that $\tilde{\phi}(x, \tilde{U}) = \text{co } \phi(x, U)$ for each x. (The new problem is a "relaxation" of the original.) One useful way to explicitly construct a relaxation without introducing too many technicalities is to define the new control set $\tilde{U}$ to consist of all points $\tilde{u} = (\lambda_0, \lambda_1, \ldots, \lambda_{2n}, u_0, u_1, \ldots, u_{2n})$, where $\lambda_i \geq 0, \sum_{i=0}^{2n} \lambda_i = 1$, and each u_i belongs to U. Set

$$\tilde{\phi}(x, \tilde{u}) = \sum_{i=0}^{2n} \lambda_i \phi(x, u_i),$$

and observe that $\tilde{\phi}(x, \tilde{U}) = \text{co } \phi(x, U)$ as a result of Carathéodory's Theorem. Note that all our assumptions on ϕ and U are inherited by $\tilde{\phi}$ and $\tilde{U}$. Although now the control set is in a higher dimensional space $R^{\tilde{m}}$ ($\tilde{m} = (2n + 1)(m + 1)$), the states x still lie in R^n.

The *relaxed problem*, denoted $\tilde{P}_M$, is the problem P_M in which the dynamics (ϕ, U) are replaced by $(\tilde{\phi}, \tilde{U})$. Note that $\tilde{P}_M$ admits "more" controls and (possibly) state arcs than does P_M. Nonetheless (because of the approximation results cited above) it is often (but not always) the case that an arc x solving P_M also solves $\tilde{P}_M$. When the state constraint is inactive, this is true if P_M is calm (see Section 4.2) and in particular if $x(b)$ lies in the interior of C_1 (i.e., if the endpoint constraint is inactive). In general, the relaxed problem $\tilde{P}_M$ is the only one for which existence theorems can be proved. For this reason, there are many who deem it the only reasonable problem to consider in practice. Because the attainable set for the relaxed dynamics is the closure of that for the original dynamics, it is also the case that numerical approximation algorithms converge to the relaxed value of the problem. Finally, let us note that if min $\tilde{P}_M$ < min P_M, then by violating the state constraint and the constraint $x(b) \in C_1$ to an arbitrarily small extent, we can find an "almost admissible" arc x for the *original* problem which returns a value for the cost functional lower than min P_M. In practice, then, the relaxed problem is often more appropriate than the original. An analogy in differential equations would be that only stable equilibria are physically meaningful.

Normality

Recall that in Section 3.4 we defined multiplier sets $M_r^\lambda(x)$ corresponding to any arc x admissible for P_D; these multipliers involve a Hamiltonian inclusion for the (true) Hamiltonian H of Section 3.2. The arc x was termed normal if the only multiplier in $M_r^0(x)$ (the set of "abnormal" multipliers) is the trivial one ($\mu = 0$, $p = 0$). We have pointed out above that $\tilde{P}_M$ and its associated P_D are equivalent (with the same value function V). Although the results of Chapter 3 are thereby available for $\tilde{P}_M$, our purpose here is to establish a connection with multipliers in the sense of the maximum principle; these multipliers involve the pseudo-Hamiltonian H_P of Section 5.2. (The Hamiltonian formulation appears to be more natural, however, since P_M and $\tilde{P}_M$ generate the *same* Hamiltonian $H(x, p)$, whereas the conditions of the maximum principle are different for the two problems.)

Pseudonormality

Given a control problem P_C and an arc x feasible for P_C, x is said to be *pseudonormal* for P_C provided that, for any control u corresponding to x, there are no p, γ, μ satisfying the conclusions of the maximum principle (i.e., Theorem 5.2.1(i)–(v)) with $\lambda = 0$. (That is, x is pseudonormal when there are no nontrivial abnormal multipliers for (x, u) whenever u is a control for x.)

5.5.1 Theorem (Controllability about Pseudonormal Trajectories)

Let the arc x be admissible and pseudonormal for $\tilde{P}_M$. Then there exist constants m and $\varepsilon > 0$ such that, for all α and β in εB, there is a state/control pair (y, v) for $(\tilde{\phi}, \tilde{U})$ satisfying the state constraint $g(y(t)) \le 0$, the conditions $y(a) \in C_0 + \alpha$, $y(b) \in C_1 + \beta$, and the inequality

$$\int_a^b |x(t) - y(t)|\, dt \le m|(\alpha, \beta)|.$$

Proof. Taking P_D to be the differential inclusion problem 3.4.1 generated by $\tilde{P}_M$, we need only show that x is normal and then apply Theorem 3.5.3. Accordingly, let $[p, \gamma, \mu, \zeta, 0]$ belong to $M_r^0(x)$; we wish to show that p and μ are zero. This follows from the hypothesis of the theorem if p, γ, μ satisfy (for $\tilde{\phi}, \tilde{U}$, and for some $\tilde{u}$) conditions 5.2.1(i)–(iv) (with $\lambda = 0$) (for then (v) must

5.5 The Relaxed Control Problem

fail). Note that (iii) and (iv) certainly hold, so we turn our attention to (i) and (ii).

In view of Theorem 2.8.2, Corollary 2, Condition (iv) of Definition 3.2.5 implies

$$(-\dot{p}, \dot{x}) = \sum_{i=0}^{2n} \lambda_i [\nabla_x \phi(x, u_i)^* q, \phi(x, u_i)]$$

(where $q = p + \int_{[a,t)} \gamma \, d\mu$), where each u_i satisfies $\langle q, \phi(x, u_i) \rangle = H(x, q)$. It follows that the relaxed control $\tilde{u} = (\lambda_0, \lambda_1, \ldots, \lambda_{2n}, u_0, \ldots, u_{2n})$ satisfies $-\dot{p} = \nabla_x \tilde{\phi}(x, \tilde{u})^* q, \langle q, \tilde{\phi}(x, \tilde{u}) \rangle = H(x, q)$, so that conditions 5.2.1(i) and (ii) hold, and the proof is complete. □

The Value Function

As in Chapter 3, Y signifies the set of solutions x to P_D (in our present setting, the arcs x which solve $\tilde{P}_M$ for some corresponding control $\tilde{u}$), and V signifies its value function (thus $V(\beta)$ is the value of $\tilde{P}_M$ when the constraint $x(b) \in C_1$ is changed to $x(b) \in C_1 + \beta$). As long as $\tilde{P}_M$ admits at least one feasible $(x, \tilde{u})$ (which we assume), then Y is nonempty by Proposition 3.2.3.

5.5.2 Theorem (Value Function)

Suppose that every x in Y is pseudonormal. Then V is Lipschitz near 0, and one has

(1) $$\partial V(0) \subset \mathrm{co}\left\{ \zeta + p(b) + \int_{[a,b]} \gamma(s) \mu(ds) \right\},$$

where the convex hull is taken over all p, γ, μ, ζ satisfying the conditions of the maximum principle (i.e., conditions 5.2.1(i)–(iv)) with $\lambda = 1$, for some solution $(x, \tilde{u})$ of $\tilde{P}_M$.

Proof. In the proof of the preceding theorem we showed that (in the notation of Section 3.4) $E[M_r^1(Y)]$ is contained in the right-hand side of (1). Since we also have $E[M_r^0(Y)] = \{0\}$, the assertion of the theorem follows from Theorem 3.4.3. □

The Hamilton–Jacobi Equation

An arc x is a *local solution* to $\tilde{P}_M$ if it solves the problem P_D relative to a tube $T(x; \varepsilon)$ for some $\varepsilon > 0$. We now suppose that all local solutions x to $\tilde{P}_M$ are such that the state constraint is inactive (i.e., such that $g(x(t)) < 0$), and we also suppose that C_0 is a singleton $\{x_0\}$.

In our present context, the Hamiltonian H is given by $H(x, p) = \max\{\langle p, \phi(x, u)\rangle : u \in U\}$. With this in mind, the (extended) Hamilton–Jacobi equation of Section 3.7 for a Lipschitz function $W(t, x)$ assumes the form

$$\min\{\alpha + \langle \beta, \phi(x, u)\rangle : u \in U, (\alpha, \beta) \in \partial W(t, x)\} = 0.$$

The following is an immediate consequence of Theorem 3.7.6.

5.5.3 Theorem
Suppose that all local solutions to $\tilde{P}_M$ are pseudonormal. Then a feasible arc x is a local solution to $\tilde{P}_M$ iff there exists a tube $T(x; \varepsilon)$ about x of positive radius ε upon which is defined a Lipschitz solution W of the Hamilton–Jacobi equation satisfying the boundary conditions

$$W(a, x_0) = f(x(b)), \qquad W(b, y) = f(y) \quad \text{for } y \text{ in } C_1 \cap \{x(b) + \varepsilon B\}.$$

As explained in Section 3.7, this result can be interpreted as affirming that for all "reasonable" problems, the Hamilton–Jacobi equation is necessary and sufficient for optimality.

Chapter Six

Mathematical Programming

But he was not permitted to leave it at that, for in universities nobody is ever fully satisfied with somebody else's definition.

ROBERTSON DAVIES, *The Rebel Angels*

We now focus upon the abstract optimization problem posed by the general mathematical program. The common thread running through the chapter is the concept of multiplier, which is defined in terms of a Lagrangian related to the data of the problem. We shall discover and explore an interesting interplay between the multipliers and the necessary conditions, the value function, the stability and sensitivity of the problem, and the feasible points under perturbation. We begin with a general form of the familiar and very useful mathematical technique known as the Lagrange multiplier rule. As is often the case, the basic idea seems to have originated with Euler.

6.1 THE LAGRANGE MULTIPLIER RULE

The Problem

Let X be a Banach space. The general mathematical programming problem, denoted P, is that of minimizing a given function $f(x)$ on X subject to three types of constraints:

(i) Inequality constraints: $g_i(x) \leq 0$, $i = 1, 2, \ldots, n$, where each g_i is a real-valued function on X.
(ii) Equality constraints: $h_j(x) = 0$, $j = 1, 2, \ldots, m$, where each h_j is a real-valued function on X.
(iii) An abstract constraint: $x \in C$, where C is a given subset of X.

Of course, constraints of types (i) and (ii) can always be absorbed into one of type (iii) by suitably defining C, but our goal is rather to exploit the explicit nature of (i) and (ii).

Basic Hypothesis

We assume throughout the chapter that C is closed, and that each function f, g_i, h_j is Lipschitz near any given point of C. (As seen in Chapter 2, this Lipschitz hypothesis would result from more familiar assumptions of convexity or differentiability.)

The Lagrangian

Let g and h be the functions $[g_1, g_2, \ldots, g_n]$ and $[h_1, h_2, \ldots, h_m]$ mapping X to R^n and R^m, respectively. Generally one has $n \geq 1$, $m \geq 1$, but we allow n or $m = 0$ to signify the case in which there are no explicit inequality or equality constraints. In these cases it is clear below that certain references to such constraints are simply to be deleted.

The *Lagrangian* is the function $L(x, \lambda, r, s, k): X \times R \times R^n \times R^m \times R \to R$ defined by

$$L(x, \lambda, r, s, k) := \lambda f(x) + \langle r, g(x) \rangle + \langle s, h(x) \rangle + k|(\lambda, r, s)|d_C(x),$$

where d_C denotes as usual the distance function associated with C (see Section 2.4). This reduces to the familiar expression in the classical theory when $C = X$ (so that d_C is identically zero).

6.1.1 Theorem (Lagrange Multiplier Rule)

Let x solve P. Then for every k sufficiently large there exist $\lambda \geq 0$, $r \geq 0$, and s, not all zero, such that $\langle r, g(x) \rangle = 0$ and $0 \in \partial_x L(x, \lambda, r, s, k)$.

6.1.2 Remark

(i) The proof will show that the conclusion is valid for any $k > \hat{k}$, where $\hat{k}$ is a Lipschitz constant for the function $[f, g, h]$ on a neighborhood of x.
(ii) The notation $r \geq 0$ means that each component r_i of r is nonnegative. The condition $\langle r, g(x) \rangle = 0$ is sometimes called "complementary slackness." It is equivalent to saying that $r_i = 0$ when $g_i(x) < 0$ (i.e., when the constraint $g_i(x) \leq 0$ is (locally) inactive).
(iii) Of course the conclusions of the theorem hold for local minima, since they are unchanged when C is replaced by $C \cap \{x + \varepsilon \bar{B}\}$ for any positive ε.
(iv) The stationarity condition $0 \in \partial_x L$ implies the "separated" condition

$$0 \in \lambda \partial f(x) + \sum_i r_i \partial g_i(x) + \sum_j s_j \partial h_j(x) + k|(\lambda, r, s)|\partial d_C(x)$$

6.1 The Lagrange Multiplier Rule

(as well as the variant in which the last term is replaced by $N_C(x)$). This condition is in general weaker than the Lagrangian form (see Proposition 2.3.3), although the two are equivalent for "smooth (or convex) data".

(v) The vector (λ, r, s) above is termed a multiplier. Note that if (λ, r, s) is a multiplier (i.e., if it satisfies the conclusions of the theorem for some k), then so is $(t\lambda, tr, ts)$ for any $t > 0$. Thus the theorem could also stipulate $|(\lambda, r, s)| = 1$ (say).

Proof of Theorem 6.1.1. Let T be the set

$$\{t = (\lambda, r, s) \in R^{1+n+m} : \lambda \geq 0, r \geq 0, |(\lambda, r, s)| = 1\},$$

let any positive ε be given, and define $F: X \to R$ by

$$F(y) = \max_T \{(\lambda, r, s) \cdot (f(y) - f(x) + \varepsilon, g(y), h(y))\}.$$

Note that F is Lipschitz near x, and that $F(x) = \varepsilon$. We claim that F is positive on C. Indeed, if $F(y) \leq 0$, then it follows easily that $g_i(y) \leq 0$, $h_j(y) = 0$ (i.e., y is feasible for P), and $f(y) \leq f(x) - \varepsilon$, a contradiction. Therefore x satisfies

$$F(x) \leq \inf_C F + \varepsilon,$$

and by Theorem 7.5.1, there is a point u in $x + \sqrt{\varepsilon}\,\overline{B}$ such that, for all y in C,

$$F(y) + \sqrt{\varepsilon}\,\|y - u\| \geq F(u).$$

If $\hat{k}$ is the Lipschitz constant of Remark 6.1.2(i), then it is easy to see that any $k > \hat{k}$ is a local Lipschitz constant (when ε is sufficiently small) for the function $F(y) + \sqrt{\varepsilon}\,\|y - u\|$ near the point $y = u$. By Proposition 2.4.3, u therefore also minimizes, over some neighborhood of u, the function

$$y \to F(y) + \sqrt{\varepsilon}\,\|y - u\| + kd_C(y)$$

$$= \max_T \{L(y, \lambda, r, s, k) - \lambda f(x) + \varepsilon\lambda\} + \sqrt{\varepsilon}\,\|y - u\|$$

$$= G(y) + \sqrt{\varepsilon}\,\|y - u\|,$$

where $G(y)$ is defined as the first term in the preceding expression.

For ε sufficiently small, then, one has

(1) $$0 \in \partial G(u) + \sqrt{\varepsilon}\,\overline{B}_*.$$

We intend to estimate ∂G by Theorem 2.8.2. To this end, we claim first that the mapping

(2) $$(t, y) \to \partial_x L(y, t, k)$$

is closed, in the sense of Definition 2.8.1. To see this, note that for any pair t_1, t_2 in T, the function

$$y \to L(y, t_1, k) - L(y, t_2, k) = (t_1 - t_2) \cdot (f, g, h)(y)$$

is Lipschitz near x of rank $k|t_1 - t_2|$; thus,

$$\partial_x L(y, t_1, k) \subset \partial_x L(y, t_2, k) + k|t_1 - t_2|\bar{B}_*$$

by Propositions 2.3.3 and 2.1.2(a). Combined with the closure property 2.1.5(b) of the generalized gradient, this observation implies that the map (2) is closed as claimed.

Since $F(u)$ is positive, there is a unique t_u in T at which the maximum defining F (and hence G) is attained. Theorem 2.8.2 now applies, and we conclude from (1):

(3) $$0 \in \partial_x L(u, t_u, k) + \sqrt{\varepsilon}\,\bar{B}_*.$$

Note that if $g_i(u) \leq 0$, then the maximizing t_u has $r_i = 0$ necessarily.

If we do the above for a sequence ε_i decreasing to 0, then the corresponding u_i converge to x, and a subsequence of t_{u_i} converges to an element of T. The assertions of the theorem then follow from (3) together with the observation that the map (2) is closed. □

Pareto Optima

Some mathematical programming problems have several objective functions, say $f_1, f_2, \ldots, f_\beta$. Let f be the vector function $[f_1, f_2, \ldots, f_\beta]$, and consider the problem of "optimizing" $f(x)$ over the points x satisfying the same constraints as in the problem P above. The feasible point x is termed (weakly) *Pareto optimal* for this multiobjective problem if there is no feasible point y for which $f_\alpha(y) < f_\alpha(x)$, $\alpha = 1, 2, \ldots, \beta$.

Let us extend the definition of the Lagrangian L by interpreting λ as a vector in R^β:

$$L(x, \lambda, r, s, k) := \langle \lambda, f(x) \rangle + \langle r, g(x) \rangle + \langle s, h(x) \rangle + k|(\lambda, r, s)|d_C(x).$$

6.1.3 Theorem

Let x be Pareto optimal for the multiobjective problem. Then the conclusions of Theorem 6.1.1 hold.

Proof. The technique used to prove Theorem 6.1.1 goes through without change. It is only necessary to interpret the ε occurring in the definition of F as $[\varepsilon, \varepsilon, \ldots, \varepsilon]$ in R^β. □

6.2 AN ALTERNATE MULTIPLIER RULE

We shall now digress from the main developments to derive a multiplier rule distinct from that of Theorem 6.1.1, one which is motivated by computational considerations. For the purposes of this section, we assume that X is finite-dimensional. We require prior to the statement the following concept.

The Relative Generalized Gradient

Let S be a subset of X. The S-relative generalized gradient of f at x, denoted $\partial|_S f(x)$, consists of the cluster points of ∂f evaluated at points in S approaching x. Formally,

$$\partial|_S f(x) := \{\zeta : \zeta \text{ is a limit point of } \zeta_i, \text{ where } \zeta_i \in \partial f(y_i), y_i \in S, y_i \to x\}.$$

The following properties of $\partial|_S f$ are immediate:

6.2.1 Proposition

Let f be Lipschitz near x. Then
(a) $\partial|_S f(x)$ *is a closed subset of* $\partial f(x)$.
(b) $\partial|_S f(x) = \partial f(x)$ *if x lies in int S;* $\partial|_S f(x) = \emptyset$ *if* $(x + \varepsilon B) \cap S = \emptyset$ *for some* $\varepsilon > 0$; $\partial|_S f(x)$ *is nonempty if* $x \in cl(S)$.
(c) $\partial|_S f(x)$ *is upper semicontinuous at x.*

We define sets G_i ($i = 1, 2, \ldots, n$), H_j ($j = 1, 2, \ldots, m$), and D via

$$G_i = \{x' : g_i(x') > 0\}$$

$$H_j = \{x' : h_j(x') \neq 0\}$$

$$D = \{x' : d_C(x') > 0\}.$$

As always, the basic hypotheses of Section 6.1 are in force.

6.2.2 Theorem
Let x solve P. Then

$$0 \in \text{co}\{\partial f(x), \partial|_{G_i} g_i(x), \pm \partial|_{H_j} h_j(x), \partial|_D d_C(x)\}.$$

6.2.3 Remarks

(i) Note that some of the sets appearing in the statement of the theorem may be empty. For example, $\partial|_{G_i} g_i(x)$ is empty precisely when $g_i \leq 0$ in a neighborhood of x. This merely serves to eliminate irrelevant constraints from the necessary conditions. When g_i is smooth and $\partial|_{G_i} g_i(x)$ is nonempty, the latter reduces to $\{Dg_i(x)\}$; it is easy to see that we recover the classical Lagrange multiplier rule for smooth data (in the absence of an abstract constraint).

(ii) The suggestion that there could be a multiplier rule with relative gradients generated by infeasible points is due to E. Polak; this property is useful for developing algorithms for nondifferentiable optimization. There are other potential sharpenings of the multiplier rule, of course. Suppose, for example, that x minimizes a smooth function f over a set C. Then x also minimizes f over any subset of C containing x; thus,

$$-\nabla f(x) \in \cap \{N_S(x) : x \in S \subset C\}.$$

(iii) It is always possible to reformulate a constraint $h_j = 0$ as two inequalities: $h_j \leq 0$, $-h_j \leq 0$. Similarly, $x \in C$ can be expressed as $d_C(x) \leq 0$. Generally, however, this will lead to a problem in which the necessary conditions hold trivially at any feasible point (since, e.g., $0 \in \partial d_C(x)$, the conditions of Theorem 6.1.1 hold automatically if one of the inequality constraints is $d_C \leq 0$). The conclusion of Theorem 6.2.2, however, continues to be nontrivial in general under this transformation (in fact, unaffected).

Proof of Theorem 6.2.2. As explained in Remark 6.2.3(iii), it suffices to prove the theorem in the case in which all the constraints are inequalities, in which case the required conclusion is that

(1) $$0 \in \text{co}\{\partial f(x), \partial|_{G_1} g_1(x), \ldots, \partial|_{G_n} g_n(x)\}.$$

To derive (1), fix any $\varepsilon > 0$, and consider the function F defined by

$$F(y) := \max\{f(y) - f(x) + \varepsilon, g_1(y), \ldots, g_n(y)\}.$$

Since x solves P (relative to $\tilde{C} := x + \delta \bar{B}$, say), it follows that F is positive on $\tilde{C}$. Because $F(x) = \varepsilon$, we have

$$F(x) \leq \inf_{\tilde{C}} F + \varepsilon,$$

so that by Theorem 7.5.1 there exists a point u in $x + \sqrt{\varepsilon}\bar{B}$ such that u minimizes over $\tilde{C}$ the function

$$y \to F(y) + \sqrt{\varepsilon}\|y - u\|.$$

We derive (for $\varepsilon < \delta^2$) that 0 belongs to $\partial F(u) + \sqrt{\varepsilon}\bar{B}$. Invoking Theorem

2.8.2 leads to (recall that $F(u) > 0$):

$$\operatorname{co}\{\partial f(u), \partial|_{G_1} g_1(u), \ldots, \partial|_{G_n} g_n(u)\} \cap \sqrt{\varepsilon}\,\bar{B} \neq \emptyset.$$

We proceed to obtain this conclusion for a sequence ε_j decreasing to 0. The corresponding u_j converge to x, and the required property (1) is then seen to follow from the upper semicontinuity of Proposition 6.2.1(c). □

Directions of Complete Descent

The proof of Theorem 6.2.2 is linked to a certain algorithmic approach to solving the mathematical program. Anticipating somewhat the notation of the following sections, let $V(0)$ denote the value of the program (i.e., inf P), and consider the function

$$F(x) := \max\{f(x) - V(0), g_1(x), \ldots, g_n(x)\}.$$

(Recall that in the proof of Theorem 6.2.2 it was shown that, because *relative* generalized gradients are involved, it is only necessary to consider inequality constraints $g_i \leq 0$.) Observe that F is nonnegative by construction, and that x solves P iff $F(x) = 0$. Thus solving P is equivalent to finding a zero of F. Note that we do not know $V(0)$ until the problem is solved, however. Nonetheless, as we are about to see, we can still produce "descent directions" based on $\partial F(x)$.

Suppose that 0 fails to belong to the set

$$\Gamma(x) := \operatorname{co}\{\partial f(x), \partial|_{G_1} g_1(x), \ldots, \partial|_{G_n} g_n(x)\}.$$

Then (by Theorem 6.2.2) x cannot be a solution to P, whence $F(x) > 0$ necessarily. We would now like to find a nonzero vector v such that, for some $T > 0$, one has

$$F(x + tv) < F(x) \quad \text{for } 0 < t < T.$$

(Such a v is what we term a "complete descent direction," since a shift in the direction of v decreases F, which measures both the cost function f as well as the degree of infeasibility.)

6.2.4 Proposition

Let γ be the point in $\Gamma(x)$ having minimum norm. Then $v = -\gamma$ is a direction of complete descent.

Proof. By assumption we have $v \neq 0$. (Note that γ exists and is unique because $\Gamma(x)$ is closed and convex.) Then v is a perpendicular to $\Gamma(x)$ at γ (see Proposition 2.5.5), so that v belongs to $N_{\Gamma(x)}(\gamma)$, which coincides with the cone of normals in the sense of convex analysis (by Proposition 2.4.4). We deduce

$$\langle v, \gamma' \rangle \leq \langle v, \gamma \rangle = -|v|^2 \quad \text{for all } \gamma' \in \Gamma(x).$$

Now let t be a positive scalar. By the mean-value theorem, one has

$$F(x + tv) - F(x) \in \langle \partial F(x^*), tv \rangle,$$

where x^* lies between x and $x + tv$.

When x^* is such that $F(x^*)$ is positive (which is certainly the case for t sufficiently small, since $F(x) > 0$ and F is continuous), one has

$$\partial F(x^*) \subset \Gamma(x^*).$$

In turn, since Γ is upper semicontinuous, for x^* sufficiently near x, one can guarantee

$$\Gamma(x^*) \subset \Gamma(x) + \frac{|v|}{2} B.$$

It follows under these conditions that $\langle \zeta, v \rangle$ is bounded above by $-|v|^2/2$ for any ζ in $\partial F(x^*)$, which combined with the preceding yields

$$F(x + tv) - F(x) \leq -t \frac{|v|^2}{2}$$

for all t sufficiently small, and confirms the desired property for v. □

When an appropriate method is specified for determining the length of the step in the descent direction, one has the core of a conceptual algorithm for solving P. Note that in general the algorithm would terminate at a point x such that $0 \in \Gamma(x)$, so that one is computing not so much a solution to P as a "stationary point." This is a common feature of many numerical procedures. As E. Polak has stressed, the development of an implementable algorithm requires more study. Such issues as the (approximate) calculation of the set $\Gamma(x)$ and the vector γ often require that one impose somewhat stronger hypotheses on the data or exploit some special structure of the functions involved. In particular, the derivation of a formula for the generalized gradient (such as Proposition 2.3.12 or Example 2.8.7) has been an important step in developing implementable algorithms.

6.3 CONSTRAINT QUALIFICATIONS AND SENSITIVITY

The necessary conditions of Theorems 6.1.1 can be viewed as being degenerate when the multiplier corresponding to f (which we have labelled λ) vanishes, since then the function being minimized is not involved. Various supplemen-

6.3 Constraint Qualifications and Sensitivity

tary conditions have been proposed under which it is possible to assert that the multiplier rule holds in "normal" form (i.e., with $\lambda = 1$). These conditions are called "constraint qualifications." Together with their bearing upon the sensitivity of the mathematical program, they are the object of study in this section.

The Multiplier Sets

Let x be feasible for P (i.e., satisfy the constraints) and let λ and k be non-negative numbers. We define $M_k^\lambda(x)$, the index λ multiplier set corresponding to x (cf. Section 3.4) to be the set of vectors (r, s) in $R^n \times R^m$ satisfying the conclusions of Theorem 6.1.1 (together with λ); that is, the vectors (r, s) such that

$$0 \in \partial_x L(x, \lambda, r, s, k), \qquad r \geq 0, \qquad \langle r, g(x) \rangle = 0.$$

Note that if (λ, r, s, k) satisfy these conditions, and if $\lambda \neq 0$, then so do $(1, r/\lambda, s/\lambda, k)$. Consequently there are two essentially different situations: $\lambda = 0$ and (by normalization) $\lambda = 1$. Theorem 6.1.1 can be rephrased as follows. If x solves P, then for k sufficiently large one has

$$M_k^1(x) \cup \left[M_k^0(x) \setminus \{0\} \right] \neq \emptyset.$$

Observe that $M_k^0(x)$, the set of "abnormal" multipliers, always contains 0.

It is possible to classify constraint qualifications into two categories: the type that makes structural assumptions about the data of the problem so that the set $M_k^0(x)$ of abnormal multipliers necessarily reduces to 0, and the type that simply assures that $M_k^1(x)$ is nonempty (even though $M_k^0(x)$ may not reduce to 0). The primary examples of the first type are the so-called Mangasarian–Fromowitz and Slater conditions and their extensions, which we study in this section. In the second category falls the constraint qualification called "calmness," which we study in the next section. It turns out that calmness can lay claim to being the weakest of these conditions, since it is implied by all the others.

The *Mangasarian–Fromowitz condition* in the classical setting, when each of the functions g_i, h_j is continuously differentiable and $C = X$, is the postulate that the vectors $\nabla h_j(x)$ ($j = 1, 2, \ldots, m$) are linearly independent, and that there is a vector v such that

$$\langle \nabla h_j(x), v \rangle = 0, \qquad j = 1, 2, \ldots, m$$

$$\langle \nabla g_i(x), v \rangle < 0 \quad \text{if } g_i(x) = 0, i = 1, 2, \ldots, n.$$

Let this condition hold, and suppose that the nonzero vector (r, s) belongs to $M_k^0(x)$. Then (since the generalized gradient reduces to the derivative) one has

$$\sum_i r_i \nabla g_i(x) + \sum_j s_j \nabla h_j(x) = 0.$$

Since the vectors $\nabla h_j(x)$ are independent, at least one r_i is nonzero. From the conditions $r \geq 0$, $\langle r, g(x) \rangle = 0$, and $g(x) \leq 0$ (recall that x is a feasible point) one deduces $r_i > 0$, $g_i(x) = 0$ for every i such that r_i is nonzero. Armed with this fact, we take the inner product with v of both sides in the equation above; a contradiction results, so that $M_k^0(x)$ reduces to $\{0\}$.

It is not difficult to formulate analogues to the Mangasarian–Fromowitz condition in the more general setting of this chapter; however, we prefer to use directly the condition $M_k^0(x) = \{0\}$.

Slater-Type Conditions

Suppose that P has no equality constraints ($m = 0$), that the functions g_i are convex, and that C is a convex set. The *Slater condition* in this context is the postulate that there exists a point $\hat{x}$ in C such that $g_i(\hat{x}) < 0$, $i = 1, 2, \ldots, n$. ($\hat{x}$ is called a "strictly feasible" point.)

6.3.1 Proposition

If the Slater condition holds in the problem described above, then $M_k^0(x)$ reduces to $\{0\}$.

Proof. To see this, suppose that $M_k^0(x)$ contained a nonzero element r. Then one has

$$0 \in \partial\{\langle r, g(x) \rangle + k|r|d_C(x)\}.$$

Since the function $y \to \langle r, g(y) \rangle + k|r|d_C(y)$ is convex (see the lemma of Proposition 2.4.4), it attains a minimum at x. Since x is feasible and $\langle r, g(x) \rangle = 0$, the minimum is zero. But the function in question is negative at $\hat{x}$, which is a contradiction. □

We shall see that conditions like the ones above do more than simply guarantee the nonexistence of nontrivial abnormal multipliers; they have implications concerning the sensitivity of the problem to perturbations in the constraints. As a preliminary step we adduce the following result due to Robinson (1976a) and to Ursescu (1975), for which we indicate a particularly simple proof.

6.3.2 Theorem

Let U be a Banach space, and let Γ be a multifunction mapping U to the subsets of X. Suppose that for some positive M and ε, one has

$$MB_X \supset \Gamma(u) \neq \varnothing \quad \text{for } u \text{ in } 2\varepsilon B_U,$$

and suppose that the graph of Γ is convex. Then, for all u, u' in εB_U, for all γ in $\Gamma(u)$, there exists γ' in $\Gamma(u')$ such that $\|\gamma - \gamma'\|_X \leq K\|u - u'\|_U$, where $K = 5M/\varepsilon$.

6.3 Constraint Qualifications and Sensitivity

Proof. In the language of Chapter 3, we are seeking to prove that Γ satisfies a certain Lipschitz condition. Fix u and γ as in the theorem, and define the function f on U by $f(u') := \inf\{\|\gamma' - \gamma\| : \gamma' \in \Gamma(u')\}$. It suffices to prove the bound $f(u') \leq (4M/\varepsilon)\|u' - u\|_U$ (for u in εB_U). We wish to observe that f is (finite and) convex on $2\varepsilon B_U$. This is implied by the following lemma, since we may write $f(u') = \inf_{\gamma'} \{\|\gamma' - \gamma\| + \psi_G(u', \gamma')\}$, where ψ_G is the indicator of the convex set $G := \text{graph}(\Gamma)$.

Lemma. *Let $\phi: U \times X \to R \cup \{+\infty\}$ be convex, and set $f(u') = \inf_x \phi(u', x)$. If f is finite on a convex set S, then f is convex on S.*

Let u_1 and u_2 belong to S, and let x_1, x_2 satisfy $\phi(u_i, x_i) < f(u_i) + \delta$, where δ is an arbitrary positive number. Then, for any λ in $[0, 1]$, one has

$$f(\lambda u_1 + (1 - \lambda)u_2) \leq \phi(\lambda u_1 + (1 - \lambda)u_2, \lambda x_1 + (1 - \lambda)x_2)$$

$$\leq \lambda \phi(u_1, x_1) + (1 - \lambda)\phi(u_2, x_2)$$

$$\leq \lambda f(u_1) + (1 - \lambda)f(u_2) + \delta.$$

Since δ is arbitrary, it follows that f is convex.

Returning now to the proof of the theorem, we observe that f is a convex function on $2\varepsilon B_U$, and that f is bounded by $2M$ on an ε-neighborhood of εB_U. It follows from the corollary to Proposition 2.2.6 that f satisfies a Lipschitz condition of rank $r = 4M/\varepsilon$ on εB_U. Noting that $f(u) = 0$ leads to $f(u') \leq r\|u' - u\|_U$, which completes the proof. $\square$

We now proceed to use the theorem to deduce stability results about parametrized optimization problems; we begin with a result due to Aubin and Clarke (1977b):

Corollary 1

Let S and T be convex subsets of the Banach spaces X and Y, and let A be a linear mapping from S to Y. Let f be a real-valued Lipschitz function defined on S, and assume that S is bounded. Let $V: Y \to R \cup \{+\infty\}$ be defined by

$$V(y) = \inf\{f(x) : x \in S, Ax \in T + y\}.$$

If 0 belongs to the interior of the set $AS - T$, then V is Lipschitz near 0.

Proof. Let the multifunction Γ be defined by

$$\Gamma(y) := \{x \in S : Ax \in T + y\}.$$

Then Γ is bounded and nonempty for y near 0, and it is easy to verify that the graph of Γ is convex. It follows from the theorem that Γ is Lipschitz near 0; this easily implies that V is Lipschitz near 0. □

Let us return now to the special case of the problem P described in Proposition 6.3.1. For p in R^n, let $V(p)$ designate the infimum in the problem P in which the constraints $g_i(x) \leq 0$ are replaced by $g_i(x) + p_i \leq 0$. (V is a special case of what is termed the "value function" in the next section.)

Corollary 2

If, in addition to the hypotheses of Proposition 6.3.1, C is bounded and f is Lipschitz on C, then the Slater condition (i.e., the existence of a strictly feasible point) implies that V is Lipschitz near 0.

Proof. Apply the theorem to deduce that the multifunction

$$\Gamma(p) := \{x \in C : g(x) + p \leq 0\}$$

is Lipschitz near 0. It follows that V is Lipschitz near 0. □

It will follow from the developments of Section 6.5 that when X is finite-dimensional, the condition that $M_k^0(x)$ reduces to $\{0\}$ for every optimal x implies that the value function satisfies a Lipschitz condition. Thus any constraint qualification which rules out abnormal multipliers (such as the Slater or the Mangasarian–Fromowitz conditions) also guarantees stability of the problem. The next constraint qualification we discuss does not have this property.

6.4 CALMNESS

Although our loyalty remains pledged to the problem P of Section 6.1, we find it necessary to imbed P in a parametrized family $P(p, q)$ of mathematical programs. The points p and q lie in R^n and R^m, respectively, and $P(p, q)$ is the problem of minimizing $f(x)$ over the points x in C which satisfy the constraints $g(x) + p \leq 0$, $h(x) + q = 0$. (If v lies in R^n, the notation $v \leq 0$ signifies that each coordinate of v is nonpositive.)

6.4.1 Definition

Let x solve P. The problem P is *calm* at x provided that there exist positive ε and M such that, for all (p, q) in εB, for all x' in $x + \varepsilon B$ which are feasible for $P(p, q)$, one has

$$f(x') - f(x) + M|(p, q)| \geq 0.$$

The *value function* $V: R^n \times R^m \to R \cup \{\pm\infty\}$ is defined via $V(p,q) = \inf\{P(p,q)\}$ (i.e., the value of the problem $P(p,q)$). Note that if there are no feasible points for $P(p,q)$, then the infimum is over the empty set, and $V(p,q)$ is assigned the value $+\infty$.

6.4.2 Proposition

Let $V(0,0)$ be finite, and suppose that one has

$$\liminf_{(p,q)\to(0,0)} \frac{V(p,q) - V(0,0)}{|(p,q)|} > -\infty$$

(*this is true in particular if V is Lipschitz near* 0). *Then, for any solution x to P, P is calm at x.*

Proof. There is an $\varepsilon > 0$ and a positive number M such that all (p,q) in εB satisfy

$$V(p,q) - V(0,0) \geq -M|(p,q)|.$$

If x' lies in $x + \varepsilon B$ and is feasible for $P(p,q)$, then one has $f(x') \geq V(p,q)$; of course one has $V(0,0) = f(x)$. The defining inequality of calmness at x results. □

In analogy to Section 4.2, the problem P is said to be *calm* provided V satisfies the hypothesis of Proposition 6.4.2. In view of Corollary 2 to Theorem 6.3.2, it follows that the special case of P described in connection with Proposition 6.3.1 is calm when the Slater condition holds. We shall prove in the following section that (when X is finite-dimensional) the absence of nontrivial abnormal multipliers implies that V is Lipschitz, and so once again that P is calm (see Theorem 6.5.2, Corollary 1). A local version of this will be proven also: if $M_k^0(x) = \{0\}$, then P is calm at x.

The concept of calmness is closely linked to a numerical technique called "exact penalization."

6.4.3 Proposition

Let x solve P, where P is calm at x. Then, for some $M > 0$, x provides a local minimum in C for the function

$$y \to f(y) + M \max\{g_i(y), |h_j(y)|, 0 : 1 \leq i \leq n, 1 \leq j \leq m\}.$$

Proof. If the assertion is false, then for each integer k there exists a point x_k in $C \cap (x + (1/k)B)$ such that

$$f(x_k) + k \max\{g_i(x_k), |h_j(x_k)|, 0\} < f(x).$$

Necessarily, the maximum above is strictly positive. It follows that $\max\{g_i(x_k), 0\} \to 0$ for each i, and $|h_j(x_k)| \to 0$ for each j. Let us define $p_k = (p_{k,1}, \ldots, p_{k,n})$ by $p_{k,i} = \max\{g_i(x_k), 0\}$, and q_k by $q_{k,j} = |h_j(x_k)|$. Then one has

$$\frac{f(x_k) - f(x)}{\max\{|p_k|, |q_k|\}} < -k,$$

and this contradicts the hypothesis that P is calm at x. Thus, for some M and $\delta > 0$, x must minimize over $C \cap (x + \delta B)$ the function defined in the statement of the proposition. □

We now note that calmness also has the attribute of a constraint qualification. The following result will be seen to be a consequence of Theorem 6.5.2:

6.4.4 Proposition

Suppose that X is finite-dimensional. Let x solve P, where P is calm at x. Then the multiplier rule, Theorem 6.1.1, holds with $\lambda = 1$.

We conclude the discussion by giving a precise meaning to the assertion "almost all problems are calm." Note that the result concerns problems without explicit equality constraints. It is not known whether the analogous result is true for problems with equality constraints.

6.4.5 Proposition

Let P incorporate only inequality constraints $g_i(x) \leq 0$ and the abstract constraint $x \in C$, and suppose that the value function $V(p)$ is finite for p near 0. Then for almost all p in a neighborhood of 0, the problem $P(p)$ is calm.

Proof. By definition, in a neighborhood of 0 the function $V(p) = V(p_1, p_2, \ldots, p_n)$ is finite and nondecreasing as a function of each component p_i. Such functions are differentiable almost everywhere with respect to Lebesgue measure (see Ward, 1935). Clearly $P(p)$ is calm whenever V is differentiable at p, so the result follows. □

6.4.6 Remark

In the general case of P, in which equality constraints exist, it is an easy consequence of Theorem 7.5.1 that $P(p, q)$ is calm for all (p, q) in a dense subset of any open set upon which V is bounded and lower semicontinuous.

6.5 THE VALUE FUNCTION

Unless rather severe restrictions are placed upon the problem P, the value function V (see Section 6.4) is in general neither convex nor differentiable at $(0,0)$ (nor even finite in a neighborhood of $(0,0)$). It is possible, however, to establish a useful formula for the generalized gradient of V under the following mild hypothesis:

6.5.1 Hypothesis
X is finite-dimensional, $V(0,0)$ is finite, and there exist a compact subset Ω of C and a positive number ε_0 such that, for all (p,q) in $\varepsilon_0 B$ for which $V(p,q) < V(0,0) + \varepsilon_0$, the problem $P(p,q)$ has a solution which lies in Ω.

Observe that Hypothesis 6.5.1 by no means requires $P(p,q)$ to admit feasible points for all (p,q) near $(0,0)$ (i.e., $V(p,q)$ can equal $+\infty$ arbitrarily near $(0,0)$), nor does it impose that V be lower semicontinuous near $(0,0)$. A simple growth condition that implies the existence of Ω and ε_0 is the following: for every r in R, the set $\{x \in C : f(x) \leq r\}$ is compact.

We denote by Σ the set of solutions to P lying in Ω. It follows from the basic hypothesis of Section 6.1 that the function $[f, g, h]$ satisfies a Lipschitz condition of some rank $\hat{k}$ on a neighborhood of Ω. Recall that a set is *pointed* if it is not possible to express 0 as a finite sum of its nonzero elements. The multiplier sets $M_k^\lambda(x)$ were defined in Section 6.3; the notation $M_k^\lambda(\Sigma)$ signifies $\cup_{x \in \Sigma} M_k^\lambda(x)$. The following involves as well as ∂V the asymptotic generalized gradient $\partial^\infty V$ defined in Section 2.9.

6.5.2 Theorem
Under Hypothesis 6.5.1, for any $k > \hat{k}$, one has

$$\partial V(0,0) = \overline{\operatorname{co}}\{M_k^1(\Sigma) \cap \partial V(0,0) + M_k^0(\Sigma) \cap \partial^\infty V(0,0)\},$$

$$\partial^\infty V(0,0) \supset \overline{\operatorname{co}}\{M_k^0(\Sigma) \cap \partial^\infty V(0,0)\}.$$

If $M_k^0(\Sigma)$ is pointed, then equality holds in the last relation, and the closure operation is superfluous in both.

We postpone the proof of the theorem until after some of its consequences are drawn.

Corollary 1

Under the hypothesis of the theorem one has

$$\partial V(0,0) \subset \overline{\text{co}} \{ M_k^1(\Sigma) + M_k^0(\Sigma) \}$$

$$\partial^\infty V(0,0) \subset \overline{\text{co}}\, M_k^0(\Sigma).$$

If $M_k^0(\Sigma)$ reduces to $\{0\}$, then V is Lipschitz near $(0,0)$, and $\partial V(0,0)$ is contained in co $M_k^1(\Sigma)$.

Proof. The bounds are evident from the theorem while the final assertion follows from Proposition 2.9.7. (We shall see in the course of the proof of the theorem that epi V is locally closed near $[0, 0, V(0,0)]$.) □

Differentiability of V

Corollary 2

Under the hypothesis of the theorem, if Σ is a singleton $\{x\}$, and if for this x one has $M_k^0(x) = \{0\}$, $M_k^1(x) = \{(r, s)\}$, then V is strictly differentiable at $(0,0)$ with $D_s V(0,0) = (r, s)$.

Proof. In view of Corollary 1, one has $\partial V(0,0) = \{(r, s)\}$. The result follows from Proposition 2.2.4. □

Existence of Normal Multipliers

Corollary 3

Under the hypothesis of the theorem, if $\partial V(0,0)$ is nonempty (in particular, if P is calm), then there exists a solution x to P lying in Σ such that $M_k^1(x) \cap \partial V(0,0) \neq \emptyset$.

Proof. The assertion is evident from Corollary 1, except for the statement that $\partial V(0,0)$ is necessarily nonempty if P is calm. But if P is calm, then it follows immediately from the definition (because V is lower semicontinuous at $(0,0)$, as will be shown in the proof of the theorem) that $V^\circ(0,0; 0,0)$ is greater than $-\infty$; this is equivalent to $\partial V(0,0)$ being nonempty (see the corollary to Theorem 2.9.1). □

Directional Derivative Bounds for V

In order to relate our results to the work of other authors, let us define the upper and lower Dini derivates V^+ and V_+:

$$V^+(0,0; u, v) = \limsup_{t \downarrow 0} \frac{V(t(u, v)) - V(0,0)}{t}$$

6.5 The Value Function

$$V_+(0,0; u, v) = \liminf_{t \downarrow 0} \frac{V(t(u,v)) - V(0,0)}{t}.$$

In the above, and in what follows, the bounds and estimates are given in terms of M_k^1 and M_k^0 as defined in Section 6.3. In all cases, they remain true (but less sharp) if M_k^1, M_k^0 are replaced by larger sets. For example, one might replace $M_k^1(x)$ by the (larger) set of multipliers (r, s) satisfying

$$0 \in \partial_x\{(1, r, s) \cdot (f, g, h)(x)\} + N_C(x),$$

as well as $r \geq 0$, $r \cdot g(x) = 0$.

Corollary 4

Under the hypothesis of the theorem, suppose that $M_k^0(x) = \{0\}$ for each x in Σ. Then one has

$$V^+(0,0; u, v) \leq \inf_{x \in \Sigma} \sup_{(r,s) \in M_k^1(x)} (r, s) \cdot (u, v)$$

$$V_+(0,0; u, v) \geq \inf_{x \in \Sigma} \inf_{(r,s) \in M_k^1(x)} (r, s) \cdot (u, v).$$

If $M_k^1(x)$ is a singleton $\{(r(x), s(x))\}$ for each x in Σ, then $V'(0,0; u, v)$ exists for each (u, v), and one has

$$V'(0,0; u, v) = \inf_{x \in \Sigma} \{(r(x), s(x)) \cdot (u, v)\}.$$

Proof. To prove the first inequality, fix any x in Σ, and define a new problem $\tilde{P}$ by simply replacing $f(y)$ by $\tilde{f}(y) = f(y) + |y - x|^2$. Note that $\tilde{P}$ continues to satisfy the hypothesis of the theorem, and that the corresponding $\tilde{\Sigma}$ for $\tilde{P}$ is $\{x\}$. Since the functions

$$y \to L(y, 1, r, s, k), \quad y \to L(y, 1, r, s, k) + |y - x|^2$$

have the same generalized gradient at x, one has $\tilde{M}_k^1(x) = M_k^1(x)$. Also, $\tilde{M}_k^0(x) = M_k^0(x)$. Applying Corollary 1 to $\tilde{P}$ therefore yields

$$\tilde{V}^+(0,0; u, v) \leq \tilde{V}^\circ(0,0; u, v) \leq \sup\{(r, s) \cdot (u, v) : (r, s) \in M_k^1(x)\}.$$

Since $\tilde{V} \geq V$ everywhere, and $\tilde{V}(0,0) = V(0,0) = f(x)$, it is clear from the definition that $V^+(0,0; u, v) \leq \tilde{V}^+(0,0; u, v)$. We derive

$$V^+(0,0; u, v) \leq \sup\{(r, s) \cdot (u, v) : (r, s) \in M_k^1(x)\}.$$

Since x in Σ is arbitrary, the first inequality is proven.

To prove the second, observe that by Corollary 1 one has

$$\limsup_{\substack{(\alpha,\beta)\to(0,0)\\ t\downarrow 0}} \frac{V((\alpha,\beta)-t(u,v))-V(\alpha,\beta)}{t} \leq \sup_{x\in\Sigma} \sup_{M_k^1(x)} -(u,v)\cdot(r,s),$$

since the left-hand side is $V^\circ(0,0;-u,-v)$ by definition. Now, $-V_+(0,0;u,v)$ may be written as

$$\limsup_{t\downarrow 0} \frac{V(0,0)-V(t(u,v))}{t}.$$

If one regards $t(u,v)$ as a possible (α,β) in the previous inequality, it follows that

$$-V_+(0,0;u,v) \leq \sup_{x\in\Sigma} \sup_{M_k^1(x)} -(u,v)\cdot(r,s),$$

which is the second inequality to be proven.

Finally, if $M_k^1(x) = \{(r(x),s(x))\}$ for each x in Σ, then the two inequalities together give $V^+ = V_+$, so that V' exists, and equals $V^+ = V_+$. $\square$

6.5.3 Remark

There are alternate Dini derivates for which corresponding bounds can be derived, as for example the left upper Dini derivate

$$V^-(0,0;u,v) = \limsup_{t\uparrow 0} \frac{V(t(u,v))-V(0,0)}{t}.$$

We omit the details.

Local Calmness

We are now able to prove a result mentioned in Section 6.4. Hypothesis 6.5.1 is not posited, although X is assumed finite-dimensional.

Corollary 5

Let x be a local solution to P. Suppose that $M_k^0(x) = \{0\}$, where $k > K$ and K is a Lipschitz constant for $[f, g, h]$ on a neighborhood of x. Then P is calm at x.

Proof. To say that P is not calm at x is to say that there is a sequence $(p_i, q_i) \to (0,0)$ and $x_i \to x$ such that x_i is feasible for $P(p_i, q_i)$ and

$$\lim_{i\to\infty} \frac{f(x_i)-f(x)}{|(p_i,q_i)|} = -\infty.$$

6.5 The Value Function

(We may assume $f(x_i) < f(x)$ for each i.) If such is the case, we claim that there is a locally Lipschitz nonnegative function θ vanishing only at x such that

$$\lim_{i \to \infty} \frac{f(x_i) + \theta(x_i) - f(x)}{|(p_i, q_i)|} = -\infty.$$

Indeed, it is easy to verify that the following θ is such a function for any $\alpha \in (0, 1)$:

$$\theta(y) = \alpha \min\{|y - x|^2, \max[f(x) - f(y), d_E(y)]\},$$

where E is the set $\{x_i\}_1^\infty$. Consider the problem $\tilde{P}$ corresponding to P with f replaced by $\tilde{f} = f + \theta$, and C replaced by $\tilde{C} = C \cap (x + \varepsilon \bar{B})$, where $\varepsilon > 0$ is such that x solves P relative to $x + \varepsilon \bar{B}$. Since $\tilde{C}$ is compact, $\tilde{P}$ satisfies Hypothesis 6.5.1, and clearly $\tilde{\Sigma} = \{x\}$. By picking α small enough in defining θ, and ε small enough, one may arrange for $[\tilde{f}, g, h]$ to have Lipschitz constant $\tilde{k} < k$ on $x + \varepsilon \bar{B}$. Note that $\tilde{M}_k^0(x) = \{0\}$, so that Corollary 1 applies to $\tilde{P}$, which is therefore calm at x. But for i large, $\tilde{V}(p_i, q_i) \leq f(x_i) + \theta(x_i)$, and of course $\tilde{V}(0, 0) = f(x)$. In consequence,

$$\lim_{i \to \infty} \frac{\tilde{V}(p_i, q_i) - \tilde{V}(0, 0)}{|(p_i, q_i)|} = -\infty,$$

a contradiction. □

Some attention has been paid to the following apparently more general way of perturbing P:

$$P_\alpha: \min\{f(x, \alpha) : g(x, \alpha) \leq 0, h(x, \alpha) = 0, (x, \alpha) \in D\},$$

where α is a vector of k real parameters. The value function V would then be a function of α: $V(\alpha) = \inf P_\alpha$. The case we have been discussing corresponds to taking $k = n + m$, $\alpha = (p, q)$, $f(x, \alpha) = f(x)$, $g(x, \alpha) = g(x) + p$, $h(x, \alpha) = h(x) + q$, $D = C \times R^{n+m}$. We owe to Rockafellar the observation that, at least when the dependence of f, g, and h on α is locally Lipschitz, there is no actual gain in generality from considering P_α rather than $P(p, q)$. This is due to the fact that $V(\alpha)$ can equally be regarded as the value of the following problem over (x, y) in $X \times R^k$:

$$\min\{f(x, y) : g(x, y) \leq 0, h(x, y) = 0, -y + \alpha = 0, (x, y) \in D\},$$

which is of the form we have been considering, with (some of) the equality constraints being varied additively. Thus the methods and results of this section can be brought to bear upon the perturbed family P_α.

Proof of Theorem 6.5.2

Lemma 1. Let $\{(p_i, q_i)\}$ be a sequence converging to $(0,0)$, with $\limsup_{i\to\infty} V(p_i, q_i) \leq V(0,0)$, and let x_i in Ω solve the problem $P(p_i, q_i)$. Then a subsequence of $\{x_i\}$ converges to an element x of Σ.

Since Ω is compact, a subsequence of $\{x_i\}$ converges to an element x of Ω. The conditions $g(x_i) + p_i \leq 0$, $h(x_i) + q_i = 0$, $x_i \in C$ imply that x is feasible for P. We have $f(x) \leq \limsup_{i\to\infty} f(x_i) \leq V(0,0)$, so that x solves P. Thus x belongs to Σ.

Note that we also derive $f(x) = \limsup_{i\to\infty} f(x_i) = V(0,0)$, so that we have also proved that V is lower semicontinuous at $(0,0)$. It also follows much as above that for some $\delta > 0$, the set $\mathrm{epi}\, V \cap \{[0,0,V(0,0)] + \delta \bar{B}\}$ is closed.

In view of how ∂V and $\partial^\infty V$ were defined, the theorem is ultimately a statement about normal cones to $\mathrm{epi}\, V$. We proceed to define

$$N_1 := \{\alpha(r, s, -1) : \alpha > 0, (r, s) \in M_k^1(\Sigma) \cap \partial V(0,0)\}$$

$$N_2 := \{(r, s, 0) : (r, s) \in M_k^0(\Sigma) \cap \partial^\infty V(0,0)\}.$$

As in Step 2 of the proof of Theorem 3.4.3, to establish the theorem it suffices to prove

Lemma 2.

$$N_{\mathrm{epi}\, V}(0, 0, V(0,0)) = \overline{\mathrm{co}}\{N_1 \cup N_2\}.$$

Now $N_1 \cup N_2$ is contained in $N_{\mathrm{epi}\, V}(0,0,V(0,0))$ by construction, so it suffices to prove that every element of the latter belongs to $\overline{\mathrm{co}}\{N_1 \cup N_2\}$. In view of Theorem 2.5.7, it suffices to prove that any point $(u_0, v_0, -\beta_0)$ of the form

$$(u_0, v_0, -\beta_0) = \lim_{i\to\infty} \frac{(u_i, v_i, -\beta_i)}{|(u_i, v_i, -\beta_i)|}$$

belongs to $N_1 \cup N_2$, where $(u_i, v_i, -\beta_i)$ is perpendicular to $\mathrm{epi}\, V$ at (p_i, q_i, α_i), and where $(u_i, v_i, -\beta_i) \to 0$, $(p_i, q_i, \alpha_i) \to (0, 0, V(0,0))$.

The following is the key to the proof:

Lemma 3. Let $(u, v, -\beta)$ be perpendicular to $\mathrm{epi}\, V$ at (p, q, α), where (p, q) belongs to $\varepsilon_0 B$ and $\alpha < V(0,0) + \varepsilon_0$ (see Hypothesis 6.5.1). Then $u \geq 0$, $\beta \geq 0$, and there exists a solution x to $P(p, q)$ which lies in Ω, and such that $u \cdot (g(x) + p) = 0$ and $0 \in \partial_x L(x, \beta, u, v, k)$.

6.5 The Value Function

For any point y in C, for any vector $\sigma \geq 0$ in R^n, for any τ in $[0, \infty)$, one has

$$[-g(y) - \sigma, -h(y), f(y) + \tau] \in \text{epi } V$$

by definition of V. The inequality of Proposition 2.5.5 translates here as:

(1) $\beta f(y) + \beta\tau + u \cdot g(y) + u \cdot \sigma + v \cdot h(y) + u \cdot p + v \cdot q - \beta\alpha$

$\quad + \frac{1}{2}\{|p + g(y) + \sigma|^2 + |q + h(y)|^2 + |\alpha - f(y) - \tau|^2\} \geq 0.$

Let x be a point in Ω which solves $P(p, q)$ (x exists by Hypothesis 6.5.1). Then note that equality holds in (1) when $y = x$, $\tau = \alpha - f(x)$, $\sigma = -g(x) - p$. (Note that $\alpha - f(x)$ is nonnegative because (p, q, α) in epi V implies $\alpha \geq V(p, q) = f(x)$; $-g(x) - p$ is nonnegative because x is feasible for $P(p, q)$.) It follows that the derivatives from the right of the left-hand side of (1) (when $y = x$) with respect to τ and with respect to σ are nonnegative when evaluated at $\alpha - f(x)$ and $-g(x) - p$, respectively. This gives $\beta \geq 0$, $u \geq 0$. If $g_i(x) + p_i < 0$ for some i, then the derivative with respect to σ_i actually vanishes at $-g_i(x) - p_i$ (since a local minimum occurs there). We derive $u_i = 0$ in this case. Thus $u \cdot (g(x) + p) = 0$.

Now put $\tau = \alpha - f(x)$ and $\sigma = -g(x) - p$ in inequality (1), and observe that we obtain the conclusion that for any y in C, one has (after replacing q by $-h(x)$),

$$\beta f(y) + u \cdot g(y) + v \cdot h(y) + \frac{1}{2}\{|f(y) - f(x)|^2 + |g(y) - g(x)|^2$$

$$+ |h(y) - h(x)|^2\} \geq \beta f(x) + u \cdot g(x) + v \cdot h(x).$$

Note that if y is sufficiently near x, the left-hand side satisfies a Lipschitz condition of rank $|(u, v, \beta)|k$. We deduce from Proposition 2.4.3 that x provides a local minimum for the function

$$\beta f(y) + u \cdot g(y) + v \cdot h(y) + |(u, v, \beta)|kd_C(y) + \frac{1}{2}\{|f(y) - f(x)|^2$$

$$+ |g(y) - g(x)|^2 + |h(y) - h(x)|^2\}.$$

Notice that this function is the sum of $L(y, \beta, u, v, k)$ plus a function whose generalized gradient at $y = x$ is $\{0\}$ (since it satisfies a Lipschitz condition of arbitrarily small rank sufficiently near x). One deduces from Proposition 2.3.3 that

$$0 \in \partial_x L(x, \beta, u, v, k),$$

and the lemma is proven.

We are now ready to complete the proof of the theorem. Let $(u_0, v_0, -\beta_0)$ be as described following Lemma 2; we wish to show that $(u_0, v_0, -\beta_0)$ belongs to $N_1 \cup N_2$. By Lemma 3, there exists for each i sufficiently large a solution x_i to $P(p_i, q_i)$ lying in Ω such that

$$\langle u_i, g(x_i) + p_i \rangle = 0, \qquad 0 \in \partial_x L(x_i, \beta_i, u_i, v_i, k),$$

and one also has $u_i \geq 0$. By Lemma 1 we may assume that x_i converges to an element x of Σ.

Lemma 4. $0 \in \partial_x L(x, \beta_0, u_0, v_0, k)$, $u_0 \geq 0$, $\langle u_0, g(x) \rangle = 0$.

The last two assertions follow immediately from taking limits. Let $\gamma_i = |(u_i, v_i, -\beta_i)|$. The function θ_i defined by

$$\theta_i(y) := L\left(y, \frac{\beta_i}{\gamma_i}, \frac{u_i}{\gamma_i}, \frac{v_i}{\gamma_i}, k\right) - L(y, \beta_0, u_0, v_0, k)$$

$$= \left\{\frac{1}{\gamma_i}(\beta_i, u_i, v_i) - (\beta_0, u_0, v_0)\right\} \cdot [f, g, h](y)$$

is easily shown to be Lipschitz of rank $k\varepsilon_i$ near x, where $\varepsilon_i = |(\beta_i, u_i, v_i)/\gamma_i - (\beta_0, u_0, v_0)|$. Note $\varepsilon_i \to 0$. We have $0 \in \partial_x L(x_i, \beta_i/\gamma_i, u_i/\gamma_i, v_i/\gamma_i, k)$ from the above. Writing $L(\cdot, \beta_0, u_0, v_0, k)$ as $L(\cdot, \beta_i/\gamma_i, u_i/\gamma_i, v_i/\gamma_i, k) + \theta_i(\cdot)$ and noting that $\partial \theta_i(x) \subset k\varepsilon_i \overline{B}$ (by Proposition 2.1.2(a)) yields

$$0 \in \partial_x L(x_i, \beta_0, u_0, v_0, k) + k\varepsilon_i \overline{B}.$$

Now the assertion of the lemma is seen to follow from the upper semicontinuity of the generalized gradient (Proposition 2.1.5(d)).

Because $(u_0, v_0, -\beta_0)$ is a limit of normalized perpendiculars, it follows from Theorem 2.5.7 that $(u_0, v_0, -\beta_0)$ belongs to $N_{\text{epi } V}(0, 0, V(0, 0))$. Thus (u_0, v_0) belongs to $\partial^\infty V(0,0)$ if β_0 is 0, and $(u_0/\beta_0, v_0/\beta_0)$ belongs to $\partial V(0,0)$ if β_0 is positive. If β_0 is zero, then (u_0, v_0) belongs to $M_k^0(x)$ as a consequence of Lemma 4, so that $(u_0, v_0, -\beta_0) = (u_0, v_0, 0)$ belongs to N_2. If β_0 is positive, then the lemma shows that $(u_0, v_0)/\beta_0$ belongs to $M_k^1(x)$, whence $(u_0, v_0, -\beta_0) = \beta_0(u_0/\beta_0, v_0/\beta_0, -1)$ belongs to N_1. In either case, then, one has $(u_0, v_0, -\beta_0) \in N_1 \cup N_2$ and the proof is complete. □

6.6 SOLVABILITY AND SURJECTIVITY

The results of the previous section have application to issues not involving optimization per se. To illustrate, let us consider the existence of solutions y to the following system of inequality, equality, and abstract constraints:

(1) $\qquad g(y) + p \leq 0, \qquad h(y) + q = 0, \qquad y \in C.$

6.6 Solvability and Surjectivity

We continue to assume throughout this section that X is finite-dimensional. Let x satisfy the constraints (1) when $p = 0$, $q = 0$ (i.e., x is feasible for P), and let $\hat{k}$ be a Lipschitz constant for $[g, h]$ on some neighborhood of x. Recall that $M_k^0(x)$ is the set of vectors (r, s) satisfying

$$r \geq 0, \quad r \cdot g(x) = 0, \quad 0 \in \partial_x\{r \cdot g + s \cdot h + k|(r, s)|d_C\}(x)$$

(i.e., the necessary conditions in abnormal form).

6.6.1 Theorem

Let $M_k^0(x)$ reduce to $\{0\}$ for some $k > \hat{k}$. Then there are constants m and $\varepsilon > 0$ such that, for all (p, q) in εB, there exists y satisfying the constraints (1) as well as

$$|y - x| \leq m|(p, q)|.$$

Proof. Let us set $f(y) = \delta|y - x|$, where $\delta > 0$ is small enough so that k exceeds some Lipschitz constant for $[f, g, h]$ on the set $x + \delta \bar{B}$. Set $\tilde{C} = C \cap \{x + \delta \bar{B}\}$, and let $\tilde{P}$ be the version of P with data f, g, h, and $\tilde{C}$. Note that $\tilde{\Sigma} = \{x\}$, and that $\tilde{P}$ conforms to Hypothesis 6.5.1. Since $d_{\tilde{C}}$ and d_C agree near x, it follows that $M_k^0(\tilde{\Sigma}) = M_k^0(x) = \{0\}$, so that by Theorem 6.5.2, Corollary 1, $\tilde{V}$ is Lipschitz near 0. Thus for some K, for all (p, q) near 0, one has

$$\tilde{V}(p, q) = \tilde{V}(p, q) - \tilde{V}(0, 0) \leq K|(p, q)|.$$

Consequently, there is a point y feasible for $\tilde{P}(p, q)$ (and hence satisfying (1)) such that $\delta|y - x| \leq K|(p, q)|$. The assertion of the theorem is seen to hold for $m = K/\delta$. □

The following is related to our proof in Chapter 7 of the inverse function theorem for Lipschitz functions.

Corollary

Let F be a map from R^n to R^n which is Lipschitz near a point x, and suppose that every matrix in the generalized Jacobian $\partial F(x)$ is nonsingular. Then there exist a constant m and neighborhoods Y and Q of x and 0, respectively, such that, for every q in Q there exists a point y in Y such that

$$F(y) = F(x) - q, \quad |y - x| \leq m|q|.$$

Proof. Apply the theorem with $h(y) = F(y) - F(x)$, g absent (or identically equal -1), and $C = X$. That $M_k^0(x)$ reduces to $\{0\}$ follows from the fact that $\partial_x\{v \cdot h\}(x)$ coincides with $v^*\partial h(x) = v^*\partial F(x)$ (by Theorem 2.6.6). □

It is possible to use the proof technique of Theorem 6.5.2 to derive information about V beyond what is inherent in the multipliers. The key role is played by the following concept, which is defined within the context of the problem P.

6.6.2 Definition

Let x belong to C, let β be a real number, and let u and v be vectors in R^n and R^m, respectively. Then (x, β, u, v) is said to satisfy the *penalty property* provided that $\beta \geq 0$, $u \geq 0$, $|(\beta, u, v)| = 1$, and provided that, for some $m > 0$, the function

$$y \to \beta f(y) + u \cdot g(y) + v \cdot h(y)$$
$$+ m\{|f(y) - f(x)|^2 + |g(y) - g(x)|^2 + |h(y) - h(x)|^2\}$$

attains a minimum over C at $y = x$.

The proof of Theorem 6.5.2 established the following:

6.6.3 Proposition

Under the hypothesis of Theorem 6.5.2, $N_{epi\ V}(0, 0, V(0, 0))$ is contained in the closed convex cone generated by the set

$$\{\left(0, \lim[u_i, v_i, -\beta_i]\right) : (x_i, \beta_i, \mu_i, v_i) \text{ satisfies the penalty property}, d_\Sigma(x_i) \to 0\}.$$

Corollary

Under the hypothesis of Theorem 6.5.2, suppose that there exists an $\varepsilon > 0$ such that whenever (x, β, u, v) satisfies the penalty property for some x in $\Sigma + \varepsilon B$, one has $\beta \geq \varepsilon$. Then V is (finite and) Lipschitz near $(0, 0)$.

Proof. It suffices to show that $\partial^\infty V(0, 0)$ reduces to $\{0\}$, in view of Proposition 2.9.7. This is equivalent to the condition that there be no vectors $[u, v, 0]$ in $N := N_{epi\ V}(0, 0, V(0, 0))$ with $[u, v]$ nonzero. Clearly this would result from the condition that, for some $\delta > 0$, every vector $[u, v, -\beta]$ in N satisfies $\beta \geq \delta|(u, v)|$. By assumption this is true for the set in the statement of the proposition, and the property is inherited by the closed convex cone generated by this set. Thus N has the property too, and the proof is complete. □

By using Proposition 6.6.3 one can derive (under appropriate smoothness hypotheses) formulas for ∂V analogous to that in Theorem 6.5.2, but with multipliers which involve *second-order* information about the data of the problem. As an illustration, we derive the following surjectivity result:

6.6.4 Theorem

Let $F: R^n \to R^n$ be twice continuously differentiable on a neighborhood of a compact set C, and suppose that for every y in C, the conditions

(i) $DF(y)^* v = 0$ and
(ii) $\sum_{i=1}^n v_i \langle w, F_i''(y)w \rangle \geq 0$ for every w such that $DF(y)w = 0$

together imply that $v = 0$. If x is any point in C such that $\{y \in C: F(y) = F(x)\} \subset int(C)$, then $F(C)$ contains a neighborhood of $F(x)$.

Proof. (Note that F_i'' refers to the $n \times n$ Hessian matrix of second partial derivatives of the ith component function of F.) Let f be identically 0, and let $h(y) = F(y) - F(x)$. Note that the resulting P (g is absent) conforms to Hypothesis 6.5.1, with $\Sigma = C \cap F^{-1}(F(x))$. We claim that the hypothesis of the Corollary to Proposition 6.6.3 is satisfied. Indeed, let (y, β, v) satisfy the penalty property for some y (in C). For ε small, $\Sigma + \varepsilon B$ is contained in the interior of C, so that the function

$$\theta(x') := \langle v, F(x') - F(x) \rangle + m|F(x') - F(y)|^2$$

attains a local minimum at $x' = y$. It follows that $\nabla \theta(y) = 0$ and that $\theta''(y)$ is positive semidefinite. These conditions imply (i) and (ii), whence $v = 0$. Since $|(\beta, v)| = 1$, we derive $\beta = 1$. The Corollary to Proposition 6.6.3 is therefore applicable here. It follows that $V(q)$ is finite for q near 0, so that for every q near 0 there is a solution y in C of the equation $F(y) = F(x) - q$, as asserted. □

The usual condition yielding surjectivity is nonsingularity of the Jacobian matrix. Note, for example, that when $DF(x)$ is nonsingular, the theorem above follows from the corollary to Theorem 6.6.1. As the following illustrates, however, it is possible to guarantee surjectivity even when $DF(x)$ is the zero matrix if appropriate second-order behavior is postulated.

Corollary

Let x be an isolated 0 of a function $F: R^n \to R^n$ which is twice continuously differentiable near x. Suppose that $DF(y)$ is nonsingular for y near x and different from x, and that $DF(x) = 0$. If the matrices $F_i''(x)$ are linearly independent and have trace 0, then there is a neighborhood of x which is mapped under F to a neighborhood of 0.

Proof. Let C be a compact neighborhood of x containing no other zeros of F, on a neighborhood of which F is C^2, and such that $DF(y)$ is nonsingular for every y in $C \setminus \{x\}$. If v is a vector satisfying condition (i) of Theorem 6.6.4 for some y in C different from x, then v is necessarily 0. If v satisfies condition (ii) for $y = x$, then the symmetric matrix $M = \Sigma v_i F_i''(x)$ is positive semidefinite on R^n. Since it has trace 0, the matrix M must be 0. Since the matrices $F_i''(x)$ are independent, we deduce $v = 0$. We have verified the hypothesis of Theorem 6.6.4, which gives the required conclusion. □

6.6.5 Remark

An example to which the preceding applies is provided by the function $F(x, y) = [x^2 - y^2, 2xy]$ at $(0, 0)$.

Chapter Seven

Topics in Analysis

You know my methods, Watson.

ARTHUR CONAN DOYLE, *Memoirs of Sherlock Holmes*

You have delighted us long enough.

JANE AUSTEN, *Pride and Prejudice*

The final chapter applies the results of the previous ones to a variety of topics in which optimization is not explicitly involved. The first three sections employ the analytical and geometrical machinery developed in Chapter 2 to treat inverse and implicit functions, epi-Lipschitzian sets, and Aumann's Theorem. The next section uses Chapter 3 to extend the classical theory of the resolvent, and the fifth and sixth sections present a theorem of Ekeland and illustrate its use in nonsmooth analysis. The final sections deal with boundary-value problems, the theory of which is a traditional application of the calculus of variations. The methods of Chapters 3 and 4 are combined with a dual variational principle to shed new light on Hamiltonian boundary-value problems of classical origin.

7.1 INVERSE AND IMPLICIT FUNCTIONS

The classical inverse-function theorem gives conditions under which a C^1 function admits (locally) a C^1 inverse. We shall give here conditions under which a Lipschitzian (not necessarily differentiable) function admits (locally) a

7.1 Inverse and Implicit Functions

Lipschitzian inverse. Let F be a given function mapping R^n to itself, and let x_0 be a point near which F is Lipschitz. The generalized Jacobian $\partial F(x_0)$ (see Section 2.6) is said to be of *maximal rank* provided every matrix M in $\partial F(x_0)$ is of maximal rank (i.e., nonsingular).

7.1.1 Theorem (Inverse Functions)

If $\partial F(x_0)$ is of maximal rank, then there exist neighborhoods U and V of x_0 and $F(x_0)$, respectively, and a Lipschitz function $G\colon V \to R^n$ such that

(i) $G(F(u)) = u$ for every $u \in U$
(ii) $F(G(v)) = v$ for every $v \in V$.

7.1.2 Remarks

(i) When F is C^1, $\partial F(x_0)$ reduces to $\{JF(x_0)\}$, and the function G above is necessarily C^1 as well; thus we recover the classical theorem.
(ii) It is not enough to assume merely that the Jacobian $JF(x)$ is of maximal rank whenever it exists at points x near x_0. A counterexample is provided by $F(x) = |x|$ with $n = 1$, $x_0 = 0$.
(iii) A simple example to which the theorem applies ($n = 2$) is the following:

$$F(x, y) = [|x| + y, 2x + |y|].$$

We find

$$\partial F(0,0) = \left\{ \begin{bmatrix} s & 1 \\ 2 & t \end{bmatrix} : -1 \leq s \leq 1,\ -1 \leq t \leq 1 \right\},$$

which clearly has maximal rank.

Proof of Theorem 7.1.1 The proof is in lemma-size steps.

Lemma 1. There are positive numbers r and δ with the following property. Given any unit vector v in R^n, there is a unit vector w in R^n such that, whenever x lies in $x_0 + rB$ and M belongs to $\partial F(x)$, one has

(1) $$\langle w, Mv \rangle \geq \delta.$$

In the following, as in Section 2.6, $B_{n \times n}$ denotes the open unit ball in the space of $n \times n$ matrices, and we consider vectors in R^n to be columns. Note that the vector w is to depend only on v and not on x. Let S signify the unit sphere in R^n. The subset $\partial F(x_0)S$ of R^n is compact and does not contain 0, since $\partial F(x_0)$ is of maximal rank. Hence for some $\delta > 0$, $\partial F(x_0)S$ is distance 2δ from 0. For ε sufficiently small, ΩS is distance at least δ from 0, where $\Omega := \partial F(x_0) + \varepsilon B_{n \times n}$. Since ∂F is upper semicontinuous (see Proposition 2.6.2), it follows that, for some positive r,

(2) $\qquad x \in x_0 + rB \quad \text{implies} \quad \partial F(x) \subset \Omega.$

We may suppose r chosen so that F satisfies a Lipschitz condition on $x_0 + r\bar{B}$.

Now let any unit vector v be given. It follows from the above that the convex set Ωv is distance at least δ from 0. By the separation theorem for convex sets, there is a unit vector w such that $\langle w, \Omega v \rangle \geq \delta$. Relation (1) follows from this together with relation (2).

Lemma 2. If x and y lie in $x_0 + r\bar{B}$, then

$$|F(x) - F(y)| \geq \delta|x - y|.$$

We may suppose $x \neq y$ and (in view of the continuity of F) $x, y \in x_0 + rB$. Set

$$v = \frac{y - x}{|y - x|}$$

$$\lambda = |y - x|,$$

so that $y = x + \lambda v$.

Let π be the plane perpendicular to v and passing through x. The set P of points x' in $x_0 + rB$ where $DF(x')$ fails to exist is of measure 0, and hence by Fubini's theorem, for almost every x' in π, the ray

$$x' + tv, \quad t \geq 0$$

meets P in a set of 0 one-dimensional measure. Choose an x' with the above property and sufficiently close to x so that $x' + tv$ lies in $x_0 + rB$ for every t in $[0, \lambda]$. Then the function

$$t \to F(x' + tv)$$

is Lipschitzian for t in $[0, \lambda]$ and has a.e. on this interval the derivative $JF(x' + tv)v$. Thus

$$F(x' + \lambda v) - F(x') = \int_0^\lambda JF(x' + tv)v \, dt.$$

Let w correspond to v as in Lemma 1. We deduce

$$w \cdot [F(x' + \lambda v) - F(x')] = \int_0^\lambda w \cdot [JF(x' + tv)v] \, dt \geq \int_0^\lambda \delta \, dt = \delta \lambda.$$

Recalling the definition of λ, we arrive at:

$$|F(x' + \lambda v) - F(x')| \geq \delta|x - y|.$$

7.1 Inverse and Implicit Functions

This may be done for x' arbitrarily close to x. Since F is continuous, the lemma follows.

Lemma 3. $F(x_0 + rB)$ contains $F(x_0) + (r\delta/2)B$.

Let y be any point in $F(x_0) + (r\delta/2)B$, and let the minimum of $|y - F(\cdot)|^2$ over $x_0 + r\overline{B}$ be attained at x. We claim first that x belongs to $x_0 + rB$. Otherwise, by Lemma 2 and the triangle inequality, one has

$$\frac{r\delta}{2} > |y - F(x_0)| \geq |F(x) - F(x_0)| - |y - F(x)|$$

$$\geq \delta|x - x_0| - |y - F(x)|$$

$$\geq \delta r - |y - F(x_0)| \quad \text{(by the minimality of } x\text{)}$$

$$> \delta r - \frac{\delta r}{2} = \frac{r\delta}{2},$$

which is a contradiction. Thus x yields a local minimum for the function $|y - F(\cdot)|^2$, and consequently

$$0 \in \partial |y - F(x)|^2.$$

By Theorem 2.6.6, this implies that 0 belongs to the set $2(y - F(x))^*\partial F(x)$. But Lemma 1 implies that every element of $\partial F(x)$ is invertible, whence $F(x) = y$ necessarily.

To prove the theorem, we now set $V = F(x_0) + (r\delta/2)B$, and we define G on V as follows: $G(v)$ is the unique x in $x_0 + rB$ such that $F(x) = v$. We choose U as any neighborhood of x_0 satisfying $F(U) \subset V$. The theorem now follows, since Lemma 2 implies that G is Lipschitz of rank $1/\delta$. □

7.1.3 Remark

The theorem is true with only minor modifications in the proof if ∂F is calculated while avoiding points in a given set S of measure 0 (i.e., if the "constrained" $\partial_S F$ of Proposition 2.6.4 is of maximal rank at x_0).

The Implicit Function Theorem

Let H be a function mapping $R^m \times R^k$ to R^k, and let us consider the equation

$$H(y, z) = 0$$

where y lies in R^m and z in R^k. We wish to "solve" this equation for z as a function of y around a point $(\hat{y}, \hat{z})$ at which the equation holds. When H is Lipschitz near $(\hat{y}, \hat{z})$, the following gives a sufficient condition for this to be possible. The notation $\pi_z \partial H(\hat{y}, \hat{z})$ signifies the set of all $k \times k$ matrices M such that, for some $k \times m$ matrix N, the $k \times (k + m)$ matrix $[N, M]$ belongs to $\partial H(\hat{y}, \hat{z})$.

Corollary

Suppose that $\pi_z \partial H(\hat{y}, \hat{z})$ is of maximal rank. Then there exists a neighborhood Y of $\hat{y}$ and a Lipschitz function $\zeta(\cdot): Y \to R^k$ such that $\zeta(\hat{y}) = \hat{z}$ and such that, for every y in Y,

$$H(y, \zeta(y)) = 0.$$

Proof. Let $n = k + m$, and define F from R^n to itself by

$$F(y, z) := [y, H(y, z)].$$

When the Jacobian matrix DF exists, it is of the form

$$\begin{bmatrix} I & 0 \\ D_y H & D_z H \end{bmatrix}.$$

It follows that $\partial F(\hat{y}, \hat{z})$ is of maximal rank. Applying the theorem to F immediately gives the desired result. In fact, one obtains a Lipschitz function $\zeta(y, u)$ such that, for every y near $\hat{y}$ and u near 0, one has

$$H(y, \zeta(y, u)) = u. \quad \square$$

7.2 AUMANN'S THEOREM

We shall now give a simple variational proof of a result which, in one form or another, lies at the heart of many qualitative issues in the calculus of variations and optimal control, notably existence and relaxation theory and the bang-bang theorem. (It is also well known in connection with Lyapounov's theorem on vector measures.) The approach, which relies upon the results of Chapter 2, has the added feature of implying a constructive procedure which we shall make explicit later on.

Let F be a multifunction mapping a real interval $[a, b]$ to R^n, and suppose that F is measurable, closed and nonempty, and integrably bounded (see Section 3.1). By co F we mean the multifunction whose value at t is co $F(t)$, the convex hull of the set $F(t)$. Recall that $\int F$ signifies the set of points of the form $\int_a^b f(t)\, dt$, where f is a measurable selection for F (and similarly for $\int \mathrm{co}\, F$). Aumann's Theorem is the following assertion:

7.2.1 Theorem

$$\int F = \int \mathrm{co}\, F.$$

7.2 Aumann's Theorem

Proof.

Step 1. It suffices to show that any point ξ lying in $\int \mathrm{co}\, F$ also lies in $\int F$; we may take $\xi = 0$ without loss of generality. The set of all measurable selections for $\mathrm{co}\, F$ is a nonempty (Theorem 3.1.1) convex weakly compact subset of L^1. (It is weakly relatively compact by the Dunford–Pettis criterion (Edwards 1965, Theorem 4.21.2), strongly closed since F is closed-valued, and thus weakly closed because convex strongly closed sets are weakly closed.) It follows that $\int \mathrm{co}\, F$ is convex, compact, and nonempty. Its dimension K is defined to be the dimension of the minimal subspace S of R^n containing it. If $K = 0$, $\mathrm{co}\, F$ is a point and must coincide with $\int F$, so we may assume $K \geq 1$, and that the theorem is true for multifunctions G such that $\dim \int \mathrm{co}\, G \leq K - 1$.

Step 2. Suppose now that 0 does not lie in the relative interior of $\int \mathrm{co}\, F$ (i.e., interior relative to S). Then there is a nonzero vector d in S normal to $\int \mathrm{co}\, F$ at 0; that is, such that

$$\text{(1)} \qquad d \cdot w \leq 0 \quad \text{for all } w \text{ in } \int \mathrm{co}\, F.$$

Let us define $g: R^n \to R$ as follows:

$$\text{(2)} \qquad g(x) := \sup\left\{ \int_a^b x \cdot f(t)\, dt : f(\cdot) \text{ is a selection for } \mathrm{co}\, F \right\}.$$

We pause to show that, in Eq. (2), we may "take the maximum under the integral." For any subset C of R^n and any point x in R^n, we use the notation $\sigma(C, x)$ to denote the quantity

$$\sup\{\langle x, c\rangle : c \in C\}$$

(i.e., the support function of C evaluated at x). Note that $\sigma(C, x) = \sigma(\mathrm{co}\, C, x)$. It is easy to see that the function $\sigma(F(t), x)$ is measurable in t and continuous in x. As a consequence of this fact and Proposition 3.1.2, it follows that, for given x, the multifunction M defined as follows is measurable:

$$M(t) = \{f \in \mathrm{co}\, F(t) : \sigma(F(t), x) = \langle f, x\rangle\},$$

and hence (Theorem 3.1.1) admits a measurable selection.

But any such selection must provide the supremum in Eq. (2), which is therefore attained. We have shown, in fact:

$$\text{(3)} \qquad g(x) = \int_a^b \sigma(F(t), x)\, dt.$$

It follows that any selection $f(\cdot)$ yielding the maximum in Eq. (2) must satisfy $\sigma(F(t), x) = \langle f(t), x\rangle$ a.e.

Since $0 \in \int \operatorname{co} F$, we have $g(\cdot) \geq 0$. On the other hand, condition (1) yields $g(d) = 0$. Thus g attains a minimum at d, and by Theorem 2.8.2 and the above, there is a selection $f_0(\cdot)$ for co F such that $\int_a^b f_0(t)\, dt = 0$ and $\sigma(\operatorname{co} F(t), d) = f_0(t) \cdot d$ a.e. If we define a new multifunction $\tilde{F}$ via

$$\tilde{F}(t) = \{f \in F(t) : \langle f, d \rangle = \sigma(F(t), d)\},$$

then $\tilde{F}$ satisfies the same hypotheses as F (Proposition 3.1.2), and $0 \in \int \operatorname{co} \tilde{F}$ (since $f_0(t) \in \operatorname{co} \tilde{F}(t)$ a.e.). Further, $\langle d, \int \operatorname{co} \tilde{F} \rangle = \langle d, \int_a^b f_0(t)\, dt \rangle = 0$, and so $\dim \int \operatorname{co} \tilde{F} \leq K - 1$. Thus we are finished by the inductive hypothesis, in view of the fact that $\int \tilde{F}$ is contained in $\int F$.

Step 3. The remaining case is the one in which 0 belongs to the relative interior of $\int \operatorname{co} F$; that is, $\delta B \cap S \subset \int \operatorname{co} F$ for some $\delta > 0$. Let us choose any nonzero vector d_1 and define

(4) $\quad g_1(x) := \max\left\{ \int_a^b \langle d_1 t + x, f(t) \rangle\, dt : f(\cdot) \text{ is a selection for co } F \right\}.$

(As in Step 2, this is equivalent to a pointwise maximization problem.) We proceed to show that g_1 attains a minimum. Let any x in R^n be expressed in the form $s + c$, where s belongs to S and c lies in its orthogonal complement. Then

$$g_1(x) = g_1(s + c) = g_1(s)$$
$$= \max\left\{ \left\langle s, \int_a^b f(t)\, dt \right\rangle + \int_a^b \langle d_1 t, f(t) \rangle\, dt : f \text{ is a selection for co } F \right\}$$
$$\geq \delta |s| + k,$$

for δ and k independent of x. It follows that g_1 attains a minimum, say at x_1. Invoking Theorem 2.8.2, we deduce the existence of a selection f_0 for co F such that $\int_a^b f_0(t)\, dt = 0$ and

(5) $\quad \langle d_1 t + x_1, f_0(t) \rangle = \sigma(F(t), d_1 t + x_1) \quad$ a.e.

Let us now define

(6) $\quad F_1(t) := \{f \in F(t) : \sigma(F(t), d_1 t + x_1) = \langle f, d_1 t + x_1 \rangle\}.$

Then F_1 satisfies the same hypotheses as F, and $0 \in \int \operatorname{co} F_1$.

We now recommence with F_1 replacing F. If Step 2 applies to F_1, we are finished (since $\int F_1 \subset \int F$). If not, we perform Step 3 again, this time with a vector d_2 linearly independent of d_1. This generates a new function g_2 attaining a minimum at x_2, and a new multifunction $F_2 \subset F_1$ all of whose points f satisfy

$$\langle d_2 t + x_2, f \rangle = \sigma(F_1(t), d_2 t + x_2) \quad \text{a.e.}$$

and such that $0 \in \int \text{co} \, F_2$.

The process continues until either Step 2 has applied (i.e., the dimension has been reduced and the inductive hypothesis can be used) or else Step 3 has been performed n times. In the latter case, we will have defined a multifunction $F_n \subset F$ such that $0 \in \int \text{co} \, F_n$ and such that for almost all t, every f in $F_n(t)$ satisfies (we let F_0 stand for F):

(7) $\qquad \langle d_i t + x_i, f \rangle = \sigma(F_{i-1}(t), d_i t + x_i), \qquad i = 1, 2, \ldots, n.$

This may be written in matrix notation as

(8) $\qquad\qquad (Dt + X)f = \Sigma(t),$

where the $n \times n$ matrix D is invertible. But since $Dt + X$ is invertible whenever t is not an eigenvalue of $-D^{-1}X$, it follows that $F_n(t)$ is a singleton a.e. Consequently, we have $0 \in \int \text{co} \, F_n = \int F_n \subset \int F$. $\square$

7.2.2 Remark

Although the above context is the one within which the applications of Aumann's Theorem to optimization take place, the theorem remains true if Lebesgue measure on $[a, b]$ is replaced by a nonatomic finite positive measure space (T, μ). Such a space admits a bounded measurable function $m: T \to R$ such that, for each t_0 in T, the set $\{t \in T : m(t) = m(t_0)\}$ has μ-measure 0. Our proof then goes through as is if in Step 3 we replace the terms $d_i t$ (as in Eq. (4)) by $d_i m(t)$.

A Constructive Approximation Procedure

Were one always able to apply Step 3 of the proof n times, it would be possible in all cases to arrive at the system of equations (8), and hence the proof would be completely constructive (rather than inductive in part). The difficulty is that when 0 does not lie in the relative interior of $\int \text{co} \, F$, the function g_1 may not attain a minimum (although it is certainly bounded below). It may be possible to circumvent this difficulty by a more careful choice of the vectors d_i (our only restriction was independence); we cannot say. However, it is possible to obtain a constructive approximation procedure entailing n minimizations, as we now show. As before, let $\{d_i\}$ be a set of n independent vectors, and define g_1 by Eq. (4). Let x_1 minimize $g_1(x) + \varepsilon |x|$, and then define F_1 again by Eq. (6). We claim that

(9) $\qquad\qquad \varepsilon \overline{B} \cap \int \text{co} \, F_1 \neq \emptyset.$

By the way x_1 is defined, we have

$$\varepsilon \overline{B} \cap \partial g_1(x_1) \neq \emptyset,$$

where ∂g_1 is the subdifferential of g_1. This implies by Theorem 2.8.2 the existence of a selection $f_0(\cdot)$ for co F such that $|\int_a^b f_0(t)\, dt| \leq \varepsilon$ and such that Eq. (5) holds. This establishes (9), which yields in turn an estimate for g_2 (which is again defined as in Eq. (4), with d_1 and F replaced by d_2 and F_1, respectively):

(10) $$g_2(x) \geq -\varepsilon|x| + k.$$

It follows that $g_2(x) + 2\varepsilon|x|$ attains a minimum, at a point we label x_2. We proceed in this way, defining g_i via d_i and F_{i-1}, letting $g_i(x) + i\varepsilon|x|$ attain a minimum at x_i, and defining by means of d_i and x_i a new multifunction F_i having the property that $i\varepsilon \overline{B} \cap \int \text{co } F_i \neq \emptyset$, and also such that a.e. every f in $F_i(t)$ satisfies

$$\langle d_i t + x_i, f \rangle = \sigma(F_{i-1}(t), d_i t + x_i).$$

If D, X are the $n \times n$ matrices whose ith rows are d_i and x_i, respectively, and if $\Sigma(t)$ is the vector whose ith component is $\sigma(F_{i-1}(t), d_i t + x_i)$, then for almost every t, every point f of $F_n(t)$ satisfies Eq. (8). It follows that $F_n(t)$ is a singleton a.e., and defines a selection for F whose integral lies in $n\varepsilon \overline{B}$. We summarize:

7.2.3 Theorem
Suppose 0 belongs to $\int F$, and let $\varepsilon > 0$ be given. Then if X and Σ are calculated as above, the system of equations

$$(Dt + X)f = \Sigma(t), \quad t \in [a, b]$$

uniquely determines a measurable selection $f(\cdot)$ for F such that

$$\left| \int_a^b f(t)\, dt \right| \leq n\varepsilon.$$

7.3 SETS WHICH ARE LIPSCHITZ EPIGRAPHS

Let C be a subset of R^n. It is useful in a variety of situations to know whether C can be viewed (locally) as the epigraph of a Lipschitz function. Formally, if x is a point in C, one would like a condition assuring the existence of an invertible linear transformation $A: R^n \to R^{n-1} \times R$ such that, for some neighborhood U of x, one has

$$C \cap U = U \cap A^{-1}(\text{epi } \phi),$$

where ϕ is a function on R^{n-1} that is Lipschitz near ξ, ξ being the R^{n-1} component of $A(x)$. When this holds, C is said to be *epi-Lipschitzian* near x. Note that in this case the part of the boundary of C in U is represented by the graph of ϕ and is a "Lipschitzian surface." The following characterization of sets C having this property (and its proof) is due to Rockafellar (1979a).

7.3.1 Theorem

A closed subset C of R^n is epi-Lipschitzian at a point x in C iff int $T_C(x) \neq \emptyset$.

Proof. Let us begin by observing that the epigraph of a Lipschitz function f has tangent cones with nonempty interior. One way to see this from results already proved is to note that f is directionally Lipschitz by Theorem 2.9.4, so that epi f admits a hypertangent (at the appropriate point) by Proposition 2.9.3, and consequently the tangent cone to epi f has nonempty interior by Theorem 2.4.8. It follows from this fact that if C can be represented near x as the epigraph of a Lipschitz function, in the sense defined above, then int $T_C(x) \neq \emptyset$. We turn now to the converse.

If $x \in \text{int } C$, the conclusion that C is epi-Lipschitzian near x is trivial. Suppose therefore that x is a boundary point of C. Then $T_C(x)$ is not all of R^n (by Corollary 2 to Theorem 2.5.6). Let $y \in \text{int } T_C(x)$, $y \neq 0$, and let H be the hyperplane through the origin orthogonal to y. Each $x' \in R^n$ can be expressed uniquely in the form $\xi' + \alpha' y$, where $\xi' \in H$, $\alpha' \in R$; the mapping $A: x' \to (\xi', \alpha')$ is a nonsingular linear transformation from R^n onto $H \times R$. Let $(\xi, \alpha) = A(x)$. Since $y \in \text{int } T_C(x)$, y is hypertangent to C at x by Theorem 2.5.8, Corollary 1, and in this property the ball B can be replaced equally well by the product of its intersection B' with H and the interval $\{ty | -1 \leq t \leq 1\}$. In terms of $A(C)$, the property is that for some $\varepsilon > 0$, $\delta > 0$, $\lambda > 0$, one has

(1) $\quad (\xi', \alpha') + t(\eta, \beta) \in A(C)$ for all $t \in [0, \lambda]$ when $(\xi', \alpha') \in A(C) \cap [(\xi, \alpha) + \delta(B' \times [-1, 1])]$ and $(\eta, \beta) \in [(0, 1) + \varepsilon(B' \times [-1, 1])]$.

For all $\xi' \in H$ define

$$\phi(\xi') = \inf\{\alpha' | \alpha' \geq \alpha - \delta, (\xi', \alpha') \in A(C)\} \geq \alpha - \delta$$

(where the convention inf $\emptyset = +\infty$ is implicit). Since C is closed, ϕ is a lower semicontinuous function with values in $(-\infty, +\infty]$. Taking $(\xi', \alpha') = (\xi, \alpha)$ in (1), one sees that

$$\phi(\xi + t\eta) \leq \alpha + t(1 - \varepsilon) \quad \text{when } \eta \in \varepsilon B', t \in [0, \lambda].$$

Hence there exists $\delta' \leq \delta$ such that for all $\xi' \in (\xi + \delta'B)$ one has $\phi(\xi') \leq \alpha + \delta$, and (consequently)

$$(\xi', \phi(\xi')) \in A(C) \cap [(\xi, \alpha) + \delta(B' \times [-1, 1])].$$

Then (1) implies (with $\beta = 1$)

$$(\xi' + t\eta, \phi(\xi') + t) \in A(C) \quad \text{when } t \in [0, \lambda], \eta \in \varepsilon B'.$$

Therefore, for all $\xi' \in (\xi + \delta'B)$, one has

$$\phi(\xi' + t\eta) \leq \phi(\xi') + t \quad \text{when } t \in [0, \lambda], \eta \in \varepsilon B',$$

so that ϕ is Lipschitzian on a neighborhood of ξ. Furthermore

$$(\xi', \phi(\xi') + t) \in A(C) \quad \text{for all } t \in [0, \lambda]$$

by (1). For $t < 0$, of course, one has $(\xi', \phi(\xi') + t) \notin A(C)$ by the definition of ϕ. Thus there is a neighborhood of $(\xi, \alpha) = (\xi, \phi(\xi))$ in which $A(C)$ coincides with the epigraph of ϕ. This proves that C is epi-Lipschitzian at x. □

7.4 DEPENDENCE ON INITIAL VALUES

A fundamental tool in differential equations is the theory of the resolvent, which describes the behavior of the solution to an ordinary differential equation as the initial condition is varied. With the help of Chapter 3, we can now extend the theory to the case in which the underlying function is Lipschitz, a setting which has been viewed as a natural one for existence and uniqueness theory.

Consider the initial-value problem

(1) $$\dot{x}(t) = f(t, x(t)) \quad \text{a.e., } a \leq t \leq b, x(a) = u,$$

where f maps $[a, b] \times R^n$ to R^n, and where (as is customary) solutions x are absolutely continuous functions on $[a, b]$ (which we have become used to calling *arcs*). Suppose that $\hat{x}$ is a solution to Eq. (1) for $u = \hat{u}$. Our goal is to determine whether solutions continue to exist for u near $\hat{u}$, and to investigate how $x(b)$ varies as u varies (near $\hat{u}$). We assume that, near $\hat{x}$, f is measurable in t and Lipschitz in x. (To be precise, we assume that f is *measurably Lipschitz* in the sense of 3.1.4.) The main result below asserts in part that a solution to Eq. (1) exists for each u near $\hat{u}$; we denote it x_u. It is well known (but true) that under the Lipschitz hypothesis solutions to Eq. (1) are unique, so that the following map F from R^n to R^n is well defined near $\hat{u}$: $F(u) = x_u(b)$.

The *linearization* about $\hat{x}$ of Eq. (1) is the differential inclusion

(2) $$\dot{y}(t) \in \partial \hat{f}(t) y(t),$$

7.4 Dependence on Initial Values

where $\partial \hat{f}(t)$ signifies $\partial_x f(t, \hat{x}(t))$ (the generalized Jacobian of f with respect to x, see Section 2.6). We define $\Phi(t, \tau)$ to be the set of all $n \times n$ matrices $Y(t)$, where $Y(\cdot)$ is a matrix solution of the initial-value problem

$$(3) \qquad \dot{Y}(s) = M(s)Y(s), \qquad Y(\tau) = I$$

for some measurable selection $M(\cdot)$ of $\partial \hat{f}(\cdot)$. (I denotes the $n \times n$ identity matrix.) The *resolvent* R is the matrix-valued multifunction on $[a, b] \times [a, b]$ whose value at (t, τ) is the *plenary hull* of $\Phi(t, \tau)$; that is, $R(t, \tau)$ is the set

$$\{M : \langle v, Mw \rangle \leq \max[\langle v, Nw \rangle : N \in \Phi(t, \tau)] \quad \text{for all } v, w \text{ in } R^n\}.$$

Note that $R(t, \tau)$ contains the convex hull $\operatorname{co} \Phi(t, \tau)$ of $\Phi(t, \tau)$. In general, $R(t, \tau)$ may contain matrices not in $\operatorname{co} \Phi(t, \tau)$, although one always has, for any v and w,

$$\max \langle v, R(t, \tau)w \rangle = \max \langle v, \operatorname{co} \Phi(t, \tau)w \rangle = \max \langle v, \Phi(t, \tau)w \rangle.$$

7.4.1 Theorem

A solution $x_u(\cdot)$ of Eq. (1) exists on $[a, b]$ for all u near $\hat{u}$. If F is the map $F(u) = x_u(b)$, then F is Lipschitz near $\hat{u}$ and one has $\partial F(\hat{u}) \subset R(b, a)$. If $\partial_x f(t, \hat{x}(t))$ reduces to a singleton a.e. on $[a, b]$, then so does $R(b, a)$, and F is then strictly differentiable at $\hat{u}$ with $DF(\hat{u}) = R(b, a)$.

Proof. Consider the matrix differential inclusion

$$\dot{P}(s) \in -\partial \hat{f}(s)^* P(s),$$

where * signifies transpose, and let $S(t, \tau)$ be its resolvent. Then $S(t, \tau)$ is the plenary hull of all elements of the form $S_M(t, \tau)$, where S_M is the resolvent in the classical sense of the linear matrix differential equation

$$\dot{P}(s) = -M^*(s)P(s),$$

and where $M(\cdot)$ is a measurable selection of $\partial \hat{f}(\cdot)$. It is known from the classical theory that $S_M(t, \tau) = R_M^*(\tau, t)$, where R_M is the (classical) resolvent of the equation

$$\dot{Y}(s) = M(s)Y(s).$$

But $R(t, \tau)$ is itself the plenary hull of all matrices $R_M(t, \tau)$ as $M(\cdot)$ varies over the measurable selections of $\partial \hat{f}(\cdot)$. We have proven:

Lemma 1. $S(t, \tau) = R(\tau, t)^*$.

We now define a differential inclusion problem P_D as in Problem 3.4.1. We denote by (x, y) points in $R^n \times R^n$, and, for any fixed unit vector v in R^n, we set

$$\tilde{F}(t, x, y) = \{[f(t, x), 0]\}$$

$$\tilde{f}(x, y) = \langle v, x \rangle$$

$$C_0 = \{(x, y) : x = y, |x - \hat{u}| \leq \varepsilon\}$$

$$C_1 = R^n \times \{\hat{u}\}$$

$$g(t, x, y) = |x - \hat{x}(t)| - \varepsilon,$$

where f is measurably Lipschitz on the tube $T(\hat{x}; 2\varepsilon)$. Note that the Hamiltonian $H(t, x, y, p, q)$ (see Section 3.2) is given by

$$H(t, x, y, p, q) = \langle p, f(t, x) \rangle.$$

Lemma 2. The problem described above is normal.

The unique feasible trajectory for the problem is the arc $(\hat{x}(t), \hat{u})$, so we need only verify that it is normal in the sense of Definition 3.5.1. A multiplier of index 0 for the arc in question turns out to be an arc p and a constant q such that

(4) $$-\dot{p} \in \partial \hat{f}(t)^* p \quad \text{a.e.}$$

$$p(a) + q = 0, \quad [p(b), q] \in \{0\} \times R^n.$$

Because p satisfies a "linear differential inclusion" and vanishes at b, it follows easily that p is identically 0. Then q is 0 also, so there exist no nontrivial abnormal multipliers, as claimed.

We are now able to invoke Corollary 1 of Theorem 3.4.3, which asserts that the value function V of the problem is Lipschitz near 0. Note that in this case

(5) $$V(u_1, u_2) = \langle v, F(\hat{u} + u_2) \rangle,$$

so that we deduce the fact that $x_u(\cdot)$ exists for all u near $\hat{u}$. Since v is an arbitrary unit vector, it also follows that F is Lipschitz near $\hat{u}$.

Theorem 3.4.3 itself yields the estimate

$$\partial V(0, 0) \subset \text{co}\{[v, 0] + [p(b), q]\},$$

where (p, q) is a multiplier of index 1, which means here that p is an arc satisfying the differential inclusion (4) and q is a constant such that

$$p(a) + q = 0, \quad [v, 0] + [p(b), q] \in \{0\} \times R^n.$$

This gives, in light of Eq. (5),

$$\partial \langle v, F \rangle(\hat{u}) \subset \mathrm{co}\{-p(a)^*\},$$

where we have now made explicit the fact that $\partial \langle v, F \rangle$ is $1 \times n$ rather than $n \times 1$ (see Remark 2.6.3). The arcs p above satisfy (4) and the condition $p(b) = -v$, from which it follows (from the measurable selection theorem, Theorem 3.1.1) that $p(t)$ is of the form $-P(t)v$, where P is a matrix solution of $\dot{P} = -M^*P$, $P(b) = I$, for some $M(\cdot)$ assuming values in $\partial \hat{f}(\cdot)$. It follows that $-p(a) = P(a)v$ belongs to $S(a, b)v$. By Lemma 1 and the preceding, we derive

$$\partial \langle v, F \rangle(\hat{u}) \subset v^*R(b, a).$$

The left-hand side coincides with $v^*\partial F(\hat{u})$ by Theorem 2.6.6. For any M in $\partial F(\hat{u})$, therefore, and for any vector w in R^n, one has

$$v^*Mw \leq \max v^*R(b, a)w \leq \max v^*\Phi(b, a)w,$$

and so M belongs to $R(b, a)$ by definition. This proves the estimate for ∂F stated in the theorem. The final assertion follows from the fact that if $\partial \hat{f}(t)$ reduces to a singleton a.e., then $\Phi(b, a)$ contains a single element. Then so does $R(b, a)$ (as is easily proven), and so $\partial F(\hat{u})$ reduces to one element also. Then each component function F_i of F (and hence F itself) is strictly differentiable at $\hat{u}$ by Proposition 2.2.4. □

7.5 EKELAND'S THEOREM

There would seem to be little that one can say about a point which *almost* minimizes a given function (it does not follow, for example, that the derivative is "almost zero"). In this section we present a theorem of Ekeland (1974) that makes a nontrivial assertion about such points. Roughly speaking, it says that there is a "nearby point" which actually minimizes a slightly perturbed functional. We include this result here for several reasons. It has played an important role in the proofs of the necessary conditions in Chapters 3, 5, and 6, and in a variety of other applications (see the survey by Ekeland, 1979a). Nonetheless, it is elementary, requiring in its proof only a familiar property of complete metric spaces. The result is also a natural part of nonsmooth analysis, since the perturbed functionals that it produces are necessarily nondifferentiable. The use of the theorem will be illustrated again in a simple setting in the next section, where we derive an extension of the contraction mapping principle.

Let V be a complete metric space with associated metric Δ, and let $F: V \to R \cup \{+\infty\}$ be a lower semicontinuous function which is bounded below.

7.5.1 Theorem
If u is a point in V satisfying

$$F(u) \leq \inf F + \varepsilon$$

for some $\varepsilon > 0$, then, for every $\lambda > 0$ there exists a point v in V such that
(i) $F(v) \leq F(u)$.
(ii) $\Delta(u, v) \leq \lambda$.
(iii) *For all $w \neq v$ in V, one has $F(w) + (\varepsilon/\lambda)\Delta(w, v) > F(v)$.*

Proof. The crux of the matter will be seen to be a judicious application of a partial ordering introduced by Bishop and Phelps (1963). For any $\alpha > 0$, define a partial ordering $\leq_\alpha$ on $V \times R$ by:

$$(v_1, r_1) \leq_\alpha (v_2, r_2) \quad \text{iff } r_2 - r_1 + \alpha\Delta(v_1, v_2) \leq 0.$$

This relation is easily seen to be reflexive, antisymmetric and transitive. Further, it is easy to verify that for every (v_1, r_1) in $V \times R$, the set

$$\{(v, r) : (v_1, r_1) \leq_\alpha (v, r)\}$$

is closed in $V \times R$.

Lemma. Let S be a closed subset of $V \times R$ such that, for some scalar m, every element (v, r) of S satisfies $r \geq m$. Then, for every (v_1, r_1) in S, there exists an element $(\bar{v}, \bar{r})$ of S satisfying $(v_1, r_1) \leq_\alpha (\bar{v}, \bar{r})$ which is maximal in S for the ordering $\leq_\alpha$.

We define a sequence $\{(v_n, r_n)\}$ in S by induction, starting with (v_1, r_1). Let (v_n, r_n) be known. Define

(1) $$S_n := \{(v, r) \in S : (v_n, r_n) \leq_\alpha (v, r)\}$$

$$m_n := \inf\{r : (v, r) \in S_n \text{ for some } v\}.$$

(Clearly $m_n \geq m$.) Now define (v_{n+1}, r_{n+1}) to be any point of S_n such that

(2) $$r_n - r_{n+1} \geq \tfrac{1}{2}(r_n - m_n).$$

7.5 Ekeland's Theorem

The sets S_n are closed and nested: $S_{n+1} \subset S_n$. (The situation is illustrated in Figure 7.1.) Moreover, one derives from (2) the inequality

$$|r_{n+1} - m_{n+1}| \leq \tfrac{1}{2}|r_n - m_n| \leq 2^{-n}|r_1 - m|.$$

Hence, for every (v, r) in S_{n+1}, one deduces from (1) that

$$|r_{n+1} - r| \leq |r_{n+1} - m_{n+1}| \leq 2^{-n}|r_1 - m|$$

$$\Delta(v_{n+1}, v) \leq \frac{2^{-n}}{\alpha}|r_1 - m|.$$

This shows that the diameter of S_n goes to zero as n goes to infinity. It follows (since $V \times R$ is metric complete) that the sets S_n have one point $(\bar{v}, \bar{r})$ in common:

$$\{(\bar{v}, \bar{r})\} = \bigcap_{n \geq 1} S_n.$$

By definition, $(v_n, r_n) \leq_\alpha (\bar{v}, \bar{r})$ for every n, in particular for $n = 1$. Now suppose that $(\tilde{v}, \tilde{r})$ is a point in S such that $(\bar{v}, \bar{r}) \leq_\alpha (\tilde{v}, \tilde{r})$. By transitivity, one gets $(v_n, r_n) \leq_\alpha (\tilde{v}, \tilde{r})$ for every n, whence $(\tilde{v}, \tilde{r}) \in \bigcap_{n \geq 1} S_n$. It follows that $(\tilde{v}, \tilde{r})$ is equal to $(\bar{v}, \bar{r})$, which is indeed maximal.

To prove the theorem, take $S = \text{epi } V$ and apply the lemma with $\alpha = \varepsilon/\lambda$ and $(v_1, r_1) = (u, F(u))$. We obtain a maximal element (v, r) in S satisfying

(3) $$(u, F(u)) \leq_\alpha (v, r).$$

Since (v, r) lies in S, we also have $(v, r) \leq_\alpha (v, F(v))$, which implies $r = F(v)$

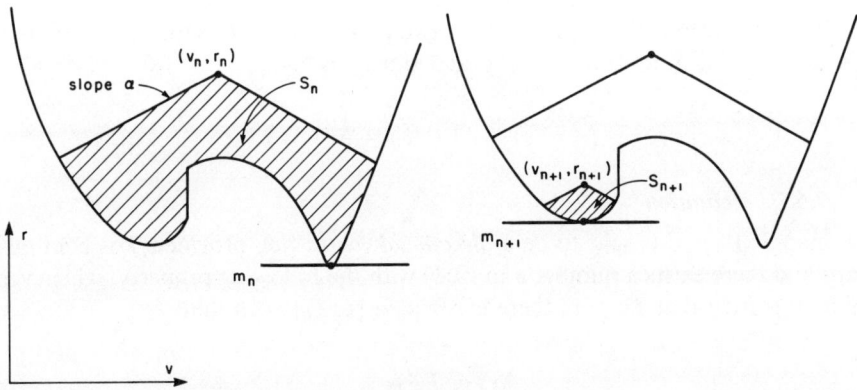

Figure 7.1 Proof of Ekeland's Theorem.

by maximality. Now (3) implies

(4) $$F(v) - F(u) + \alpha\Delta(v, u) \leq 0,$$

which gives (i) of the theorem. The maximality of $(v, F(v))$ in S implies that for any w in V different from v, whenever $F(w)$ is finite, the relation $(v, F(v)) \leq_\alpha (w, F(w))$ fails; this gives (iii). Finally, because one has $F(u) \leq \inf F + \varepsilon$, one must have $F(v) \geq F(u) - \varepsilon$. This gives (ii) when combined with (4). □

7.6 DIRECTIONAL CONTRACTIONS AND FIXED POINTS

Let T be a continuous mapping of a complete metric space (V, Δ) to itself. The celebrated contraction principle of Banach asserts that if there exists a number σ in $(0, 1)$ such that

(1) $$\Delta(Tx, Ty) \leq \sigma\Delta(x, y) \quad \text{for all } x, y \in V$$

(T is then said to be a *contraction*), then T has a (unique) fixed point; that is, a point x such that $Tx = x$. The uses of this principle are many and varied.

Our purpose is to establish a "pointwise" version of the result. One can imagine various weakened forms of condition (1). For example, suppose that for each x in X there is some neighborhood $N(x)$ of x such that

$$\Delta(Tx, Ty) \leq \sigma\Delta(x, y) \quad \text{for all } y \in N(x).$$

Must T have a fixed point? That the answer is negative follows from the fact that *any* function T satisfies this condition when Δ is the discrete metric. It is clear therefore that any pointwise criterion must be accompanied in some way by at least an indirect hypothesis concerning the metric structure.

The *open segment* $]x, y[$ defined by two points x, y in V is the set of points z (if any) in V distinct from x and y and satisfying

$$\Delta(x, z) + \Delta(z, y) = \Delta(x, y).$$

7.6.1 Definition

A map $T: V \to V$ is said to be a *directional contraction* provided T is continuous and there exists a number σ in $(0, 1)$ with the following property: whenever v in V is such that $Tv \neq v$, there exists w in $]v, Tv[$ such that

$$\frac{\Delta(Tv, Tw)}{\Delta(v, w)} \leq \sigma.$$

7.6.2 Theorem
Every directional contraction of a complete metric space admits a fixed point.

7.6.3 Remark
A *convex* metric space is one for which $]x, y[$ is nonempty whenever $x \neq y$. For such spaces, the theorem implies the Banach contraction principle, and may apply when the latter does not. Consider the case in which V is R^2, with the metric Δ induced by the norm $\|(x, y)\| = |x| + |y|$. The open segment determined by any two distinct points (x_1, y_1), (x_2, y_2) consists of the closed rectangle having the two points as diagonally opposite corners, with those two points deleted. (This reduces to a line segment in the usual sense if x_1 and x_2 or y_1 and y_2 coincide.) Define T as follows:

$$T(x, y) := \left(\frac{3x}{2} - \frac{y}{3}, x + \frac{y}{3}\right).$$

It is easily seen that T is not a contraction even in a local sense. However, T is a directional contraction. For let $T(x, y) \neq (x, y)$. Then (setting $T(x, y) = (a, b)$) it follows that $b \neq y$ (for otherwise $a = x$ also), so that the open segment between (x, y) and $T(x, y)$ contains points of the form (x, z) with z arbitrarily close to y. But for such points we have

$$\frac{\Delta(T(x, z), T(x, y))}{\Delta((x, z), (x, y))} = \frac{2}{3},$$

so that the theorem applies. Note that the fixed points of T are all points of the form $(x, 3x/2)$, so that the hypotheses of the theorem do not imply the uniqueness of the fixed point.

Proof of Theorem 7.6.2. Let $F: V \to R$ be defined by

$$F(v) := \Delta(v, Tv).$$

Note that F is continuous and bounded below. The idea would be to find a point v minimizing F, and to show that $F(v) = 0$, so that v would be the desired fixed point. The problem is that in the absence of compactness, there is no reason a priori why F should attain a minimum. This is where Ekeland's Theorem comes in. By Theorem 7.5.1 there is a point v such that, for all w in V, one has

(2) $$\Delta(w, Tw) + \frac{1-\sigma}{2}\Delta(w, v) \geq \Delta(v, Tv)$$

(we have applied the theorem with $\varepsilon = (1 - \sigma)/2$, $\lambda = 1$). If $v = Tv$, we are done, so supposing the contrary we know by assumption that a point w exists such that $w \neq v$ and

(3) $$\Delta(v, w) + \Delta(w, Tv) = \Delta(v, Tv)$$

(i.e., w lies in $]v, Tv[$), as well as

(4) $$\frac{\Delta(Tw, Tv)}{\Delta(w, v)} \leq \sigma.$$

We calculate

$$0 \leq \sigma\Delta(w, v) - \Delta(Tw, Tv) \qquad \text{(by (4))}$$

$$\leq \sigma\Delta(w, v) - \Delta(Tw, w) + \Delta(w, Tv) \qquad \text{(triangle inequality)}$$

$$= (\sigma - 1)\Delta(w, v) - \Delta(Tw, w) + \Delta(v, Tv) \qquad \text{(by (3))}$$

$$\leq \frac{\sigma - 1}{2}\Delta(w, v) \qquad \text{(by (2))}$$

Since $\sigma < 1$, this implies $w = v$. This contradiction proves the theorem. □

7.7 HAMILTONIAN TRAJECTORIES AND BOUNDARY-VALUE PROBLEMS

The evolution of virtually every physical system is describable in terms of some *variational principle* (i.e., an axiom to the effect that the laws of motion of the system coincide with the necessary conditions of a certain problem in the calculus of variations). The prototypical situation is that of classical mechanics, where (through such physical laws as the principle of least action) the laws governing a physical system are expressible as a system of Hamiltonian equations

$$-\dot{p} = \nabla_x H(x, p), \qquad \dot{x} = \nabla_p H(x, p).$$

The more complete our understanding of Hamiltonian systems, the more complete will be our knowledge of the underlying physical reality.

The analysis of Hamiltonian systems is for this reason a long-standing and ongoing area of research in mathematics. Many of the results in the field are inspired by the work of Poincaré and Birkhoff. Being nonlinear, the Hamiltonian equations are not easily analyzed. Poincaré referred to them as an "impregnable fortress" with, however, one possible weak point: the periodic solutions. The hope is that by combining knowledge about critical points, periodic solutions, and global behavior, it will eventually be possible to piece together a complete picture of the behavior of Hamiltonian systems.

As mentioned above, the (classical) calculus of variations has played a central role in the study of Hamiltonian systems (and other boundary-value problems as well). In this section (and the next) we shall see that more modern

7.7 Hamiltonian Trajectories and Boundary-Value Problems

optimization techniques can also be of use. The main tools will be those of the preceding chapters, along with a new "dual form" of the principle of least action introduced by the author (Clarke, 1978c).

We shall now write Hamiltonian systems in the form

$$J\dot{z} = \nabla H(z),$$

where $z = (x, p)$ is an amalgam of the state x and the adjoint variable p, and where J is the following $2n \times 2n$ matrix:

$$J = \begin{bmatrix} 0 & -I \\ I & 0 \end{bmatrix}.$$

(I signifies the $n \times n$ identity matrix.) Note that the transpose J^* and the inverse J^{-1} both coincide with $-J$.

As we shall see, solutions $z(\cdot)$ of Hamiltonian systems such as the above evolve on a level surface of H; i.e., $H(z(t)) = c$. We begin with a study of certain boundary-value problems in which the constant c (called the "energy" because the Hamiltonian is often interpreted as such) is specified. In the next section we focus upon the issue of periodic trajectories, in the case when the period of the motion is prescribed.

Convex Energy Surfaces

We wish to study the trajectories z of a Hamiltonian system evolving on a given energy level: $H(z(t)) = c$. Let S be the surface $\{z : H(z) = c\}$. Of course there are any number of functions H giving rise to S (i.e., whose c-level set is S). We plan to focus attention upon S, which we assume to be the boundary of a compact convex set containing 0 in its interior. (S is what we call a *convex energy surface*.)

One of the functions representing S is its (Minkowski) gauge g_S, the function defined by

$$g_S(z) = \text{the (unique) nonnegative } \lambda \text{ such that } z \in \lambda S.$$

Clearly, S coincides with the set $\{z : g_S(z) = 1\}$. We leave as an exercise the verification that g_S is convex, in fact subadditive and positively homogeneous. It does not follow under our hypotheses that g_S is differentiable (this is known to be true, except at the origin, when S is smooth, which we do not assume). Associated with g_S is the Hamiltonian system (inclusion)

(1) $$J\dot{z}(t) \in \partial g_S(z(t)).$$

Let us pause to confirm that a solution z of this relation remains on a level surface of g_S. In fact, we prove a somewhat more general result.

7.7.1 Proposition

If for almost all t in $[a, b]$ the arc z satisfies $J\dot{z} \in \partial H(z)$, where, for each t, H is Lipschitz near $z(t)$ and regular at $z(t)$, then $H(z(t))$ is constant on $[a, b]$.

Proof. Let θ be the (Lipschitz) function $\theta(t) = H(z(t))$. It suffices to prove that $\theta'(t) = 0$ a.e. Let t be any point such that $\theta'(t)$ and $\dot{z}(t)$ exist and such that $J\dot{z}(t)$ belongs to $\partial H(z(t))$. Because $\langle w, Jw \rangle = 0$ for any w in R^{2n}, one has

$$0 = \langle J\dot{z}(t), \dot{z}(t) \rangle \leqslant H^0(z(t); \dot{z}(t)) \quad \text{(by Proposition 2.1.2)}$$

$$= H'(z(t); \dot{z}(t)) \quad \text{(since } H \text{ is regular at } z(t))$$

$$= \lim_{\delta \downarrow 0} \frac{H(z(t) + \delta \dot{z}(t)) - H(z(t))}{\delta}$$

$$= \lim_{\delta \downarrow 0} \frac{H(z(t + \delta)) - H(z(t))}{\delta} \quad \text{(because } H \text{ is Lipschitz)}$$

$$= \theta'(t).$$

Thus $\theta'(t)$ is nonnegative. To see that $\theta'(t)$ is 0, we calculate, much as above,

$$\theta'(t) = \lim_{\delta \downarrow 0} \frac{H(z(t) - \delta \dot{z}(t)) - H(z(t))}{-\delta}$$

$$= -H'(z(t); -\dot{z}(t)) = -H^0(z(t); -\dot{z}(t))$$

$$\leqslant -\langle -\dot{z}(t), J\dot{z}(t) \rangle = 0. \quad \square$$

Note that g_S, being convex, is regular (Proposition 2.3.6), so that any solution z of (1) on some interval remains in a fixed level surface of g_S. Given any point s in S, it is known (since ∂g_S is upper semicontinuous, convex and bounded on S; see for example Aubin, Cellina, and Nohel (1977)), that some (not necessarily unique) arc z satisfying (1) exists, with z defined on $(-\infty, \infty)$ and satisfying $z(0) = s$. Thus $z(t)$ lies in S for all t. The arc z is said to be a *trajectory* passing through s.

Geodesics

A *geodesic* on S is defined to be any curve lying on S which admits as a parametrization an arc z defined on an interval $[0, T]$ (for $T > 0$) which satisfies on this interval the Hamiltonian inclusion (1). Note that a geodesic is an oriented curve. If its associated parametrization z has $z(0) = s$, $z(T) = s'$, then we say the geodesic *joins* s to s'.

7.7 Hamiltonian Trajectories and Boundary-Value Problems

The concept of geodesic has been defined through the Hamiltonian flow induced on S via g_S and (1). However, geodesics are actually intrinsic to S, as we now show by verifying that any other suitable Hamiltonian H representing S gives rise to the same geodesics. A "suitable" representation is defined to be a function H which is Lipschitz near S and regular at every point of S, and which satisfies co $S = \{z : H(z) \leq 1\}$, $0 \notin \partial H(S)$. (The last condition assures "nondegeneracy"; it is satisfied by g_S, since otherwise g_S, being convex, would attain a minimum at a point of S.) Note that g_S satisfies these requirements.

7.7.2 Proposition
All Hamiltonians H representing the convex energy surface S in the sense defined above give rise to the same geodesics on S.

Proof. Let a given curve C on S be a geodesic relative to the representation of S given by H. Then C admits a parametrization $z(\cdot)$ on an interval $[0, T]$, where $J\dot{z} \in \partial H(z)$ a.e. By Theorem 2.4.7, Corollary 1, the normal cone to the (convex) set co S at $z(t)$ is given by $\cup_{\lambda \geq 0} \lambda \partial H(z(t))$, or again by $\cup_{\lambda \geq 0} \lambda \partial g_S(z(t))$. It follows that $J\dot{z}(t)$ belongs to $\lambda(t) \partial g_S(z(t))$ for some λ bounded away from zero (we may suppose that $\lambda(\cdot)$ is measurable). If we define a change of variables by

$$\tau(t) = \int_0^t \lambda(s)\, ds, \quad 0 \leq t \leq T,$$

then the arc $\tilde{z}(\cdot)$ defined on $[0, \tilde{T}]$ (where $\tilde{T} := \int_0^T \lambda(s)\, ds$) by $\tilde{z}(\tau) = z(t(\tau))$ is a parametrization for C and satisfies

$$J \frac{d}{d\tau} z(\tau) \in \partial g_S(z(\tau)) \quad \text{a.e.}$$

Thus C is a geodesic relative to g_S.

A similar argument shows that geodesics relative to g_S are also geodesics relative to H, so the proposition follows. □

We remark that although different Hamiltonian representations generate the same geodesics, the trajectories z differ (they are different parametrizations of the underlying curve).

Symplectic Matrices

A $2n \times 2n$ matrix M is said to be *symplectic* if

$$M^* J M = J$$

(where * denotes transpose). Note that I, $-I$, and J are symplectic. Symplectic transformations (which can be shown to form a group) are of great importance in classical mechanics, where "the motion of a mechanical system corresponds to the continuous evolution or unfolding of a canonical (i.e., symplectic) transformation" (Goldstein, 1950, Section 8-6). The following can be viewed as a kind of converse: any (linear) symplectic transformation is realized on some orbit of any convex energy surface.

7.7.3 Theorem

Let S be a convex energy surface and let M be a symplectic matrix. Then, for some s in S, there is a geodesic on S joining s to Ms.

Proof. The proof, which hinges upon a direct variational principle, achieves the desired result by producing a solution z on S of (1) on an interval $[0, T]$ ($T > 0$) having the property that $z(T) = Mz(0)$. We shall need the *polar* Σ of S. This is defined as the unique convex set Σ whose support function is g_S; that is, Σ is the convex set satisfying, for every s in R^{2n},

$$\max\{\langle s, \sigma \rangle : \sigma \in \Sigma\} = g_S(s).$$

It follows (see Proposition 2.1.4) that Σ is a convex compact set containing the origin in its interior. We denote by Y the set of all absolutely continuous functions $y: [0, 1] \to R^{2n}$ satisfying $J\dot{y}(t) \in \Sigma$ a.e., and we let $A: Y \to R^{2n}$ be defined by

$$A(y) = y(1) - My(0).$$

Finally, let E_1 be the eigenspace of M for eigenvalue 1 ($E_1 = \{0\}$ if 1 is not an eigenvalue), and let π and ρ be the projections onto E_1 and $E_1^\perp$, respectively. ($E_1^\perp$ is the orthogonal complement to E_1.)

Lemma 1. There exist positive constants c and k such that, for every y in Y,

$$|\rho(y(0))| \leq c + k|A(y)|.$$

There is a constant k such that, for every v in $E_1^\perp$, $|(I - M)v| \geq |v|/k$ and a constant b such that $|y(1) - y(0)| \leq b$ for all y in Y (since Σ is bounded). Let any y in Y be given. Since $(I - M)v = (I - M)\rho(v)$ for any v in R^n, we have

$$(I - M)\rho(y(0)) = (I - M)y(0) = y(0) - y(1) + A(y)$$

and consequently

$$|\rho(y(0))| \leq k(b + |A(y)|).$$

The lemma follows by taking $c = bk$.

7.7 Hamiltonian Trajectories and Boundary-Value Problems

We denote by f the functional on Y defined by

$$f(y) = -\frac{1}{2}\int_0^1 \langle J\dot{y}, y\rangle\, dt.$$

We now define a parametrized family $P(\alpha)$ of problems: to minimize $f(y)$ over the elements y of Y which satisfy $|y(0)| \leq c + k$ and $A(y) = \alpha$. We denote by $V(\alpha)$ the minimum in problem $P(\alpha)$.

Lemma 2. The minimum $V(\alpha)$ is finite and attained for all α near 0; the function $V(\cdot)$ is Lipschitz in a neighborhood of 0.

Since Σ contains a neighborhood of 0, it follows that the feasible set for $P(\alpha)$ is nonempty for small α. To deduce the existence of a solution to $P(\alpha)$ for such α, note that $P(\alpha)$ can be viewed as a problem of Bolza with

$$L(t, y, \dot{y}) = -\tfrac{1}{2}\langle J\dot{y}, y\rangle + \psi_\Sigma(J\dot{y})$$

$$l(y_0, y_1) = \psi_{(c+k)\bar{B}}(y_0) + \psi_{\{0\}}(y_1 - My_0 - \alpha).$$

The corresponding Hamiltonian H for this L is

$$H(t, y, p) = \sup\{\langle p, \dot{y}\rangle + \tfrac{1}{2}\langle J\dot{y}, y\rangle : J\dot{y} \in \Sigma\}$$

$$= \sup_{\sigma \in \Sigma}\{\langle \sigma, Jp + \tfrac{1}{2}y\rangle\}$$

$$= g_S(Jp + \tfrac{1}{2}y).$$

Since g_S satisfies a bound of the form $g_S(q) \leq K|q|$ for all q, it follows that H satisfies the growth condition of Theorem 4.1.3, as does l. The theorem implies the existence of a solution to $P(\alpha)$ for all α near 0.

The fact that $V(\cdot)$ is Lipschitz near 0 is an immediate consequence of Corollary 1, Theorem 6.3.2.

Lemma 3. There is a nonconstant solution $\hat{y}$ of $P(0)$ such that $|\hat{y}(0)| \leq c$.

That there is a solution $\hat{y}$ follows from the preceding lemma. If $\hat{y}$ is constant, then $V(0) = 0$. But this cannot be, since it is easy to produce y with zero endpoints such that $f(y) < 0$, and for sufficiently small ε we will have εy feasible for $P(0)$ with $f(\varepsilon y) < 0$. Now let us observe that whenever $A(y) = 0$ and u is any element of E_1, we have $A(y - u) = 0$ and

$$f(y - u) = f(y) + \frac{1}{2}\int \langle J\dot{y}, u\rangle\, dt = f(y) + \frac{1}{2}\langle J(M - I)y(0), u\rangle$$

$$= f(y) + \tfrac{1}{2}\langle y(0), (I - M^*)Ju\rangle = f(y),$$

in view of the fact that, when M is symplectic, $Mu = u$ is equivalent to $M^*Ju = Ju$. Thus by "subtracting off" we may choose the solution $\hat{y}$ to satisfy $\pi(\hat{y}(0)) = 0$, and the required result now follows from Lemma 1.

It follows from the definition of V that for any y in Y satisfying $|y(0)| \leqslant c + k$, we have

(2) $$f(y) \geqslant V(A(y)),$$

and we have equality for any y solving $P(\alpha)$ for any α, in particular for $\hat{y}$. We may interpret (2) as follows: the function $\hat{y}$ provides a strong local solution for the problem of minimizing

$$-\frac{1}{2}\int_0^1 \langle J\dot{y}, y\rangle\, dt - V(y(1) - My(0))$$

subject to $J\dot{y} \in \Sigma$. (The constraint $|y(0)| \leqslant c + k$ is inactive near $\hat{y}$ as a consequence of Lemma 3.)

We now apply the necessary conditions of Theorem 4.2.2 to this problem, which may be viewed as a problem of Bolza much as in the proof of Lemma 2 (note, however, that now l is Lipschitz, which assures calmness). These assert the existence of an absolutely continuous function p satisfying

(3) $$(-\dot{p}, \dot{\hat{y}}) \in \partial H(\hat{y}, p) \quad \text{a.e.}$$

(4) $$(p(0), -p(1)) \in \partial l(\hat{y}(0), \hat{y}(1)),$$

where l is the function

$$l(y, y') = -V(y' - My),$$

and where H is the Hamiltonian of the problem (calculated above):

$$H(y, p) = g_S\left(Jp + \frac{y}{2}\right).$$

Relation (3) thus reduces to

$$J\dot{\hat{y}} = -2\dot{p} \in \partial g_S\left(Jp + \frac{\hat{y}}{2}\right) \quad \text{a.e.}$$

It follows that the function Jp is a translate of $\hat{y}/2$, so that the function $\hat{z} := Jp + \hat{y}/2$ is nonconstant (Lemma 3). Note the relation

(5) $$J\dot{\hat{z}} \in \partial g_S(\hat{z}) \quad \text{a.e.,}$$

7.7 Hamiltonian Trajectories and Boundary-Value Problems

which implies by Proposition 7.7.1 that $g_S(\hat{z}(t)) = h$ for some constant h. Because $\hat{z}$ is nonconstant, h is positive.

We now turn to (4), which can be rewritten $p(1) = r, p(0) = M^*r$, where r is an element of $\partial V(0)$. Armed with this we calculate

$$M\hat{z}(0) = MJp(0) + \tfrac{1}{2}M\hat{y}(0)$$

$$= MJM^*p(1) + \tfrac{1}{2}\hat{y}(1)$$

$$= Jp(1) + \tfrac{1}{2}\hat{y}(1) = \hat{z}(1).$$

We now effect a transformation which places $\hat{z}$ on S (rather than hS). Let $z(t) = \hat{z}(th)/h$ for $0 \leq t \leq 1/h =: T$. Then $g_S(z(t)) = 1$, since g_S is positively homogeneous of degree 1, and we have (in view of (5))

$$J\dot{z} \in \partial g_S(z) \quad \text{a.e.}$$

since $\dot{z} = \dot{\hat{z}}$ and ∂g_S is positively homogeneous of degree 0. Of course, z continues to satisfy $z(T) = Mz(0)$, and so we have arrived at the z alluded to at the beginning of the proof. □

Some Consequences of the Theorem

Corollary 1

Any convex energy surface S admits at least one closed geodesic.

Of course we derive this by taking $M = I$ in the theorem. This global existence result is due to Rabinowitz (1978) and to Weinstein (1978). A proof of this corollary, simpler in this special case but similar in spirit, is given in Clarke (1978c), where the use of the polar or dual variational principle was inaugurated.

The following consequence of the theorem is reminiscent of Noether's Theorem in the calculus of variations, inasmuch as an invariance of the system implies an invariance of (in this case, some) extremals. An *orbit* on S is defined as a curve on S which is locally a geodesic; orbits need not be of finite length.

Corollary 2

Let the symplectic matrix M satisfy $MS = S$. Then there is an orbit C such that $MC = C$. If M is of finite order, there is a closed geodesic with this property.

Proof. First we apply the theorem and, as mentioned earlier, get a function $z(t)$ on $[0, T]$ satisfying (1) and also $z(T) = Mz(0)$. We now observe that $z(\cdot)$ can be extended to $[T, 2T]$ as a solution of (1) as follows. Let $z(t) = Mz(t - T)$

for $T < t \leqslant 2T$, and note: z is continuous at T, and

$$J\dot{z}(t) = JM\dot{z}(t - T)$$
$$\in -JMJ\partial g_S(z(t - T))$$
$$= -JMJM^*\partial g_S(Mz(t - T))$$
$$= \partial g_S(z(t)) \quad \text{as required.}$$

(We have used the identity $M^*\partial g(Mz) = \partial g(z)$, which follows by differentiating $g(Mz) = g(z)$.) We extend z indefinitely forward this way, and also in reverse time, to get an orbit C such that $MC = C$. If M has finite order (i.e., $M^K = I$), then of course we cycle in finitely many steps, and C is a closed geodesic. □

We now proceed to show that the hypothesis that M is symplectic cannot be deleted from the theorem. We do this by studying the case in which S is the sphere, for which a convenient Hamiltonian is $H(x, p) = \Sigma\{(x^i)^2 + (p^i)^2\}/2$ (summation 1 to m). The Hamiltonian system is easily solved, and we find

$$x^i(t) = x_0^i \cos t + x_1^i \sin t$$
$$p^i(t) = x_1^i \cos t - x_0^i \sin t.$$

It follows that $Mz(0) = z(t)$ is equivalent to

(6) $$Mu = \{(\cos t)I - (\sin t)J\}u,$$

where I is the $2n \times 2n$ identity matrix and u is the vector (x_0, x_1). For (x, p) to be nonzero, we must have a nontrivial solution u of Eq. (6), for some $t > 0$. It is not difficult to find (nonsymplectic) matrices M for which this is not possible, demonstrating that the theorem is false for general M. As a consequence of the theorem and the example, we deduce:

Corollary 3
If M is symplectic, then there exist numbers α, β such that $\alpha^2 + \beta^2 = 1$ and

$$\det(M - \alpha I - \beta J) = 0.$$

A final consequence of the theorem is the following "antipodality" result, which can also be derived from the known fact that the transformation induced by a symplectic matrix is volume-preserving.

Corollary 4
If M is symplectic, then any convex energy surface contains a point whose image under M also lies in S.

7.8 CLOSED TRAJECTORIES OF PRESCRIBED PERIOD

In the previous section we sought to find certain Hamiltonian trajectories on a given energy surface $H = c$. We now deal with the question of the existence of periodic trajectories of a given system when the *period* of the motion is prescribed rather than the energy level. It is not possible to prescribe both the energy and the period, so one cannot say a priori what energy surface will contain the trajectory in question. Because of this, the theory in this case is less geometrical in nature, and growth properties of the Hamiltonian play an important role. Nonetheless, the central idea remains the use of a dual variational principle, combined this time with the theory of existence and necessary conditions developed in Chapter 4.

A *trajectory* z is a solution of the Hamiltonian inclusion $J\dot{z} \in \partial H(z)$ on some interval $[0, T]$, $T > 0$. The trajectory z is said to have *minimal* (or true) period T if $z(T) = z(0)$ and $z(t) \neq z(0)$ for t in $(0, T)$. As in the preceding section, z in R^{2n} corresponds to (x, p) in $R^n \times R^n$.

7.8.1 Theorem
Let the Hamiltonian H be finite-valued, nonnegative, and convex on R^{2n}, with $H(0) = 0$. Suppose that $\alpha > 0$, where

$$\alpha := \liminf_{|z| \to 0} \frac{H(z)}{|z|^2},$$

and that, for constants c and k, one has

$$|\partial_x H(x, p)| \leq c + k|p| \quad \text{for all } (x, p).$$

Then for all $T > \pi/\alpha$ there exists a trajectory of minimal period T. (*Note: We allow above the case $\alpha = +\infty$.*)

Proof. Let T be as in the theorem, and let G be the convex function conjugate to H:

$$G(y, q) := \sup_{x, p} \{y \cdot x + q \cdot p - H(x, p)\}.$$

G is convex by construction; our hypotheses do not imply that G is finite everywhere. However, one has:

Lemma 1. *G is nonnegative, and equal to 0 at $(0, 0)$. There is an $\varepsilon > 0$ and a scalar λ in $(0, 1)$ such that*

$$G(y, q) \leq \frac{T\lambda}{4\pi}|(y, q)|^2 \quad \text{for } |(y, q)| \leq \varepsilon.$$

By definition, $G(y, q)$ is no less than $-H(0,0) = 0$, and if $(y, q) = (0, 0)$, then (because H is nonnegative) $G(0, 0)$ cannot exceed 0; this gives the first two assertions. To prove the third, note first that the inequality $\alpha > \pi/T$ implies the existence of a positive number η and a scalar λ in $(0, 1)$ such that

$$H(x, p) \geq \frac{\pi}{T\lambda} |(x, p)|^2 \quad \text{for } |(x, p)| \leq \eta.$$

This implies, for any such (x, p) and for any (y, q), the inequality

$$y \cdot x + q \cdot p - \frac{\pi}{T\lambda} |(x, p)|^2 \geq y \cdot x + q \cdot p - H(x, p)$$

which, upon maximizing, yields

$$\frac{T\lambda}{4\pi} |(y, q)|^2 \geq \sup\{y \cdot x + q \cdot p - H(x, p) : |(x, p)| \leq \eta\}.$$

The expression being extremized on the right attains its (unconstrained) maximum for (x, p) in $\partial G(y, q)$, and its value is then $G(y, q)$. In other words, the right-hand side coincides with $G(y, q)$ as long as $\partial G(y, q)$ is contained in the ball ηB. But $\partial G(0, 0)$ is known to reduce to $\{(0, 0)\}$ because it is the set of points at which H attains its minimum. Since ∂G is upper semicontinuous, we have $\partial G(y, q) \subset \eta B$ for (y, q) sufficiently near 0. The lemma follows.

We now define a problem of Bolza P_B (see Section 4.1) for arcs (y, q) mapping $[0, 1]$ to $R^n \times R^n$, with data

$$L(y, q, \dot{y}, \dot{q}) := G(-\dot{q}, \dot{y}) + T\dot{q} \cdot y$$

$$l(y_0, q_0, y_1, q_1) := \psi_{\langle(0,0,0,0)\rangle}(y_0, q_0, y_1, q_1).$$

Note that the arc $(y(t), q(t)) = (0, 0)$ assigns a finite value to the Bolza functional. Let us calculate the Hamiltonian $\tilde{H}$ corresponding to L. If we use (p, r) to denote the adjoint variable, then

$$\tilde{H}(y, q, p, r) := \sup_{\dot{y}, \dot{q}}\{p \cdot \dot{y} + r \cdot \dot{q} - G(-\dot{q}, \dot{y}) - T\dot{q} \cdot y\}$$

$$= \sup_{\dot{y}, \dot{q}}\{-\dot{q} \cdot [Ty - r] + \dot{y} \cdot [p] - G(-\dot{q}, \dot{y})\}$$

$$= H(Ty - r, p),$$

the latter because H is the conjugate of G.

Lemma 2. $\tilde{H}$ satisfies the growth condition of Theorem 4.1.3.

7.8 Closed Trajectories of Prescribed Period

We calculate

$$\tilde{H}(y, q, p, r) = H(Ty - r, p)$$
$$= H(0, p) + \langle \zeta, Ty - r \rangle$$

(where ζ belongs to $\partial_x H(y^*, p)$ for some y^* between 0 and $Ty - r$, by the mean value theorem)

$$\leq H(0, p) + \{c + k|p|\}|Ty - r|$$
$$\leq H(0, p) + |r|\{c + k|p|\} + T|y|\{c + k|p|\}.$$

We now observe that $\tilde{H}$ satisfies the required growth condition (with μ being the first two terms in the last expression, and with $\sigma = Tc$, $\rho = Tk$).

It follows from Theorem 4.1.3 that the problem P_B admits a solution (y_0, q_0). We now wish to apply the necessary conditions of Theorem 4.2.2. It is clear that $\tilde{H}$ satisfies the strong Lipschitz condition, so the only missing ingredient is provided by:

Lemma 3. *The problem P_B is calm.*

We shall show that the function V_1 defined in Section 4.2 satisfies the calmness condition of that section. Note that $V_1(0)$ is finite, as shown above. Rockafellar's proof (1975) of Theorem 4.1.3 actually shows that there exists a positive constant M such that any arc (y, q) satisfying

(1) $$\int_0^1 L(y, q, \dot{y}, \dot{q}) \, dt \leq V_1(0) + 1, \quad y(0) = q(0) = 0$$

also satisfies $\int_0^1 |(\dot{y}, \dot{q})| \, dt \leq M$. Now let (y, q) be any arc satisfying (1), and let $y(1) = -s_1$, $q(1) = -s_2$. The arc $(y', q')(t) = (y, q)(t) + t(s_1, s_2)$ is admissible for our problem, and we easily find

$$\int_0^1 L(y', q', \dot{y}', \dot{q}') \, dt \leq \int_0^1 L(y, q, \dot{y}, \dot{q}) \, dt + TM|(s_1, s_2)|.$$

We deduce

$$V_1(0) \leq \int_0^1 L(y, q, \dot{y}, \dot{q}) \, dt + TM|(s_1, s_2)|,$$

which gives (since (y, q) is any arc satisfying (1))

$$V_1(s_1, s_2) \geq V_1(0) - TM|(s_1, s_2)|,$$

which implies that V_1 satisfies the required condition.

We now apply the necessary conditions of Theorem 4.2.2 to deduce the existence of an adjoint arc (p_0, r_0) satisfying the Hamiltonian inclusion for $\tilde{H}$ and for (y_0, q_0). This deciphers as:

$$\begin{bmatrix} -\dot{p}_0 \\ T\dot{y}_0 \end{bmatrix} \in T\partial H(Ty_0 - r_0, p_0) \quad \text{a.e.}$$

$$\dot{r}_0 = 0, \qquad T\dot{q}_0 = \dot{p}_0 \quad \text{a.e.}$$

The last relation yields $p_0(0) = p_0(1)$. Let us set, for $0 \leqslant t \leqslant T$,

$$x(t) := Ty_0\left(\frac{t}{T}\right) - r_0, \qquad p(t) := p_0\left(\frac{t}{T}\right).$$

It follows then that (x, p) satisfies on $[0, T]$ the relation

$$\begin{bmatrix} -\dot{p} \\ \dot{x} \end{bmatrix} \in \partial H(x, p) \quad \text{a.e.}$$

and one has $x(0) = x(T)$, $p(0) = p(T)$. We need only verify that T is the minimal period of (x, p) to complete the proof of the theorem. The following will help:

Lemma 4. The minimum in P_B is negative.

We need only exhibit a feasible arc (y, q) for which the Bolza functional is negative. Let u in R^n be a unit vector, and set $y(t) = \beta u \cos(2\pi t)$, $q(t) = -\beta u \sin(2\pi t)$, where the scalar β will be specified presently. Substituting gives

$$\int_0^1 L(y, q, \dot{y}, \dot{q}) \, dt = \int_0^1 G(2\pi\beta u \cos(2\pi t), -2\pi\beta u \sin(2\pi t)) \, dt$$

$$- \int_0^1 T\beta^2 2\pi \cos^2(2\pi t) \, dt.$$

The last term is computed to be $-T\pi\beta^2$. We estimate the first by means of Lemma 1. If $0 < 2\pi\beta < \varepsilon$, we deduce

$$\int_0^1 L(y, q, \dot{y}, \dot{q}) \, dt \leqslant \int_0^1 \frac{4\pi^2\beta^2 T\lambda}{4\pi} \, dt - T\pi\beta^2$$

$$= (\lambda - 1)T\pi\beta^2 < 0.$$

We now complete the proof of the theorem by assuming that (x, p) is periodic of period T/j for some integer j greater than 1, and deducing a contradiction. Note that in this case (y_0, q_0) is $(1/j)$-periodic. Define (y', q')

7.8 Closed Trajectories of Prescribed Period

on $[0, 1]$ via

$$y'(t) = \frac{j}{T} x\left(\frac{tT}{j}\right), \qquad q'(t) = \frac{j}{T} p\left(\frac{tT}{j}\right).$$

Note that (y', q') is feasible for P_B. We compute

$$\int_0^1 L(y', q', \dot{y}', \dot{q}') \, dt = \int_0^1 L\left(\frac{j}{T} x\left(\frac{tT}{j}\right), \frac{j}{T} p\left(\frac{tT}{j}\right), \dot{x}\left(\frac{tT}{j}\right), \dot{p}\left(\frac{tT}{j}\right)\right) dt$$

$$= \int_0^{1/j} L\left(\frac{j}{T} x(sT), \frac{j}{T} p(sT), \dot{x}(sT), \dot{p}(sT)\right) j \, ds$$

$$= j \int_0^{1/j} L\left(jy_0(s), jp_0(s), \dot{y}_0(s), \frac{1}{T}\dot{p}_0(s)\right) ds$$

(since an additive constant in the y variable can be ignored)

$$= j \int_0^{1/j} L(jy_0(s), jq_0(s), \dot{y}_0(s), \dot{q}_0(s)) \, ds$$

(since L does not depend on the q variable)

$$= j^2 \int_0^{1/j} \{G(-\dot{q}_0(s), \dot{y}_0(s)) + T\ddot{q}_0(s) \cdot y_0(s)\} \, ds$$

$$- j(j-1) \int_0^{1/j} G(-\dot{q}_0(s), \dot{y}_0(s)) \, ds$$

$$\leq j \int_0^1 \{G(-\dot{q}_0(s), \dot{y}_0(s)) + T\ddot{q}_0(s) \cdot y_0(s)\} \, ds$$

(since y_0, q_0 are $1/k$ periodic, and since G is nonnegative)

$$= j \min(P_B) < \min(P_B)$$

(since the latter is negative). This implies that (y', q') assigns a lower value to the Bolza functional than does (y_0, q_0), which is a contradiction. □

Comments

CHAPTER 1

Most of the references for this chapter are cited later on when the topics are treated in detail.

Section 1. The L^1 example was inspired by Bartels and Conn (1980), and the diode example by McClamroch (1980) (who discusses many interesting models). The two problems arising in engineering design are discussed in Polak (1982) and in Polak and Wardi (1982).

Section 2. The concepts defined here were introduced by Clarke (1973); see Chapter 2.

Sections 3 and 4. As general references for the calculus of variations we suggest Akhiezer (1962), Bliss (1946), Carathéodory (1965, 1967), Hestenes (1966), Young (1969); for optimal control, Berkovitz (1974c), Ioffe and Tihomirov (1979), Lee and Marcus (1967), Pontryagin et al. (1962), and Warga (1972).

CHAPTER 2

Je me détourne avec effroi et horreur de cette plaie lamentable des fonctions qui n'ont pas de dérivées. Hermite, in a letter to Stieltjes

Hermite's aversion to nonsmoothness notwithstanding, there have of course long been efforts to deal analytically with nondifferentiable functions. Among the better known such efforts are those involving the Dini derivates and approximate derivatives (see Saks, 1937, 1964), the generalized (or distributional) derivative, and the subdifferential of convex analysis (see Rockafellar, 1970a). There have been countless extended or alternate definitions of derivatives; see, for example, Bruckner and Leonard (1966), and Averbuh and Smoljanov (1968) for surveys. As early as 1892, Péano proposed that a function $f: R \to R$ should be termed differentiable at x provided the limit

$$\lim_{y \to x, z \to x} \frac{f(y) - f(z)}{y - z}$$

exists, saying that this notion "relates much better to derivatives as utilised in the physical sciences than the usual definition." (This is the strict differentiability of Section 2.2, which is in fact what the generalized gradient generalizes!) Many approaches to certain classes of nondifferentiable functions have been motivated by optimization (two of the better known are those of Neustadt (1976) and Pshenichnyi (1971)). The generalized gradient is distinguished, in our opinion, by its generality and constructive (nonaxiomatic) nature, its rich calculus and associated geometric theory, and its broad applicability.

> *à l'horreur, qui m'obsède*
> *quelle tranquillité succède*
> *Oui, le calme rentre dans mon coeur*
>
> Nicholas-Francois Guillard, *Iphigénie en Tauride*

Sections 1-3. The generalized directional derivative and gradient, the class of regular functions, and the first elements of the calculus were introduced in Clarke (1973, 1975a); the calculus was broadened in Clarke (1976c, 1976d, 1981a). The mean-value theorem, Theorem 2.3.7, is due to Lebourg (1975) (see Hiriart-Urruty (1980) for a discussion of mean values). Formula 2.3.10 appeared in Aubin and Clarke (1979); 2.3.13, 2.3.14 are due to Hiriart-Urruty (1979a, 1979c), whose chain rule was amalgamated with an earlier one in Clarke (1981a) to get Theorem 2.3.9. More detail on the calculus is to be found in Aubin (1978b, 1980b) and in the works of Hiriart-Urruty (1978, 1979abc, 1980, 1981ab, 1982ab) (see also Spingarn (1981) for a discussion of regularity).

Section 4. Generalized tangents and normals appeared first in Clarke (1973, 1975a); more detail can be found in Aubin and Clarke (1977a), Aubin (1980b) (who also discusses the Bouligand or contingent derivate), Rockafellar (1979a), Cornet (1981) (see also Watkins, (to appear) and Treiman (to appear)). The distance function has been used in the literature: see, for example, Redheffer and Walters (1974). Formula 2.4.7 is due to Rockafellar (1979b), as well as the theory of hypertangents (the "radial tangents" of Hiriart-Urruty).

Section 5. An interesting open question is how to extend the useful characterization 2.5.7 to infinite dimensions. The formula in 2.5.1 has been extended to separable Banach spaces by Hiriart-Urruty and Thibault (1980).

Section 6. The generalized Jacobian was introduced in Clarke (1973, 1976d); a discussion appears in Hiriart-Urruty (1982b). Proposals for extending the scope of the definition have been made by Thibault (1979, 1982a, b), Papageorgiou (to appear), and Mirica (1980).

Section 7. A version of the main result appeared in Clarke (1981a); the alternate minimax proof stems from Clarke (1977b). A form of Theorem 2.7.5 appeared in Aubin and Clarke (1979).

Section 8. There is a large literature on the differential properties of max functions: see, for example, Danskin (1967), Hogan (1973), and Dem'janov and Malozemov (1971). The first version of Theorem 2.8.2 appeared in Clarke (1973, 1975a), and 2.8.6 is taken from Clarke (1976e). See Hiriart-Urruty (1978) for more developments on max functions. Example 2.8.7 is drawn from Polak and Wardi (1982).

Section 9. The generalized gradient was defined for extended-valued maps in Clarke (1973, 1975a), but most of the existing calculus is for the Lipschitz case. The results of this section are due to Rockafellar (1979b, 1980) (see also Rockafellar, 1979c, 1981); some parallel developments have been achieved by Aubin (1980b), Hiriart-Urruty (1979a, 1979b, 1982a), and Borwein and Stròjwas (to appear).

Applications of the theory of this chapter to a variety of topics have been made; many will be cited later. See also Ekeland (1979a), Lebourg (1979), Shu-Chung (1980), and Janin (1982).

"Man muss immer generalisieren" (one should always generalize), said Jacobi. True to this dictum, there have been concepts defined which generalize the generalized gradient: see Halkin (1976a, 1978), Ioffe (1981), and Warga (1976, 1978a). (These approaches are similar in spirit to the abstract nonconstructive one of Neustadt, 1976.) Morduhovic (1980, 1981) has proposed and used a modified definition of the generalized gradient. Rockafellar's work (1979b, 1980) extends the context to that of locally convex spaces. A bibliography of nonsmooth optimization has been published by Gwinner (1981).

CHAPTER 3

Section 1. The theory of multifunctions is studied in Castaing and Valadier (1977), Rockafellar (1969), and Wagner (1977). Proposition 3.1.2 is proven in Clarke (1973); Theorem 3.1.3 appears in Aumann (1965). The main results of this section parallel those in Clarke (1975c); the proof of 3.1.6 is adapted from Filippov (1967). Theorem 3.1.7 is a standard type of result proven by many authors. Differential inclusions have been studied by, for example, Antosiewicz and Cellina (1975), Aubin, Cellina, and Nohel (1977), Filippov (1967), Hermes (1970), and Roxin (1965).

Section 2. Versions of the differential inclusion problem are studied in Berliocchi and Lasry (1973), Blagodatskih (1975), Boltjanskii (1973), Clarke (1976a, 1976f), and Fedorenko (1970). (Let us note that in Clarke (1976a), the hypothesis that the multifunction be convex-valued (as in this chapter) must be added to make the proof correct, or alternatively, that the terminal endpoint constraints are given by (Lipschitz) inequalities.) The preceding treat the problem in less generality (for example, without state constraints); the present definition of multiplier is new.

Section 3. The results of this section are the joint work of Clarke and Darrough (1981).

Sections 4 and 5. The results of these sections are new; there is earlier related work in optimal control and in mathematical programming (see Chapters 5 and 6).

Section 6. The transformation technique used here stems from the calculus of variations; it was first used for the present purpose in Clarke (1980a).

Section 7. The sufficiency of a modified Hamilton–Jacobi equation was studied by the author's students Havelock (1977) and Offin (1979); the necessity (for normal problems) is a new result (announced in Clarke, 1982b); further developments appear in Clarke and Vinter (to appear). As mentioned, the basic verification technique is an old and often rediscovered one. It is known in the Soviet literature as the method of Krotov functions (see Hrustalev, 1973). Examples of the use of the method are provided by Adams and Clarke (1979), Clark, Clarke, and Munro (1979), and Clarke and Darrough (1981).

CHAPTER 4

The impetus (at least for us) to study the generalized problem of Bolza was the seminal work of Rockafellar (1970c, 1971b, 1972, 1973, 1975, 1976b), most of which is concerned with the convex case. The possibility of a general approach along these lines was known to Young, who went as far as to identify the "true Hamiltonian" for the optimal control problem; that is, the function $H(x, p) := \max\{\langle p, \phi(x, u)\rangle - F(x, u) : u \in U\}$ (see Young, 1969, p. 230). Despite his belief that "conceptually, the importance of Hamiltonians compares with that of complex numbers" (p. 46), Young fails to cross the Rubicon, but instead turns to the pseudo-Hamiltonian of Pontryagin (after remarking that H is lacking in smoothness (p. 230)). Nonsmooth analysis is the bridge that makes progress in this direction possible.

Section 1. Detailed studies of equivalence, measurability, and other matters appear in the articles of Rockafellar cited above. The transform used to define H from L is a basic tool of convex analysis; see Rockafellar (1966, 1967, 1970a, 1970b, 1974). There is an abundant literature on existence theory; see, for example, Berkovitz (1974a, 1974c), Cesari (1971), Ioffe (1976, 1977), and Olech (1969).

Section 2. Necessary conditions for the general nonconvex, nonsmooth problem of Bolza were obtained by Clarke (1973, 1975b, 1976b, 1977a); we refer the reader to Clarke (1976b) for a discussion of the necessary conditions in Lagrangian (rather than Hamiltonian) terms, as well as for results on relaxation. The calmness constraint qualification was introduced in Clarke (1973). It seems essential to have a qualification of some kind, since P_B does not lend itself to expressing the necessary conditions in abnormal form (this of course is because all the constraints of the problem are implicit).

Section 3. Zeidan uses the technique of canonical transformations to obtain stronger sufficiency theorems than the simple one we have given here; we refer to Zeidan (1982, 1983) for a thorough discussion.

Section 4. The utility of a growth condition such as the one postulated here was suggested by Jean-Michel Lasry.

Section 5. The proof of the multiplier rule was ultimately completed by McShane (1939); the results of this section appeared in Clarke (1977c).

Section 6. This result appeared in Clarke (1977b); for related work see Arnautu (1980), Barbu (1981, 1982).

CHAPTER 5

We have given general references for optimal control theory in the notes to Chapter 1.

Section 1. Controllability is an active area of research; see, for example, Brockett (1976), Hermann and Krener (1977), and Hermes (1974, 1982).

Section 2. The difficult issues in proving a general maximum principle stem from the lack of smoothness and continuity assumptions on the data, the fact that the set of state arcs is not closed (since the problem is not relaxed), and the generality of the endpoint constraints. The free-endpoint case is much simpler to treat, and the case in which the endpoint constraint is of the form $\theta(x(b)) \leq 0$ can be reduced to it by observing that (when $F = 0$) x then minimizes $\max\{f(y(b)) - f(x(b)), \theta(y(b))\}$ over the admissible state arcs y. (See Clarke (1980a) for details of this technique, which also allows analysis of Pareto optima.) An early article by Dunn (1967) foreshadowed the use of Hamiltonian level curves in phase-plane analysis. One of the early attempts to deal with nonsmooth control problems is due to Luenberger (1970) (see also Wierzbicki, 1972). A systematic theory was developed by Neustadt (1976). See Haussmann (1981) for the stochastic maximum principle.

Section 3. In applying the necessary conditions of Chapter 4, we are led to seek a Hamiltonian extremal pair (x, p) which does exist, whereas there may be no (x, p) satisfying the conditions of the maximum principle. This stems from the fact that the original and relaxed problems (only the latter of which is guaranteed to have a solution) give rise to the same Hamiltonian, but to different pseudo-Hamiltonians. A further consequence of this observation is that if there is a single solution (x, p) of the Hamiltonian conditions, and if x is feasible for the orginal problem, then x must solve both the original and the relaxed problems. The corresponding assertion with the maximum principle is untrue.

Section 4. The possibility of applying variational methods to optimal control, and the frequent equivalence of the two, has been well recognized: Young (1969, p. 218), Berkovitz (1961). See the notes to Chapter 4, Section 1, for remarks on existence, and Chapter 3, Section 7, for sufficient conditions (as well as Seierstad and Sydsaeter, 1977).

Section 5. The relaxed problem was introduced and studied by Warga (1962, 1972) (and by Young (1969) as generalized curves in the calculus of variations). For other sensitivity results in optimal control see Gollan (to appear (c)) and Maurer (1979a, 1979b).

CHAPTER 6

McCormick (1975) and Rockafellar (1976a) survey optimality criteria in mathematical programming. See also Mangasarian (1969).

Section 1. The choice of Lagrangian, and whether the conclusions of the multiplier rule are expressed in separated or partly separated form play an important role in determining the relative precision of the resulting conditions (see the discussion in Rockafellar, 1982b). From this point of view the (new) Lagrangian used in this section appears to be best. See Morduhovic (1980, 1981) for other multiplier rules with generalized gradients, and Nguyen, Strodiot, and Mifflin (1977), and Strodiot and Nguyen (1979) for related developments. Many examples and refinements of certain aspects of the theory have been developed by Hiriart-Urruty (1979a, 1979b, 1981, 1982a).

Section 2. The numerical aspects of nonsmooth optimization have been the subject of study (see, for example, Wolfe and Balinski, 1975); for methods involving generalized gradients see Auslender (1977), Feuer (1974), Goldstein (1977), Mifflin (1977), Polak (1982), Polak, Mayne, and Wardi (to appear), and Polak and Wardi (1982). Chaney (1982a, b, 1983) has studied sufficiency under nonsmoothness.

Sections 3 and 4. Robinson (1976a, 1976b) has elucidated the relationship between stability and constraint qualifications. Calmness was introduced in Clarke (1973), and some of the results in these sections in Clarke (1976c).

Section 5. There is a considerable literature on sensitivity; one of the early results in the present vein is Gauvin's (1979). We refer also to Aubin (1980a), Aubin and Clarke (1977b, 1979), Auslender (1978), Gollan (1981ab, to appear (a)), Hiriart-Urruty (1978), Lempio and Maurer (1980), Maurer (1979b), Rockafellar (1982b) (which has a thorough discussion of other related work), and Pomerol (1982).

Section 6. Robinson (1976a, 1976b, 1982) has achieved many of the results in this area; see also Borwein (to appear) and the references for Section 5.

CHAPTER 7

Section 1. The first inverse and implicit function theorems of comparable generality to appear in the literature seem to be Warga's (1976, 1978b) (see also Halkin (1976b) for the related interior mapping theorem). The main result appears in Clarke (1976d); that the implicit function theorem follows was first explicitly noted by Hiriart-Urruty (1979a, Theorem 11).

Section 2. This proof of Aumann's Theorem (Aumann, 1965) appeared in Clarke (1981c).

Section 3. This result is taken from Rockafellar (1979a).

Section 4. For the smooth case, the most general results in the vein of the one proven here seem to be those of Hestenes (1966, Appendix).

Section 5. See Ekeland (1979a) for a survey of applications of the theorem.

Section 6. This result first appeared in Clarke (1978a).

Section 7. The dual variational principle that lies at the heart of this section and the next was introduced in Clarke (1978c); see also Clarke (1979b, 1981b) for related work. The main result of the section appears in Clarke (1982a). Desolneux-Moulis (1979) surveys recent developments in the theory of periodic solutions of Hamiltonian systems. The dual variational principle has subsequently been used by Clarke and Ekeland (1980, 1982), and also by Ambrosetti and Mancini (1981), Brézis (1980), Brézis, Coron, and Nirenberg (1980), Ekeland (1979b), and Ekeland and Lasry (1980). The global existence of a closed orbit (Corollary 1) was proven by Rabinowitz (1978) and by Weinstein (1978).

Section 8. The result proven here is distinct from, but closely related to, the main result in Clarke and Ekeland (1980), in which an ad hoc approach to existence is employed.

References

Adams, R. A., and Frank H. Clarke (1979). Gross's logarithmic Sobolev inequality: A simple proof, *Am. J. Math.* **101**, 1265–1269.

Akhiezer, N. I. (1962). *The Calculus of Variations*, Aline H. Frink, Transl., Blaisdell Publishing, Boston.

Ambrosetti, A., and G. Mancini (1981). Solutions of minimal period for a class of convex Hamiltonian systems, *Math. Ann.* **255**, 405–421.

Antosiewicz, H. A., and A. Cellina (1975). Continuous selections and differential relations, *J. Differ. Eq.* **19**, 386–398.

Arnautu, V. (1980). Characterisation and approximation of optimal control of a class of nonconvex distributed control problems, *Mathematika* **22**, 189–205.

Aubin, J. P. (1978a). *Applied Functional Analysis*, Wiley Interscience, New York.

―――― (1978b). Gradients généralisés de Clarke, *Ann. Sci. Math. Quebec* **2**, 197–252.

―――― (1980a). Further properties of Lagrange multipliers in nonsmooth optimization, *Appl. Math. Optimization* **6**, 79–90.

―――― (1980b). Contingent derivatives of set-valued maps, Math. Research Center, Univ. of Wisconsin, Madison, Tech. Rep. 2044.

Aubin, J. P., A. Cellina, and J. Nohel (1977). Monotone trajectories of multivalued dynamical systems, *Ann. Mat. Pura Appl.* **115** (IV), 99–117.

Aubin, J. P., and Frank H. Clarke (1977a). Monotone invariant solutions to differential inclusions, *J. London Math. Soc.* **16** (2), 357–366.

―――― (1977b). Multiplicateurs de Lagrange en optimisation non-convexe et applications, *Co. R. Acad. Sci. Paris* **285**, 451–454.

―――― (1979). Shadow prices and duality for a class of optimal control problems, *SIAM J. Control Optim.* **17**, 567–587.

Aumann, R. J. (1965). Integrals of set-valued functions, *J. Math. Anal. Appl.* **12**, 1–12.

―――― (1967). Measurable utility and the measurable choice theorem, *La Décision*, Actes Coll. Int. du CNRS, Aix-en-Provence, pp. 15–26.

Auslender, A. (1977). Minimisation sans contraintes de fonctions localement lipschitziennes, *Co. R. Acad. Sci. Paris* **284**, 959–961.

——— (1978). Differential stability in non convex and non differentiable programming, *Mathematical Programming Study 10*, P. Huard, Ed., North-Holland, Amsterdam.

Averbuh, V. I., and O. G. Smoljanov (1968). Different definitions of derivative in linear topological spaces, *Usp. Mat. Nauk* **23**, 67–116.

Barbu, V. (1981). Necessary conditions for nonconvex distributed control problems governed by elliptic variational inequalities, *J. Math. Anal. Appl.* **80**, 566–597.

——— (1982a). Boundary control problems with nonlinear state equation, *SIAM J. Control Optim.* **20**, 125–143.

——— (1982b). Necessary conditions for multiple integral problems in the calculus of variations, *Math. Ann.* **260**, 175–289.

Bartels, R. H., and A. R. Conn (1980). Linearly constrained discrete l_1 problems, *ACM Trans. Math. Software* **6**, 594–608.

Berkovitz, L. D. (1961). Variational methods in problems of control and programming, *J. Math. Anal. Appl.* **3**, 145–169.

——— (1974a). Lower semicontinuity of integral functionals, *Trans. Am. Math. Soc.* **192**, 51–57.

——— (1974b). Existence and lower closure theorems for abstract control problems, *SIAM J. Control Optim.* **12**, 27–42.

——— (1974c). *Optimal Control Theory*, Springer-Verlag, New York.

Berliocchi, H., and J. M. Lasry (1973). Principe de Pontryagin pour des systèmes régis par une équation differentielle multivoque, *Co. R. Acad. Sci. Paris* **277**, 1103–1105.

Bishop, E., and R. R. Phelps (1963). The support functionals of a convex set, *Proceedings of the Symposium in Pure Mathematics*, Vol. 7: Convexity, American Mathematical Society, pp. 27–35.

Blagodatskih, V. I. (1975). Time optimal control problem for differential inclusions, in Banach Center Publ. 1, Warsaw.

——— (1976). On the theory of sufficient conditions for optimality, *Sov. Math. Dokl.* **17**, 1680–1683.

Bliss, G. A. (1930). The problem of Lagrange in the calculus of variations, *Am. J. Math.* **52**, 673–744.

——— (1946). *Lectures on the Calculus of Variations*, Univ. of Chicago Press, Chicago.

Boltjanskii, V. G. (1973). The maximum principle for problems of optimal steering, *Differ. Uravn.* **9**, 1363–1370.

Borwein, J. M. (to appear). Stability and regular points of inequality systems.

Borwein, J. M., and H. M. Stròjwas (to appear). Directionally Lipschitzian mappings on Baire spaces.

Brézis, H. (1980). Periodic solutions of nonlinear vibrating strings, in *Proc. AMS Symp. on the Mathematical Heritage of H. Poincaré*, Bloomington, Ind.

Brézis, H., J. M. Coron, and L. Nirenberg (1980). Free vibrations for a nonlinear wave equation and a theorem of P. Rabinowitz, *Commun. Pure Appl. Math.* **33**, 667–684.

Brockett, R. W. (1976). Nonlinear systems and differential geometry, *Proc. IEEE* **64**, 61–72.

Bruckner, A. M., and J. L. Leonard (1966). Derivatives, in *The Slaught Memorial Papers*, *Am. Math. Mon.* **73**, 24–56.

References

Campbell, H. F. (1980). The effect of capital intensity on the optimal rate of extraction of a mineral deposit, *Can. J. Econ.* **13**, 349–355.

Carathéodory, C. (1965, 1967). *Calculus of Variations and Partial Differential Equations of the First Order*, vol. 1 and 2, R. B. Dean and J. J. Brandstatter, Transl., Holden-Day, San Francisco.

Castaing, C., and M. Valadier (1977). *Convex Analysis and Measurable Multifunctions*, Lecture Notes in Mathematics, No. 580, Springer, Berlin.

Cesari, L. (1971). Closure, lower closure, and semicontinuity theorems in optimal control, *SIAM J. Control Optim.* **9**, 287–315.

Chaney, R. W. (1982a). Second-order sufficiency conditions for nondifferentiable programming problems, *SIAM J. Control Optim.* **20**, 20–33.

────── (1982b). On sufficient conditions in nonsmooth optimization, *Math. Oper. Res.* **7**, 463–475.

────── (1983). A general sufficiency theorem for nonsmooth nonlinear programming, *Trans. Am. Math. Soc.* **276**, 235–246.

Clark, C. W., Frank H. Clarke, and G. R. Munro (1979). The optimal exploitation of renewable resource stocks, *Econometrica* **47**, 25–47.

Clarke, Frank H. (1973). *Necessary Conditions for Nonsmooth Problems in Optimal Control and the Calculus of Variations*, Ph.D. thesis, Univ. of Washington.

────── (1975a). Generalized gradients and applications, *Trans. Am. Math. Soc.* **205**, 247–262.

────── (1975b). The Euler-Lagrange differential inclusion, *J. Differ. Eq.* **19**, 80–90.

────── (1975c). Admissible relaxation in variational and control problems, *J. Math. Anal. Appl.* **51**, 557–576.

────── (1976a). Necessary conditions for a general control problem, in *Calculus of Variations and Control Theory*, D. Russell, Ed., Mathematics Research Center, Pub. 36, Univ. of Wisconsin, Academic Press, New York, pp. 259–278.

────── (1976b). The generalized problem of Bolza, *SIAM J. Control Optim.* **14**, 682–699.

────── (1976c). A new approach to Lagrange multipliers, *Math. Oper. Res.* **1**, 165–174.

────── (1976d). On the inverse function theorem, *Pac. J. Math.* **64**, 97–102.

────── (1976e). The maximum principle under minimal hypotheses, *SIAM J. Control Optim.* **14**, 1078–1091.

────── (1976f). Optimal solutions to differential inclusions, *J. Optim. Theory Appl.* **19**, 469–478.

────── (1977a). Extremal arcs and extended Hamiltonian systems, *Trans. Am. Math. Soc.* **231**, 349–367. (A Russian translation of this article appears in *Differ. Uravn.* **13**(1977), 427–442.)

────── (1977b). Multiple integrals of Lipschitz functions in the calculus of variations, *Proc. Am. Math. Soc.* **64**, 260–264.

────── (1977c). Inequality constraints in the calculus of variations, *Can. J. Math.* **3**, 528–540.

────── (1978a). Pointwise contraction criteria for the existence of fixed points, *Bull. Can. Math. Soc.* **21**, 7–11.

────── (1978b). Nonsmooth analysis and optimization, in *Proc. Int. Cong. of Mathema-*

ticians (Helsinki).

——— (1978c). Solutions périodiques des équations hamiltoniennes, *Co. R. Acad. Sci. Paris* **287**, 951–952.

——— (1979a). Optimal control and the true Hamiltonian, *SIAM Rev.* **21**, 157–166.

——— (1979b). A classical variational principle for periodic Hamiltonian trajectories, *Proc. Am. Math. Soc.* **76**, 186–188.

——— (1980a). The Erdmann condition and Hamiltonian inclusions in optimal control and the calculus of variations, *Can. J. Math.* **32**, 494–509.

——— (1980b). The dual action, optimal control and generalized gradients, in *Mathematical Control Theory*, Proc. International Semester on Optimal Control Theory, S. Banach Mathematical Research Center.

——— (1981a). Generalized gradients of Lipschitz functionals, *Adv. Math.* **40**, 52–67. Appeared earlier as MRC Tech. Rep. 1687, Aug. 1976 (Madison, Wisconsin).

——— (1981b). Periodic solutions of Hamiltonian inclusions, *J. Differ. Eq.* **40**, 1–6.

——— (1981c). A variational proof of Aumann's Theorem, *Appl. Math. Optim.* **7**, 373–378.

——— (1982a). On Hamiltonian flows and symplectic transformations, *SIAM J. Control Optim.* **20**, 355–359.

——— (1982b). The applicability of the Hamilton–Jacobi verification technique, in *Proc. 10th IFIP Conf.* (New York, 1981), R. F. Drenick and F. Kozin, Eds., *System Modeling and Optimization Ser.*, no. 38, Springer-Verlag, New York, pp. 88–94.

Clarke, Frank H., and Masako Darrough (1981). Resource extraction versus capital investment under fixed proportions: a nondifferentiable control problem, IAMS Technical Report No. 81-2, UBC.

Clarke, Frank H., and I. Ekeland (1980). Hamiltonian trajectories having prescribed minimal period, *Commun. Pure Appl. Math.* **33**, 103–116.

——— (1982). Nonlinear oscillations and boundary-value problems for Hamiltonian systems, *Arch. Ration. Mech. Anal.* **78**, 315–333.

Clarke, Frank H., and Richard B. Vinter (to appear). Local optimality conditions and Lipschitzian solutions to the Hamilton–Jacobi equation, SIAM J. Control Optim.

Cornet, B. (1981). *Théorie Mathématique des Mecanismes; Dynamiques d'Allocation des Ressources*, thesis, Université de Paris IX, France.

Danskin, J. M. (1967). *The Theory of Max Min*, Springer-Verlag, New York.

Dasgupta, P. S. and G. M. Heal (1979). *Economic Theory and Exhaustible Resources*, Cambridge Univ. Press, England.

Dem'janov, V. F., and V. N. Malozemov (1971). On the theory of non-linear minimax problems, *Russ. Math. Surv.* **26**, 57–115.

Dem'janov, V. F., and A. M. Rubinov (1980). On quasidifferentiable functionals, *Sov. Math. Dokl.* **21**, 14–17.

Desolneux-Moulis, N. (1979). Orbites périodiques des systèmes hamiltoniens autonomes, *Semin. Bourbaki* **32**, No. 552.

Dunn, J. C. (1967). On the classification of singular and nonsingular extremals for the Pontryagin maximum principle, *J. Math. Anal. Appl.* **17**, 1–36.

Edwards, R. E. (1965). *Functional Analysis*, Holt, New York.

Ekeland, I. (1974). On the variational principle, *J. Math. Anal. Appl.* **47**, 324–353.
—— (1979a). Nonconvex minimization problems, *Bull. Am. Math. Soc. (N.S.)* **1**, 443–474.
—— (1979b). Periodic Hamiltonian trajectories and a theorem of Rabinowitz, *J. Differ. Eq.* **34**, 523–534.
Ekeland, I., and J. M. Lasry (1980). On the number of periodic trajectories for a Hamiltonian flow on a convex energy surface, *Ann. Math.* **112**, 283–319.
Ekeland, I., and R. Temam (1976). *Convex Analysis and Variational Problems*, North-Holland, Amsterdam.
Fedorenko, R. P. (1970). A maximum principle for differential inclusions, *USSR Comp. Math. and Math. Phys.* **10**, 57–68.
Feuer, A. (1974). Minimizing well-behaved functions, 12th Allerton Conf. on Circuit and Systems Theory, University of Illinois, Allerton, Ill.
Filippov, A. F. (1967). Classical solutions of differential equations with multivalued right-hand side, *SIAM J. Control Optim.* **5**, 609–621.
Gauvin, J. (1979). The generalized gradient of a marginal function in mathematical programming, *Math. Oper. Res.* **4**, 458–463.
Goldstein, A. A. (1977). Optimization of Lipschitz continuous functions, *Math. Program.* **13**, 14–22.
Goldstein, H. (1950). *Classical Mechanics*, Addison-Wesley, Reading, Mass.
Gollan, B. (1981a). Higher order necessary conditions for an abstract optimization problem, *Mathematical Programming Study*, **14**, pp. 69–76.
—— (1981b). Perturbation theory for abstract optimization problems, *J. Optim. Theory Appl.* **35**, 417–441.
—— (to appear (a)). On the marginal function in nonlinear programming, *Math. Oper. Res.*
—— (to appear (b)). On the optimal value functions of optimal control problems.
Gwinner, J. (1981). Bibliography on non-differentiable optimization and non-smooth analysis, *J. Comp. Appl. Math.* **7**, 277–285.
Halkin, H. (1972). Extremal properties of biconvex contingent equations, in *Ordinary Differential Equations* (NRL-MRC Conf.), Academic Press, New York.
—— (1976a). Mathematical programming without differentiability, in *Calculus of Variations and Control Theory*, D. L. Russell, Ed., Academic Press, New York, pp. 279–288.
—— (1976b). Interior mapping theorem with set-valued derivative, *J. d'Anal. Math.* **30**, 200–207.
—— (1978). Necessary conditions for optimal control problems with differentiable or nondifferentiable data, in *Mathematical Control Theory* (Proc. Conf., Australian Nat. Univ., Canberra, 1977), *Lecture Notes in Mathematics*, no. 680, Springer, Berlin, pp. 77–118.
Hartl, R. F., and S. P. Sethi (to appear). Sufficient conditions for the optimal control of a class of systems with differential inclusions and applications.
Haussmann, U. (1981). Some examples of optimal stochastic control or: the stochastic maximum principle at work, *SIAM Rev.* **23**, 292–307.
Havelock, D. (1977). *A Generalization of the Hamilton–Jacobi Equation*, M.Sc. thesis, Univ. of British Columbia, Canada.

Hermann, R., and A. J. Krener (1977). Nonlinear controllability and observability, *IEEE Trans. Autom. Control* **AC-22**, 728–740.

Hermes, H. (1970). The generalized differential equation $\dot{x} \in R(t, x)$, *Adv. Math.* **4**, 149–169.

―― (1974). On necessary and sufficient conditions for local controllability along a reference trajectory, in *Geometric Methods in Systems Theory*, R. Brockett and D. Mayne, Eds., Reidel Publishing, Holland.

―― (1982). On local controllability, *SIAM J. Control Optim.* **20**, 211–220.

Hestenes, M. R. (1966). *Calculus of Variations and Optimal Control Theory*, John Wiley, New York.

Hiriart-Urruty, J. B. (1978). Gradients généralisés de fonctions marginales, *SIAM J. Control Optim.* **16**, 301–316.

―― (1979a). Tangent cones, generalized gradients and mathematical programming in Banach spaces, *Math. Oper. Res.* **4**, 79–97.

―― (1979b). Refinements of necessary optimality conditions in nondifferentiable programming I, *Appl. Math. Optim.* **5**, 63–82.

―― (1979c). New concepts in nondifferentiable programming, *Bull. Soc. Math. France*, Memoire 60, 57–85.

―― (1980). Mean value theorems in nonsmooth analysis, *Numer. Funct. Anal. Optim.* **2**, 1–30.

―― (1981a). Optimality conditions for discrete nonlinear norm-approximation problems, in *Optimization and Optimal control* (Proc. Conf., Mathematics Research Institute, Oberwolfach, 1980), *Lecture Notes in Control and Information Science*, no. 30, Springer, Berlin, pp. 29–41.

―― (1981b). A better insight into the generalized gradient of the absolute value of a function, *Applic. Anal.* **12**, 239–249.

―― (1982a). Refinements of necessary optimality conditions in nondifferentiable programming II, *Mathematical Programming Study*, **19**, 120–139.

―― (1982b). Characterizations of the plenary hull of the generalized Jacobian matrix, *Mathematical Programming Study* **17**, 1–12.

Hiriart-Urruty, J. B., and L. Thibault (1980). Existence et caractérisation de differentielles généralisées, *Co. R. Acad. Sci. Paris* **290**, 1091–1094.

Hogan, W. (1973). Directional derivatives for extremal-value functions with applications to the completely convex case, *Oper. Res.* **21**, 188–209.

Hörmander, L. (1954). Sur la fonction d'appui des ensembles convexes dans un espace localement convexe, *Ark. Math.* **3**, 181–186.

Hotelling, H. (1931). The economics of exhaustible resources, *J. Political Economy* **39**, 137–175.

Hrustalev, M. M. (1973). Necessary and sufficient conditions for an optimal control problem, *Sov. Math. Dokl.* **14**, 983–986.

Ioffe, A. D. (1976). An existence theorem for a general Bolza problem, *SIAM J. Control Optim.* **14**, 458–466.

―― (1977). On lower semicontinuity of integral functionals I, *SIAM J. Control Optim.* **15**, 521–538.

―― (1981). Nonsmooth analysis: differential calculus of nondifferentiable mappings, *Trans. Am. Math. Soc.* **266**, 1–56.

Ioffe, A. D., and V. L. Levin (1972). Subdifferentials of convex functions, *Trans. Moscow Math. Soc.* **26**, 1–72.

Ioffe, A. D., and V. M. Tihomirov (1979). *Theory of Extremal Problems*, North-Holland, Amsterdam.

Janin, R. (1982). Sur les multiapplications qui sont des gradients généralisés, *Co. R. Acad. Sci. Paris* **294**, 115–117.

Krener, A. J. (1977). The high order maximal principle and its applications to singular extremals, *SIAM J. Control Optim.* **15**, 256–293.

Lebourg, G. (1975). Valeur moyenne pour gradient généralisé, *Co. R. Acad. Sci. Paris* **281**, 795–797.

——— (1979). Generic differentiability of Lipschitzian functions, *Trans. Am. Math. Soc.* **256**, 125–144.

Lee, E. B., and L. Markus (1967). *Foundations of Optimal Control Theory*, John Wiley, New York.

Lempio, F., and H. Maurer (1980). Differential stability in infinite-dimensional nonlinear programming, *Appl. Math. Optim.* **6**, 139–152.

Luenberger, D. G. (1970). Control problems with kinks, *IEEE Trans. Autom. Control* **AC-15**, 570–575.

Mangasarian, O. L. (1966). Sufficient conditions for the optimal control of nonlinear systems, *SIAM J. Control Optim.* **4**, 139–152.

——— (1969). *Nonlinear Programming*, McGraw-Hill, New York.

Maurer, H. (1979a). Differential stability in optimal control problems, *Appl. Math. Optim.* **5**, 283–295.

——— (1979b). First order sensitivity of the optimal value function in mathematical programming and optimal control, in *Proc. Symp. on Mathematical Programming with Data Perturbations* (Washington, D.C.).

McClamroch, N. H. (1980). *State Models of Dynamic Systems*, Springer-Verlag, New York.

McCormick, Garth P. (1975). Optimality criteria in nonlinear programming, in *Proc. Symp. on Applied Mathematics*, Vol. 9, R. W. Cottle, Ed., AMS, SIAM.

McLeod, R. M. (1965). Mean value theorems for vector-valued functions, *Proc. Edinburgh Math. Soc.* **14**(2), 197–209.

McShane, E. J. (1939). On multipliers for Lagrange problems, *Am. J. Math.* **61**, 809–819.

Mifflin, R. (1977). Semismooth and semiconvex functions in optimization, *SIAM J. Control Optim.* **15**, 959–972.

Mirica, Stefan (1980). A note on the generalized differentiability of mappings, *Nonlinear Anal.* **4**, 567–575.

Morduhovic, B. S. (1980). Metric approximations and necessary optimality conditions for general classes of nonsmooth extremal problems, *Sov. Math. Dokl.* **22**, 526–530.

——— (1981). Penalty functions and necessary conditions for an extremum for nonsmooth nonconvex optimization problems, *Usp. Mat. Nauk.* **36**, 215–216.

Morrey, Jr., C. B. (1966). *Multiple Integrals in the Calculus of Variations*, Springer-Verlag, New York.

Neustadt, L. W. (1976). *Optimization*, Princeton Univ. Press, Princeton, N.J.

Nguyen, V. Hien, J. J. Strodiot, and R. Mifflin (1980). On conditions to have bounded multipliers in locally Lipschitz programming, *Math. Program.* **18**, 100–106.

Offin, D. (1979). *A Hamilton–Jacobi Approach to the Differential Inclusion Problem*, M.Sc. thesis, Univ. of British Columbia, Canada.

Olech, C. (1966). Extremal solutions of a control system, *J. Differ. Eq.* **2**, 74–101.

——— (1969). Existence theorems for optimal control problems with vector-valued cost functions, *Trans. Am. Math. Soc.* **136**, 157–180.

Papageorgiou, N. S. (to appear). Nonsmooth analysis on partially ordered vector spaces. Part 2: nonconvex case, Clarke's theory, *Pac. J. Math.*

Péano, G. (1892). Sur la définition de la dérivée, *Mathesis* **2**(2), 12–14. Also *Opere Scelte*, Vol. 1, Edizioni Crenonese, Rome, 1957, pp. 210–212.

Polak, E. (1982). An implementable algorithm for the optimal design, centering, tolerancing, and tuning problem, *J. Optim. Theory Appl.* **37**, 45–68.

Polak, E., D. Q. Mayne, and Y. Wardi (to appear). On the extension of constrained optimization algorithms from differentiable to nondifferentiable problems, SIAM J. Control Optim.

Polak, E., and Y. Wardi (1982). A nondifferentiable optimization algorithm for the design of control systems having singular value inequalities, *Automatica, J. IFAC* **18**, 267–283.

Pomerol, J. C. (1982). The Lagrange multiplier set and the generalized gradient set of the marginal function of a differentiable program in a Banach space, *J. Optim. Theory Appl.* **38**, 307–317.

Pontryagin, L. S., V. G. Boltyanskii, R. V. Gamkrelidze, and E. F. Mischenko (1962). *The Mathematical Theory of Optimal Processes*, K. N. Trirogoff, Transl., L. W. Neustadt, Ed., John Wiley, New York.

Pshenichnyi, B. N. (1971). *Necessary Conditions for an Extremum*, Marcel Dekker, New York.

Puu, T. (1977). On the profitability of exhausting natural resources, *J. Environ. Econ. Manage.* **4**, 185–199.

Rabinowitz, P. H. (1978). Periodic solutions of Hamiltonian systems, *Commun. Pure Appl. Math.* **31**, 157–184.

Redheffer, R., and W. Walter (1974). A differential inequality for the distance function in normed linear spaces, *Math. Ann.* **211**, 299–314.

Roberts, A. W., and D. E. Varberg (1974). Another proof that convex functions are locally Lipschitz, *Am. Math. Mon.* **81**, 1014–1016.

Robinson, S. M. (1976a). Regularity and stability for convex multivalued functions, *Math. Oper. Res.* **1**, 130–145.

——— (1976b). Stability theory for systems of inequalities, part II: Differentiable nonlinear systems, *SIAM J. Numer. Anal.* **13**, 497–513.

——— (1982). Generalized equations and their solutions, part II: Applications to nonlinear programming, *Mathematical Programming Study* **19**, 200–221.

Rockafellar, R. T. (1966). Characterization of the subdifferentials of convex functions, *Pac. J. Math.* **17**, 497–510.

——— (1967). Conjugates and Legendre transforms of convex functions, *Can. J. Math.* **19**, 200–205.

—— (1968a). Integrals which are convex functionals, *Pac. J. Math.* **24**, 525–540.

—— (1968b). Duality in Nonlinear Programming, in *Mathematics of the Decision Sciences*, Part 1, G. B. Dantzig and A. F. Veinott, Eds., *Lectures in Applied Mathematics*, vol. II, American Mathematical Society, pp. 401–422.

—— (1969). Measurable dependence of convex sets and functions on parameters, *J. Math. Anal. Appl.* **28**, 4–25.

—— (1970a). *Convex Analysis*, Princeton Mathematics Ser., vol. 28, Princeton Univ. Press.

—— (1970b). Conjugate convex functions in optimal control and the calculus of variations, *J. Math. Anal. Appl.* **32**, 174–222.

—— (1970c). Generalized Hamiltonian equations for convex problems of Lagrange, *Pac. J. Math.* **33**, 411–428.

—— (1971a). Integrals which are convex functionals II, *Pac. J. Math.* **39**, 429–469.

—— (1971b). Existence and duality theorems for convex problems of Bolza, *Trans. Am. Math. Soc.* **159**, 1–40.

—— (1972). State constraints in convex problems of Bolza, *SIAM J. Control Optim.* **10**, 691–715.

—— (1973). Optimal arcs and the minimum value function in problems of Lagrange, *Trans. Am. Math. Soc.* **180**, 53–83.

—— (1974). *Conjugate Duality and Optimization*, Conference Board of Mathematical Sciences Ser., no. 16, SIAM Publications.

—— (1975). Existence theorems for general control problems of Bolza and Lagrange, *Adv. in Math.* **15**, 312–333.

—— (1976a). Lagrange multipliers in optimization, *SIAM-AMS Proc.*, vol. 9, R. W. Cottle and C. E. Lemke, Eds., pp. 145–168.

—— (1976b). Dual problems of Lagrange for arcs of bounded variation, in *Calculus of Variations and Control Theory*, D. L. Russell, Ed., Academic Press, New York.

—— (1976c). Integral functionals, normal integrands and measurable selections, in *Nonlinear Operators and the Calculus of Variations*, L. Waelbroeck, Ed., Lecture Notes in Mathematics, no. 543, Springer, Berlin, pp. 157–207.

—— (1979a). Clarke's tangent cones and the boundaries of closed sets in R^n, *Nonlinear Anal. Theor. Meth. Appl.* **3**, 145–154.

—— (1979b). Directionally Lipschitzian functions and subdifferential calculus, *Proc. London Math. Soc.* **39**, 331–335.

—— (1979c). *La Théorie des Sous-Gradients et Ses Applications à l'Optimisation: Fonctions Convexes et Non Convexes*, Collection Chaire Aisenstadt, Presses de l'Université de Montréal, 130 pp.

—— (1980). Generalized directional derivatives and subgradients of nonconvex functions, *Can. J. Math.* **32**, 157–180.

—— (1981). *The Theory of Subgradients and Its Applications to Problems of Optimization: Convex and Nonconvex Functions*, Helderman Verlag, Berlin.

—— (1982a). Proximal subgradients, marginal values, and augmented Lagrangians in nonconvex optimization, *Math. Oper. Res.* **6**, 427–437.

—— (1982b). Lagrange multipliers and subderivatives of optimal value functions in nonlinear programming, *Mathematical Programming Study* **17**, 28–66.

―――― (1982c). Favorable classes of Lipschitz continuous functions in subgradient optimization, in *Nondifferentiable Optimization*, E. Nurminski, Ed., Pergamon Press, New York.

―――― (1982d). Augmented Lagrangians and marginal values in parametric optimization problems, in *Generalized Lagrangian Methods in Optimization*, A. Wierzbicki, Ed., Pergamon Press, New York.

―――― (to appear (a)). Marginal values and second-order conditions for optimality, *Math. Program.*

―――― (to appear (b)). Directional differentiability of the optimal value function in a nonlinear programming problem, *Mathematical Programming Stud.*

Roxin, E. (1965). On generalized dynamical systems defined by contingent equations, *J. Differ. Eq.* **1**, 188–205.

Saks, S. (1937, 1964). *Theory of the Integral, Monografie Matematyczne Ser.*, no. 7 (1937); 2nd rev. ed., Dover Press, New York (1964).

Seierstad, A., and K. Sydsaeter (1977). Sufficient conditions in optimal control theory, *Int. Econ. Rev.* **18**, 367–391.

Shu-Chung, Shi (1980). Remarques sur le gradient généralisé, *Co. R. Acad. Sci. Paris* **291**, 443–446.

Spingarn, J. E. (1981). Submonotone subdifferentials of Lipschitz functions, *Trans. Am. Math. Soc.* **264**, 77–89.

Strodiot, J. J., and V. Hien Nguyen (1979). Caractérisation des solutions optimales en programmation non différentiable, *Co. R. Acad. Sci. Paris* **288**, 1075–1078.

Thibault, L. (1979). Cônes tangents et épi-différentiels de fonctions vectorielles, *Trav. Sem. Anal. Convexe* **9**(2), Exp. no. 13.

―――― (1982a). Subdifferentials of nonconvex vector-valued functions, *J. Math. Anal. Appl.* **86**, 319–344.

―――― (1982b). On generalized differentials and subdifferentials of Lipschitz vector-valued functions, Nonlinear Anal. Theor. Meth. Appl. **6**, 1037–1053.

Treiman, J. S. (to appear). Characterization of Clarke's tangent and normal cones in finite and infinite dimensions, *J. Optim. Theory Appl.*

Ursescu, C. (1975). Multifunctions with closed convex graph, *Czech. Math. J.* **25**, 438–441.

Vinter, R. B. (to appear). The equivalence of strong calmness and calmness in optimal control theory, *J. Math. Anal. Appl.*

Vinter, R. B., and R. M. Lewis (1978). A necessary and sufficient condition for optimality of dynamic programming type, making no a priori assumptions on the controls, *SIAM J. Control Optim.* **16**, 571–583.

―――― (1980). A verification theorem which provides a necessary and sufficient condition for optimality, *IEEE Trans. Autom. Control* **AC-25**, 84–89.

Vinter, R. B., and G. Pappas (1982). A maximum principle for non-smooth optimal control problems with state constraints, *J. Math. Anal. Appl.* **89**, 212–232.

Wagner, D. H. (1977). Survey of measurable selection theorems, *SIAM J. Control Optim.* **15**, 859–903.

Ward, A. J. (1935). Differentiability of vector monotone functions, *Proc. London Math. Soc.* **39**(2), 339–362.

Warga, J. (1962). Relaxed variational problems, *J. Math. Anal. Appl.* **4**, 111–128.

—— (1972). *Optimal Control of Differential and Functional Equations*, Academic Press, New York.

—— (1976). Derivate containers, inverse functions and controllability, in *Calculus of Variations and Control Theory*, D. L. Russell, Ed., Academic Press, New York, pp. 13–46.

—— (1978a). Controllability and a multiplier rule for nondifferentiable optimization problems, *SIAM J. Control Optim.* **16**, 803–812.

—— (1978b). An implicit function theorem without differentiability, *Proc. Am. Math. Soc.* **69**, 65–69.

Watkins, G. G. (to appear). Clarke's tangent vectors as tangents to Lipschitz continuous curves.

Weinstein, A. (1978). Periodic orbits for convex Hamiltonian systems, *Ann. Math.* **108**, 507–518.

Wierzbicki, A. P. (1972). Maximum principle for semiconvex performance functionals, *SIAM J. Control Optim.* **10**, 444–459.

Wolfe, P., and M. L. Balinski, Eds. (1975). "Nondifferentiable Optimization," *Mathematical Programming Study* 3, North-Holland, Amsterdam.

Yorke, J. (1971). Another proof of the Liapounov convexity theorem, *SIAM J. Control Optim.* **9**, 351–353.

Young, L. C. (1969). *Lectures on the Calculus of Variations and Optimal Control Theory*, Saunders, Philadelphia.

Zeidan, V. (1982). *Sufficient Conditions for Optimal Control and the Calculus of Variations*, Ph.D. thesis, Univ. of British Columbia, Canada.

—— (1983). Sufficient conditions for the generalized problem of Bolza, *Trans. Am. Math. Soc.* **275**, 561–586.

About the Author

Frank Clarke was born in 1948 in Montreal, where his mother, née Rita Tourville, had the goodness to raise him. He received the B.Sc. (summa cum laude) in Honours Mathematics in 1969 and the M.Sc. in 1970 from McGill University. He wrote his doctoral thesis at the University of Washington under R. T. Rockafellar. Ever since, he has been a member of the Mathematics Department at the University of British Columbia, and has been a visitor at l'Université de Paris and the University of California, Berkeley. Among his academic distinctions are the invitation to speak at the 1978 International Congress of Mathematicians, the award of a coveted Killam Research Fellowship by the Canada Council in 1979–80 and in 1980–81, and the Coxeter–James Lectureship of the Canadian Mathematical Society in 1980.

Index

Absolute value, 28
Action, 8, 271
   dual, 8, 9, 271, 279
Adjoint equation, 211
Antipodality, 278
Approximating line, 2, 49
Arc, 14, 114
   piecewise smooth, 188
Attainable set, 150, 200
Aubin, J.P., 78, 89, 95, 237
Aumann, R.J., theorem of, 112, 256
Aumann, R.J., selection theorem of, 166
Austen, J., 252
Autonomous, 152, 169, 213

Ball, unit, 25
Banach, S., 2, 88, 268
Banach space, 24
Basic hypotheses:
   control problem P_C, 200
   differential inclusion problem P_D, 120
   mathematical programming problem P, 228
   problem of Bolza P_B, 167
Birkhoff, G.D., 270
Bishop, E., 266
Blake, W., 24
Bliss, G.A., 20
Bolza, problem of, 14, 166, 219, 280
Bourbaki, N., 30

Calmness, 168, 175, 238, 240
   local, 244
Campbell, H.F., 132
Carroll, L., 199
Castaing, C., 111
Cellina, A., 272
Chain rule, 42, 45, 72, 106
Clark, C.W., 132
Complementary slackness, 228

Cone:
   contingent, 55
   normal, 11, 51, 52, 68
   tangent, 11, 51, 53
Conjugate points, 23, 179
Constraint qualification, 234
   Mangasarian-Fromowitz, 235, 238
   Slater, 236, 238
   see also Calmness
Contingent cone, 55
Continuous differentiability, 32
Contraction, 268
Control, 15, 200
Controllability, 18, 147, 200, 224
   null, 209
Convex energy surface, 271
Convex function, 34
Convexity condition, 167
Corner point, 188

Dasgupta, P.S., 132
Davies, R., 227
Dependence on initial values, 262
Derivative:
   continuous, 32
   Fréchet, 30
   Gâteaux, 30
   Hadamard, 30
   strict, 30
Descent, direction of, 233
Dido, 194
Differential inclusion, 16, 110
Dini derivate, 244
Diode, 4, 213, 221
Directional contraction, 268
Directional derivative:
   generalized, 10, 25, 96
   usual, 30
Direction of descent, 233

Distance function, 3, 8, 50
  Euclidean, 65
Doyle, A.C., 252
Dual action, 8, 9, 271, 279
Dual space, 27
Dual variational principle, 8, 277
Du Bois Reymond lemma, 127
Dunford-Pettis criterion, 119, 257

Eigenvalue, 6, 94
Ekeland, I., 197
Ekeland's theorem, 8, 265
Elastic band, 4
Energy surface, 271
Epigraph, 12, 59
  Lipschitzian, 260
Erdmann conditions, 187
Euler, L., 227
Euler-Lagrange equation, 19, 82, 187
Exact penalization, 7, 51, 239
Existence, 18, 22
  control problem P_C, 222
  differential inclusion problem P_D, 120
  problem of Bolza P_B, 167
Extended-valued functions, 13, 61

Fenchel transform, 20
Fenman's Law, 110
Fields, 23
Filippov, A.F., 17, 212, 223
Fixed points, 268
Fixed proportions, 7, 132
Fréchet derivative, 30
Free time, 151, 213

Gâteaux derivative, 30
Gauge, 271
Generalized gradient:
  of absolute value, 28
  asymptotic, 101
  of compositions, 42, 45, 72, 106
  definition, 9, 27
  of distance function, 67
  extended, 61, 95
  of the Hamiltonian, 121
  of indefinite integrals, 34, 42
  of integral functionals, 75, 79, 82
  on L_p, 82
  on L_∞, 79
  at local extrema, 38, 61
  of maximal eigenvalue, 94
  nonempty, 97
  partial, 48, 64, 65, 85, 121, 130
  of pointwise maxima, 47, 85, 92
  of products and quotients, 48
  relative, 231

  as rows or columns, 70
  of scalar multiples, 38
  of sums, 38, 102
  of supremum norm, 88
  on value function, 143, 225, 241
  of variational functionals, 81
Generalized Jacobian, 69
Geodesic, 272, 277
Goldstein, H., 274
Gradient, see Generalized gradient
Graph of multifunction, 29
Growth condition, 22, 180

Hadamard derivative, 30
Hahn-Banach theorem, 27
Hamiltonian, 8, 19, 20
  boundary-value problems, 270, 279
  constancy of, 130, 169, 213, 272
  for control problem P_C, 21, 287
  for differential inclusion problem P_D, 121
  equations, 8, 19, 270
  generalized gradient of, 121
  inclusion, 21, 122, 169, 272, 279
  level surface, 271
  multipliers, 121
  for problem of Bolza P_B, 167
  pseudo, 19, 199, 211
  for resource problem, 134
Hamilton-Jacobi equation, 23, 155, 225
  extended, 158
Heal, G.M., 132
Hestenes, M.R., 20, 127
Hiriart-Urruty, J.B., 91, 95
Hooke's Law, 4
Hotelling, H., 131
Husserl, E., 199
Hypertangents, 57, 98

Implicit function theorem, 255
Indefinite integral, 34, 42
Indicator function, 62
Initial-value problem, 3, 262
Inverse function theorem, 252
Ioffe, A.D., 77, 80, 89

Jacobian, generalized, 69, 253
Jacobi condition, 19, 179

Lagrange multiplier rule, 227, 231
Lagrangian, 19, 228
  convexity of, 22, 177, 179
  growth condition, 180
Lebourg, G., 41
Legendre transform, 19
Levin, V.L., 77, 80, 89
Linearization, 262

Linear regulator, 213
Lipschitz:
  condition, 9, 25
  directionally, 98
  epigraph, 260
  multifunction, 113
  rank, 9, 25
  strong, 168

McClamroch, N.H., 214
McShane, E.J., 20
Mangasarian, O.L., 220
Mangasarian-Fromowitz constraint qualification, 235, 238
Mathematical programming, 5, 6, 227
Maximum principle, 20, 210
  variants of, 212
Mayer, problem of, 187, 222
Mean value theorem, 41
  vector, 72
Measurable:
  $L \times B$, 166
  multifunction, 111
  selection, 111, 166
Minimax theorem, 78
Minkowski gauge, 271
Morrey, C.B., 197
Multifunction, 16, 29, 111
  associated distance function, 114
  closed, 29
  graph of, 29
  integrably bounded, 112
  integral of, 112, 256
  Lipschitz, 113
  measurable, 111
  measurably Lipschitz, 113
  monotone, 37
  selection for, 111, 166
  upper semicontinuous, 29
Multiobjective, 230
Multiple integral, 197
Multiplier, 121, 142, 235
  rule, 20, 187, 227, 231
Munro, G.R., 132

Necessary conditions, 18, 19, 20
  for basic problem, 186
  for control problem P_C, 20, 138
  for differential inclusion problem P_D, 20, 122, 147
  for inequality constraints, 187
  in Lagrangian form, 186, 288
  for mathematical programming, 228, 231
  for multiple integral problems, 197
  for problem of Bolza P_B, 20, 168, 180
Noether's theorem, 277

Nohel, J., 272
Nonsmooth analysis, 1
Norm, 25, 70
  Euclidean, 65
  supremum, 2, 88
Normal, 11, 51, 68
  cone, 11, 51, 68
  in convex analysis, 52
  in finite dimensions, 68
  to intersections, 105
  to level sets, 56, 108
  to products, 54
Normality, 147, 235
  weak, 142

Orbit, 277

Pappas, G., 119
Paradigms, 13
Pareto optima, 230
Penalization, 7, 51, 239
Penalty property, 250
Period, 279
Perpendicular, 11, 66
Perturbation, 6, 142, 148, 224
Phelps, R.R., 266
Plenary hull, 263
Poincaré, H., 270
Pointed set, 143, 241
Polak, E., 232, 234
Polarity, 11, 274
Pontryagin, L.S., 20, 199
Problem:
  autonomous, 152, 169, 213
  basic, 19, 187
  of Bolza, P_B, 14, 166, 219, 280
  control, P_C, 15, 210
  differential inclusion, P_D, 16, 110
  diode, 213, 221
  equivalence, 16, 219
  free time, 151, 213
  initial-value, 3, 262
  linear regulator, 213, 221
  mathematical programming, P, 5, 6, 227
  of Mayer, 187, 222
  multiple integral, 197
  of Queen Dido, 194
  relationships, 17
  relaxed, 175, 223
  resource, 131
Pseudo-Hamiltonian, 19, 199, 211
Pseudonormality, 224

Queen Dido's problem, 194

Rabinowitz, P.H., 277

Rademacher's theorem, 6, 63
Rank condition, 188
  maximal, 253
Regular function, 39, 59, 61, 272
  set, 55, 57, 59
Relaxed problem, 175, 222
  trajectory, 117
Resolvent, 4, 263
Resource economics, 7, 131
Roberts, A.W., 34
Robinson, S.M., 236
Rockafellar, R.T., 22, 57, 69, 95, 144, 166, 167, 219, 261
Russell, B., 1

Santayana, G., 166
Second order conditions, 250
Sensitivity, 18, 237. *See also* Calmness; Perturbation; Value function
Singular interval, 137
Slater constraint qualification, 236, 238
Solar panel, 4
Solvability and surjectivity, 248. *See also* Perturbation
State, 15
  constraint, 15, 200
Strict differentiability, 30
Subdifferential, 10, 36
Sufficient conditions, 18, 22
  for control problem P_C, 219
  for differential inclusion problem P_D, 155
  for problem of Bolza P_B, 177
  see also Hamilton-Jacobi equation
Support function, 28, 257
Symplectic matrix, 273, 278

Tangent, 3, 11, 51

cone, interior of, 57, 68, 261
  in convex analysis, 53
  to intersections, 105
  to level sets, 55, 108
  to products, 54
Temam, R., 197
Tonelli existence theorem, 22
Trajectory, 114, 200, 279
  compactness of, 118
  relaxed, 117
Transversality condition, 122, 187, 212
Tube, 113

Upper semicontinuous multifunction, 29
Ursescu, C., 236

Valadier, M., 111
Value function, 6, 142, 225, 239, 241
  differentiability of, 242
  directional derivative of, 242
  Lipschitz property for, 143, 242
Varberg, D.E., 34
Variational principle, 8, 270, 277
  dual, 8, 277
Vector mean value theorem, 72
Verification technique, *see* Hamilton-Jacobi equation
Vinter, R.B., 119

Wagner, D.H., 166
Weierstrass condition, 19, 187
Weinstein, A., 277
Wilder, T., 24

Young, L.C., 16, 23, 110

Zeidan, V., 23, 177, 219